THE PENDULUM PARADIGM

THE PENDULUM PARADIGM

Variations on a Theme and the Measure of Heaven and Earth

MARTIN BEECH

BrownWalker Press
Boca Raton

The Pendulum Paradigm: Variations on a Theme and the Measure of Heaven and Earth

BrownWalker Press
Boca Raton, Florida • USA
2014

ISBN-10: 1-61233-730-9
ISBN-13: 978-1-61233-730-2

www.brownwalker.com

Cover image © Can Stock Photo Inc./ dekay

Library of Congress Cataloging-in-Publication Data

Beech, Martin, 1959- author.
    The pendulum paradigm : variations on a theme and the measure of heaven and Earth / Martin Beech.
        pages cm
    Includes bibliographical references and index.
    Summary: This book explores the many applications of the pendulum, from its employment as a fundamental experimental device, such as in the Cavendish torsion balance for measuring the universal gravitational constant, to its everyday, practical use in geology, astronomy and horology.
    ISBN 978-1-61233-730-2 (pbk. : alk. paper) -- ISBN 1-61233-730-9 (pbk. : alk. paper)
    1. Pendulum. I. Title.
    QA862.P4B44 2014
    620.001'185--dc23

                                                    2013040494

This book is humbly dedicated to Robert Hooke (1635-1703), a man who marveled at the rhythmical vibrancy of the universe, and a man who never ceased in putting nature to the question.

TABLE OF CONTENTS

INTRODUCTION

The world is in perpetual turmoil and motion – it is a vibrant extravaganza of interactions, oscillations and pulses. Indeed, we have all witnessed the to and fro of nature's ever changing mood; the rhythmical swaying of leaf-dappled trees; the voluminous ripple of wheat stems in a gentle breeze; the furious beating of a tattered flag in a winter's gale; the stately rise and fall of ocean tides. Likewise the numinous 'tock' of a grandfather clock, repeating in a darkened room on a sleepy Sunday evening, reminds us of the steady ebb and flow of time. Movement surrounds us, and our lives dance to its ever-changing tune. Deep down, however, there is an underlying order, at least of sorts, and we can breakdown the complex motions of the everyday world into a summation of simple harmonic waves. Underpinning all this dynamical mêlée is the pendulum – perhaps the simplest experimental device ever invented. The pendulum for all its apparent simplicity, however, has been at the historical center of humanities exploration of the Earth and the sky above, and this book is about some of that remarkable history. Indeed, this book is concerned with the transformation of the pendulum from its early origins as an ungainly piece of string and an attached weight to that of a precision scientific instrument. Incredibly, for so it seems at first, by simply counting the number of oscillations, back and forth, the pendulum has acted as the key tool in the discovery of Earth's motion and its internal structure, and it has also enabled the experimental measure of the fundamental physical constants defining our universe.

More than just an intricate scientific device and timepiece regulator, however, the metaphorical pendulum is an integral part of our everyday language and lives; all politicians fear the swing of the ideological pendulum, and as Ralph Waldo Emerson so dryly put it, "Old age brings along with its ugliness the comfort that you will soon be out of it – which ought to be a substantial relief to such discontented pendulums as we are". And, indeed, we all hope that the pendulum will swing in our favor. The pendulum is a wonderful metaphor for the confined cycle of change; "what goes around comes around".

The story of the pendulum is primarily the story of observing small changes to quantify the very large – it literally displays the rift between the ideal and reality. It condenses the vastness of the Earth into laboratory-sized experiments which can be run on any weekday afternoon – as all undergraduate students of science well know. The pendulum additionally takes us from the local to the global – elevating our reach beyond the confines of the laboratory to the very edge of the universe. Indeed, as we shall see in Chapter 4 it was a pendulum experiment performed by Isaac Newton in his Cambridge University chambers that convinced him that there was no such element as the gravitational ether, or universal medium through which gravity is transmitted. Newton not only pinned down a formula to describe how gravity must work, therefore, he also showed (albeit non-rigorously by modern standards) that gravity works outside of any transmitting fluid.

And, while the exact workings of gravity are still a mystery to present day researchers, so too are the workings of time. Wryly, the composer Hector Berlioz was once heard to quip that, "time is a great teacher, but it unfortunately kills all its pupils". Sadly this is all too true, but for all this, at the core of time measurement and the pace of musical measure are the pendulum and the metronome (the latter, of course, being an inverted rigid pendulum). While less relevant to mensuration in the modern era, where atomic clocks rule supreme, the pendulum clock historically allowed astronomers to chart the heavens and measure the spin and shape of the Earth. Likewise, the innate human desire for consistency has resulted in the globe being divided into standard time zones; each swath 15 degrees wide in longitude, with every clock (pendulum regulated or otherwise) being forced to agree on the time decreed. And, all this standardization and international cooperation is centered on a concept that we still cannot explain. Time is a universal mystery; we don't know what it is, and we cannot define its origins. As Albert Einstein so rightly told us, almost a century ago now, "time is an illusion" – to which we humbly add, "but what a marvelous illusion". Indeed, through the power of commerce and the desire for order, humanity has managed to ensnare itself within the great illusion of time, and there appears to be little hope of our ever escaping from this overwhelming entrapment. Perhaps, as ever, playwright William Shakespeare found the right words to describe how it is,

I wasted time, and now doth time waste me;
For now hath time made me his numbering clock:
My thoughts are minutes; and with sighs they jar
Their watches on unto mine eyes, the outward watch,
Whereto my finger, like a dial's point,
Is pointing still, in cleansing them from tears.
Now sir, the sound that tells what hour it is
Are clamorous groans, which strike upon my heart,
Which is the bell: so sighs and tears and groans
Show minutes, times, and hours.

So speaks King Richard II, in the play of the same name. Earth has been encaged, and we are tied to time, that grand illusion which is rhythmically sliced by the scythe-like swing of the pendulum bob. And yet, time comes and goes, and according to our mood and surroundings it passes in a quick-silver shimmer or as slow as molasses – sometimes it even stops for a breath-less moment. Time may be an illusion according to physics, but the pendulum is the physical manifestation of its task master.

The mighty pendulum, for so it seems worthy of this great appellation, has accompanied humanity on its quest to discover and annotate the greater world and universe. It is a device that is easy to contemplate, but is subtle in its power and long in its reach; it is an instrument that spans the chasms between human nature, time and space – mighty indeed is the pendulum, and remarkable, as we shall see in subsequent pages, is its story of revelation.

And finally, a few comments are probably in order with respect to what this book is intended to be about and what it is not. To begin with the latter, this book is not intended to be a comprehensive tome on the history of science; nor is it intended to be a mathematical or physical treatise. My inspiration, as much as I can remember it now, has been to write the sort of book that set my mind and imagination racing when I was an undergraduate student studying mathematics and physics in the mid-1970s at the University of Sussex, in Brighton England. Accordingly it is intended that this book serves as a review of the history of science with snippets of the physical principles and some of the mathematical details included (this is in contrast to the apparent norm where they are totally excluded). The material and topics covered are intentionally eclectic, broad and wide-ranging, and they will hopefully inspire the reader's imagination as much as they have, over the years, inspired the author's. In addition to being a review of the history and science relating to pendulum physics, my aim has also been to bring the topic to the present day – the pendulum may be one of humanities oldest of inventions, but it is also one of its most enduring. Far from being a device of the ancients only, the mighty pendulum is still one of the great tools of modern science with a future that is as bright as its history is long. The layout of the text is such that it need not be read in a linear fashion, but many of the chapters are interrelated – so dip-in or dig-deep as the mood takes you, but most of all it is my hope that you will enjoy and be inspired to look further into the stories that you find.

CHAPTER 1
THE PENDULUM SWINGS

If people do not believe that mathematics is simple, it is only because they do not realize how complicated life is.

J. von Neumann

Mention the word pendulum in the modern era and the mind immediately turns to Galileo, the great renaissance philosopher who revolutionized the way in which we think about the physical word. Galileo lead a charmed life for many years, playing with great panache and skill the complex game of individual advancement within 17th Century Italian society. His inability to suffer fools lightly, however, or as some would say his very arrogance, ultimately resulted in his fall from grace, and following several misadventures with Church authorities he was eventually forced, in 1633, to publicly recant his heresy of teaching, as true, the Copernican theory of the solar system. The Inquisition argued that Galileo should be jailed for the remainder of his natural life, but his sentence was commuted to that of house arrest. It was while confined to a comfortable home penitentiary, in Arcetri, just outside Florence, however, that Galileo produced his most enduring, if not most famous, work *Dialogues Concerning Two New Sciences* (published 1638). This work was based upon many decades of diligent labor, research, experimentation and intellectual reasoning, and it revolutionized the way in which subsequent scientific investigation would be conducted. The *Dialogues* is concerned with an imagined conversation between three interlocutors: Salviate, Sagredo and Simplicio. Lasting a total of four days, these three academics explore the possible explanations for numerous physical phenomena; it is the voice of Salviate, however, that always carries the argument, for indeed, Salviate is the strident voice of anti-Aristotelian reasoning. Galileo was unabashedly taking-on the intellectual establishment in his *Dialogues*, and he was fighting against the entrenchment of ancient, un-tested ideas. His voice, the voice of Salviate, was that of the new order, and the new experimentalist who carefully measured and reasoned without resort to unjustified assumption and dogma. The pendulum, in recognition of its great physical importance, is an object of discussion on days one, three and four within the *Dialogues*, and here is how it is first introduced:

Sagredo: *Another question deals with vibrations of pendulums which may be regarded from several viewpoints; the first is whether all vibrations, large, medium, and small, are* *performed in exactly and precisely equal times: another is to find the ratio of the times of vibrations of pendulums supported by threads of unequal length.*

Salviati: *These are interesting questions: but I fear that here, as in the case of all other facts, if we take up for discussion any one of them, it will carry in its wake so many other facts and curious consequences that time will not remain to-day for the discussion of all.*

Galileo cuts to the very core of our subject within his introductory, first-day dialog. Sagredo raises a number of important queries concerning the essential motion of the pendulum, and as Galileo, under the voice of Salviati, notes they are interesting questions that will take more than one day to address. Here, then, is our starting point. While this chapter won't take more than a day to read, we shall nonetheless attempt to unravel within its compass the long and fascinating story behind the mathematics of the pendulum, and we shall also see how pendulum experiments can be used to address questions relating to fundamental physics, metrology, geophysics, astronomy and even philosophy.

The physicist and moral philosopher Stephen Toulmin was once heard to wisely argue that, definitions are like belts, the shorter they are the more flexible they need to be. And, indeed, the standard definition of the pendulum, found in any dictionary, will require a little stretching-out at times. The word pendulum is etymologically derived from the Latin *pendulus*, meaning 'hanging', which is derived from *pendrere*, meaning 'to hang'. So, a pendulum is composed of a body (called 'the bob') of mass m, suspended from a fixed support by a wire or thread of length L in such a manner that it can swing freely, backwards and forwards, under the influence of gravity.

The dictionary style definition of a pendulum seems reasonable enough and agrees well with our everyday expectations. There is, however, some necessary unpacking, as Toulmin warned us there might be, of the definition supplied. Firstly, one might ask for all its triviality, why the suspended object is called a 'bob'; well, as with many such things, the name is a matter of history. There is an old saying which sagely notes that even a stopped clock (a mechanical one with hands that is) is right twice per day. Indeed, if such statements apply to the hands of a clock, then one might further argue, with some considerable degree of confidence, that a stationary pendulum is always right. Right, that is, if it is the local vertical that one wishes to find. The simple device of a plumb-line is still employed by construction workers to this very day, and consists of a heavy weight attached to a piece of string. The name plumb-line is derived from the Latin *plumbum*, meaning 'lead', which described the weight part of the line. This word for lead was later corrupted to plum bob and then cut down to the diminutive bob, and it is this final corruption that is now given to any weight attached to the end of a line used to measure the local vertical. Since a stopped pendulum is a plumb-line, the name bob fits naturally to the hanging weight, and as we shall see in later chapters there are many different uses for a stopped pendulum.

Perhaps more importantly for the analysis to follow, the dictionary definition for a pendulum should be expanded to say that the wire or thread supporting the bob is inextensible. A bob attached to an elastic suspension wire is a mathematical beast, albeit a highly interesting one, that will not be consider here (but see Chapter 3 later). Likewise, the definition should also be expanded to explain what is meant by the term 'mass', and for that matter what exactly does 'under the influence of gravity' mean. To see why such further expansion of our definition is required, we will develop in the next section a dimensional analysis formulation for the relationship that binds together the observable and physical properties of a pendulum.

Dimensional Analysis

The idea of dimensional analysis[1] surfaced in 1687 and it was first applied by the great mathematician and alchemist Isaac Newton (1643 - 1727). The use of dimensional analysis is a standard dodge still used to this very day by physicists and engineers alike. The point of dimensional analysis is to uncover a functional relationship (i.e., an equation) between the physical variables that enter into a specific problem by making sure that all of the units agree. The term 'physical variable' should first be explained before worrying about what a 'unit' is – although the two are closely linked. A physical variable describes a quantity, such as the length of the pendulum wire L, whose numerical value depends upon the units being used. For example, the distance $D = 1$ kilometer can be written as $D = 1000$ meters, or $D = 0.6213712$ miles, or $D = 39370.079$ inches, or $D = 1.057 \times 10^{-13}$ light years. In each case the distance D is exactly the same, but its numerical value changes according to the units being employed. In the discussion that follows *SI* or *Système International* units are going to be used. Commissioned at the request of King Louis XVI of France the *SI* units were first developed in the early 1790s. Indeed, on August 1st, 1793 the meter was introduced as the standard measure of length, and on April 7th, 1795 the gram and kilogram were adopted as standard measures for mass. The standard unit of time didn't change under the *SI* scheme, and it is taken to be the second – this being said, the second was only officially defined in the 1820s when it was agreed that it should correspond to $1/86,400$th of a mean solar day. The meter is technically defined as the distance between two marks scribed on a platinum-iridium bar corresponding to one ten millionth (1 / 10,000,000) of the distance between the equator and the North Pole along the meridian passing through Paris. As is often the case with good intentions, however, there was, in spite of the heroic efforts of the surveyors and mathematicians, a miscalculation in the spacing of the two standard marks for the meter – they are in fact $1/5$th of a millimeter too close together with respect to the original definition[2]. The miscalculation, however, has been allowed to stand. Standard units, after all, are ultimately just convenient measures that people (or more specifically Governments) either agree to use or they don't.

In spite of common language usage, whereby the words weight and mass are taken to mean the same thing, they are, in fact, very different physical quantities. Mass is defined in terms of the amount of matter that constitutes a given body. Specifically, the gravitational mass is a measure of the strength with which an object interacts with a gravitational field. Indeed, once the gravitational field is 'fixed' a smaller mass will experiences a smaller gravitational force than that experienced by a larger mass. The gravitational force associated with a specific mass defines its weight, and in physical terms the weight of an object is the product of its mass times the gravitational field strength g (gravitational field strength is expressed in terms of an acceleration and correspondingly has the units of meters per second squared – m/s^2). In this fashion a mass m has an associated weight $W = m\,g$, where at the Earth's surface $g \approx 9.8$ m /s^2. Weight is often confused with mass in every day life simply because the gravitational acceleration experienced by people (living, of course, on Earth's surface) is very nearly constant, and consequently the weight of an object doesn't noticeable change as it is moved from one location to another. All this being said, an object of mass m will weigh twice as much in a gravitational field of strength $2g$; it will weigh half as much in a gravitational field of strength $g\,/\,2$.

In the *SI* scheme the mass of an object is measured in kilograms, while weight is measured in the units of Newtons (with $1\ N = 1$ kg m / s^2). Just as with the meter, the determination of a definition and the development of a standard for the kilogram has not gone entirely smoothly. Strictly speaking, the definition for the kilogram runs something like this: a kilogram is the mass of one liter of pure water at standard atmospheric pressure when the temperature is 277.16 Kelvin (3.98° Centigrade) [3]. The latter temperature condition corresponds to the so-called triple point of water, and is the temperature at which it acquires its greatest density. The actual standard kilogram mass, however, consists of a circular cylinder made of 90% platinum and 10% iridium kept at the *Bureau International des Poids et Measures*, in Paris. The first prototype cylinder was made in 1879 and after careful comparison was deemed to be equivalent in mass to the more formal pure water based definition. The one-kilogram mass cylinder was finally adopted as the standard mass unit at the first General Conference on Weights and Measures held in 1889. The saga of the kilogram definition, however, continues to this very day and recent comparison measurements indicate that the original Paris standard, in spite of careful monitoring and storage, has been loosing mass. In an attempt to remedy this situation a consortium of laboratories from around the world has formed the Avogadro Project[4], which will attempt to redefine the kilogram not in terms of a physical artifact (as it is at the present), but in terms of the equivalent number of carbon-12 atoms.

So, where does all this get us? Well, having agreed to adopt *SI* units we can now describe all physical variables in terms of the base units of mass (m), length (l) and time (s) each raised to some rational power. The standard notation is to write down $[x]$ to denote the units of any physical quantity x with respect to m, l and s. Accordingly, the units for $[x] = l^\mu\, m^\beta\, s^\gamma$, where α, β and γ are rational numbers. In terms of the three powers (α, β, γ) a physical variable relating to a length measurement will correspond to the numbers (α, β, γ) = (1, 0, 0); a physical variable relating to a velocity, which is defined as the distance traveled divided by the time taken to travel that distance, will have units described by the powers (α, β, γ) = (1, 0, −1); a variable relating to a force will be described by the numbers (α, β, γ) = (1, 1, −2).

To see how all the above discussion helps us with the pendulum, we can derive a formula that links together the period P of the pendulum, being defined as the time to complete one oscillation from the starting point and back again, to the physical quantities associated with a pendulum itself. It does not seem unreasonable to guess that the period should dependent upon $[L]$, the length of the pendulum wire, $[M]$ the mass of the pendulum bob and $[g]$ the gravitational field in which the pendulum swings. What the dimensional analysis approach now says is that the following expression must hold true

$$[P] = k\,[L]^A\,[M]^B\,[g]^C \tag{1.1}$$

where A, B, C, and k are dimensionless numerical constants. Now, in terms of the units we can write (1.1) as follows

$$
\begin{aligned}
l^0\, m^0\, s^1 &= [l^1\, m^0\, s^0]^A\ [l^0\, m^1\, s^0]^B\ [l^1\, m^0\, s^{-2}]^C \\
&= [l^{A+C}\, m^{B}\, s^{2C}] \tag{1.2}
\end{aligned}
$$

What we have to do now is make sure that the powers of the base units on both sides of (1.2) are equal – that is, we need to make sure that the units agree. If we equate the powers in length, mass and time separately we have

$$
\begin{aligned}
0 &= A + C \quad &\text{for the length unit } l \\
0 &= B \quad &\text{for the mass unit } m \\
1 &= -2C \quad &\text{for the time unit } s
\end{aligned}
$$

From these relationships we immediately find that B = 0, and consequently derive the result that the period of a pendulum is independent of the mass of its bob. We can also see that C = - ½ and that A = -C = ½. Our dimensional analysis is now complete and we can say that the period of oscillation P for a pendulum of length L in a region in which the acceleration due to gravity is g is

$$P = k\sqrt{L/g} \qquad\qquad (1.3)$$

where the square root symbol $\sqrt{}$ has been used to denote powers to one-half. From (1.3) we can deduce, for example, that the period of a pendulum with a wire of length $4L$ will be twice that of a pendulum having a support wire of length L (assuming g remains constant). While equation (1.3) tells us how the period of a pendulum changes with respect to L and g, it does not tell us what the actual period is in seconds. The problem, of course, is that we don't know what the numerical value of the dimensionless constant k is. The value of this term can only be found by a more refined analysis – which will be performed shortly. Interestingly, however, we can be reasonably sure that the value of k won't be an especially large number. No less an authority than Albert Einstein commented upon this latter point in 1911. Specifically, Einstein noted that the dimensionless numbers[5], "are generally of order magnitude one [6]. To be sure, this cannot be strictly required, because why should it not be possible for a numerical factor $(12\pi)^3$ to appear in a mathematical-physical analysis? But such cases are unquestionably rare". Indeed, writing upon the unreasonable effectiveness of dimensional analysis in their book *The Anthropic Cosmological Principle* (Oxford University Press, 1986) John Barrow and Frank Tipler suggest that the dimensionless constants are usually small because we live in a low dimensional (that is 3 spatial and one time dimension) universe – we shall pick up on this discussion later. For the moment, however, having argued that the constant in equation (1.3) should be small, what we need to do next is perform the detailed analysis which determines its actual value.

A First Approximate Equation

Sir Isaac Newton entered into our earlier discussion as being the first analyst to use the idea of dimensional analysis – or as he called it, the principle of similitude. It seems appropriate therefore that Newton also provides us with the key mathematical tool needed to determine the constant k in equation (1.3). The mathematical tool is that of calculus[7], which has two basic operations; that of differentiation and that of integration. There will be no attempt to explain the full meaning behind these two operations here, but it is perhaps worth pointing out that the so-called *fundamental theorem of calculus* tells us that the two operations are essentially the inverse of each other. In very general terms differentiation is concerned with determining the rate at which quantities change enabling, for example, a velocity to be described in terms of the time variation of distance traveled. Integration, in contrast, is about adding together infinitely small quantities to find, for example, the total distance traveled in the time interval T_1 to T_2 by an object moving with a variable velocity.

To see how the ideas of differentiation and integration can help us with respect to the pendulum, first take a look at figure 1.1. This figure shows the position of the pendulum at time T_0, when the support wire makes an angle φ_0 with the vertical and the bob is held at rest – that is, the bob's initial velocity $V_0 = 0$ m/s. After some small amount of time Δt, the pendulum bob has moved to a new position for which the support wire makes an angle φ with respect to the vertical. We can now say that in the time interval Δt, the angle describing the position of the pendulum bob with respect to the vertical has changed by an amount $\Delta \varphi = \varphi_0 - \varphi$. What we want to do next is find an expression relating the quantities $\Delta \varphi$, and Δt to the velocity and acceleration of the bob. Before we can do this, however, we must first introduce an additional unit of measure – the radian.

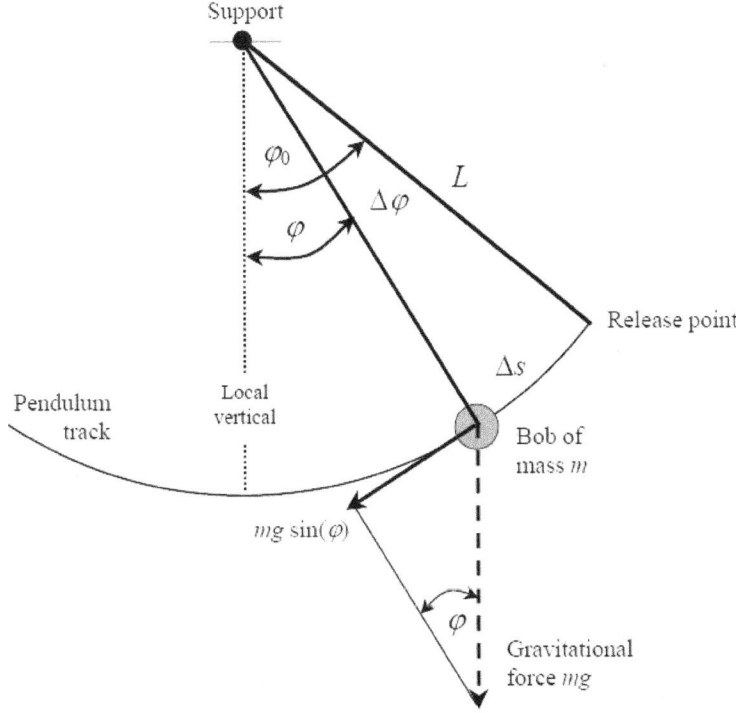

Figure 1.1. The simple pendulum.

The radian is an angular measure, just like the familiar degree unit used in everyday life, but rather than there being 360 degrees subtended in a circle there are 2π radians. The scholars of mathematical history usually tell us that the word 'radian' was first read and witnessed by students sitting an exam set by Professor James Thomas, at Queens College Belfast, on June 5th, 1873. Apparently, however, the idea of the rad, as the radian is commonly abbreviated, had been around much longer and Roger Cotes (1682 – 1716), a student of Sir Isaac Newton's and first Plumian Professor of Astronomy and Experimental Philosophy at Cambridge University in England, is credited with using the concept as early as 1714. By definition one radian is the angle subtended at the center of a circle by an arc along its circumference equal in length to its radius (see figure 1.2). In degree measure, 1 rad = 360 / 2π = 57º.29577951..., and unless otherwise stated mathematicians always assume that angles are stated in radians. The radian is the 'official' unit of angular measure in the *SI* system, although it is always taken to be a dimensionless number.

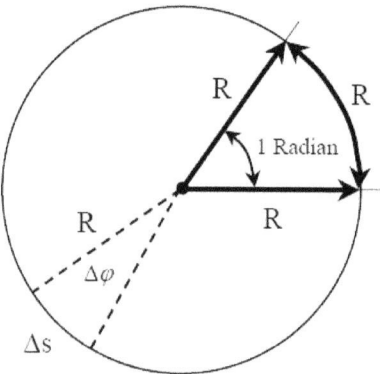

Figure 1.2. The radian is defined as the angle subtended at the center of a circle by a circumference arc equal in length to the circle's radius. One very convenient use of the radian is that the arc length, Δs, subtended by an angle $\Delta\varphi$ can be expressed as the product: $\Delta s = \Delta\varphi\, R$, where R is the radius of the circle.

With the radian now defined the distance traveled along the arc of the pendulum track Δs (see figure 1.1) can be written as $\Delta s = L\,\Delta\varphi$, and the velocity of the pendulum bob is accordingly $V = \Delta s\, /\, \Delta t = L\,\Delta\varphi\, /\, \Delta t$. The acceleration, defined as the rate of change of the velocity, experienced by the pendulum bob in time Δt can further be expressed as $a = \Delta V\, /\, \Delta t = (V - V_0)\, /\, \Delta t = L\,(\Delta\varphi\, /\, \Delta t)\, /\, \Delta t$. Now, the key point about differential calculus is that it is concerned with variations when the Δ terms approach the limit of becoming infinitely small. In this manner the differential expressions for the velocity and the acceleration are respectively $V = L\,(d\varphi\, /\, dt)$ and $a = dV\, /\, dt = L\,(d^2\varphi\, /\, dt^2)$ – the d symbol indicates the infinitely small limit of the Δ's have been taken. In words, what these two equations tell us is that the velocity is related to the first time derivative of the position angle φ, and that the acceleration is related to the second time derivative of the position angle φ. The velocity is therefore described according to the rate at which the angle φ

is swept out, and the acceleration is expressed as the rate of change at which the angle φ is swept out.

Having derived expressions for the velocity and acceleration of the pendulum bob, we now want to relate these to the gravitational force mg acting upon the bob. Here recall, m is the mass of the bob in kilograms and g is the acceleration due to gravity. The product mg is the gravitational force, acting downwards, between the pendulum bob and the Earth. We shall have much more to say about gravity and gravitational forces later on, but for the moment it can be seen from figure 1.1 that the component of the force acting to move the bob along its constrained arc has a magnitude $mg \sin(\varphi)$. It is this component of the gravitational force that initially acts to accelerate the pendulum bob away from its release point.

To complete the derivation we must now use Newton's second law of motion, which relates acceleration to an applied force. Specifically, what Newton's second law tells us is that if a force F acts upon an object of mass m, then the acceleration a produced is given by the expression: $a = F / m$. For our pendulum the force $F = -mg \sin(\varphi)$ where the minus sign has been introduced by convention since gravity acts in a downwards direction. We have already seen, however, that the acceleration is related to the second derivative of the angle φ, and we can accordingly write: $L\, d^2\varphi/dt^2 = a = F/m = -g\sin(\varphi)$. Now, this equation is all well and good, but we still haven't found our sought after expression for the pendulum period. To make progress in this direction we must first make use of a convenient simplification related to radians. Specifically, it turns out, the value of the quantity $\Psi = \sin(\varphi) / \varphi$, when φ is expressed in radians, approaches unity as φ goes to zero. Schematically we can write this as $\Psi \to 1$ as $\varphi \to 0$. Where this approximation helps us is that the differential equation for the pendulum motion can, when φ is small, be written as $L\, d^2\varphi /dt^2 = -g\, [\sin(\varphi) / \varphi]\, \varphi \approx -g\, \varphi$. The $\approx$ sign has been used in our new expression to indicate that the approximation only holds true if the angle φ through which the pendulum swings is small. The reason why this small angle approximation helps us is that we can now find an exact solution to our differential equation by substitution. Specifically, the solution follows by writing $\varphi(t)$, the angle φ at time t after the release of the pendulum bob, as $\varphi(t) = \varphi_0 \cos[\, (2\, \pi\, /\, P)\, t\,]$ where φ_0 is the initial angle at the time of release (assumed small remember), and P is the time for the pendulum to complete one oscillation. The motion is constrained to move between φ_0 and $-\varphi_0$, with the lowest point on the pendulum's track being the vertical position $\varphi = 0$ (corresponding to an instantaneous plumb-line). If we now differentiate the expression for $\varphi(t)$ with respect to time t twice and substitute the resultant terms into the differential equation for the pendulum motion[8] then the expression for the period P that we have been looking for emerges, namely, $L (2\pi / P)^2 = g$, which is more conveniently written as

$$P = 2\pi\sqrt{L/g} \qquad (1.4)$$

So, now we have our answer, a complete, first approximation result for the period of oscillation to a simple, small angle of oscillation, pendulum. Comparing equation (1.4) with equation (1.3) we find, just as the dimensional analysis predicted, the period varies according to the square root of the length of the pendulum L divided by the acceleration due to gravity g. But even more usefully, we also find that the dimensionless constant $k = 2\pi = 6.283185307...$ (a relatively small number, of order ten, as Einstein argued should be the case). This result is helpful since we can now calculate the actual period of a pendulum in seconds[9].

While equation (1.4) describes a useful relationship between the period of a pendulum, its length and the gravitational field strength[10], it is interesting to note what quantities the period doesn't depend upon. From our dimensional analysis, for example, the period was found to be independent of the mass of the bob, and from the first approximation analysis it is also found that the period is independent of the initial angle (φ_0) that it is released from (well, up to a point, as we shall see below). These are remarkable observations, and they underscore the great importance of the pendulum with respect to time keeping. For a given length of support wire the period is isochronal: the swing of a pendulum literally slices time into equal intervals irrespective of its starting conditions. In addition, it is also found that it is of no great consequence what the pendulum bob is made of; clay, wood, gold or lead it is irrelevant to the pendulum's period of swing. This latter point is, in fact, of great fundamental interest, as Isaac Newton first realized and as we shall see later in Chapter 4.

The Seconds Pendulum

The standard unit of time measurement is the second; a word derived from the ancient Greeks who called one sixtieth of a degree the 'first small part', and one sixtieth of 'the first small part' was, appropriately enough, called the 'second small part'. In Latin the 'first small part' would be written as *pars minuta prima* while the 'second small part' would be written as *pars minuta secunda*, and it is from these expressions that the words minute and second have descended. The fact that the divisions are taken as the 1/60th part is due to the ancient Babylonians, who used 60 as the base of their counting scheme (the sexagenary system) – as a side note, it was the ancient Egyptians who introduced the idea of dividing the day into 24 hours.

Since the period of the small-amplitude pendulum is constant (that is, isochronal) and independent of its starting position, it seems only natural to ask, what is the length of a pendulum that takes one second to complete one swing – that is, what is the length of a pendulum that has a period of two seconds. The point being, of course, that once calibrated such a pendulum could provide a standard for time that is easily transportable from one location to another. Not only this, the length of the Seconds Pendulum could also be adopted as a standard unit of length. Indeed, it was originally intended that the meter should be based upon the length of the Seconds Pendulum as determined at the latitude of 45 degrees near the town of Bordeaux in France. This specific proposal was first presented to the new Revolutionary Government of France by Charles-Maurice de Talleyr and in 1790, after some no doubt complicated but certainly skilful political wrangling, both Britain and the newly confederated United States of America agreed to adopt the proposed unit of length. The Commission of Weights and Measures to the French Government met, however, on March 19th 1791, and, rather surprisingly, decided to drop the idea of the Seconds Pendulum as the bases for the standard of measure, proposing instead the one ten millionth distance from the North Pole to the Equator. Neither Britain nor the United States were particularly impressed with the new scheme and accordingly declined to join in the geodetic survey to map out the meridian from which the meter would be derived. The reason why the length of the Seconds Pendulum was dropped from consideration as the standard of measure was explained by the Committee along the lines that it would be based upon the unit of time (i.e., the second), and this, they argued might change. Indeed, the Revolutionary Government at that time was seriously considering the installation of a decimalized day of 10 hours duration, with each hour being 100 minutes long and each minute being composed of 100 seconds[11]. These new hours, minutes and seconds, of course, would be of different duration to our Babylonian based measures of the same name. Additionally, the Commission argued that by basing the meter on the Earth's quarter circumference it would reflect the global nature of the new standard and would apply to all humanity. Critics of the new scheme immediately pointed out (but to no avail) that the new definition was itself entirely arbitrary and was not even a real distance, but one based upon a mathematical calculation of an imaginary arc extrapolated from a small measured segment of the whole. Well, of course, no one said that agreeing on fundamental units was going to be easy. The important point of this story, however, is that as a result of its decision the French Government set about sponsoring several scientific expeditions to carefully measure the Earth, and pendulum based experiments were an important part of the survey reductions - as we shall see later in Chapter 3.

The length of a seconds pendulum can de determined with the aid of equation (1.4). The calculation proceeds by setting the period P equal to 2 seconds, and taking g to be the standard reference acceleration due to gravity as 9.806650 m/s². Accordingly, the length of the standard Seconds Pendulum is 0.993621 meters.

At this stage, many readers will have probably realized that one of the problems associated with the use of the seconds pendulum as a 'movable' standard for time and length is that the acceleration due to gravity is not the same at all locations on Earth Indeed, this was why the original French proposal required the measurements to be made at a specific location. The suggested location was somewhere along latitude of 45 degrees, halfway between the Equator and the North Pole (and also a latitude that crosses France), and the ancient City of Bordeaux fitted the bill nicely. Importantly, it was noted, Bordeaux was located close to sea-level and well away from any mountain ranges that might bias the pendulum result (as we shall discuss later). Table 1.1 provides some examples of the measured values (how they were obtained will be described shortly) for the acceleration due to gravity at several large cities situated around the world.

The gravitational acceleration experienced doesn't vary much as we move from one location to another on Earth's surface, and the extreme values in Table 1.1, corresponding to Helsinki and Sydney, only differ by 0.1 per cent. This difference may seem small, but it is crucial, and it does indeed negate the Seconds Pendulum from being the fundamental standard for both length and time. If, for example, one compared the seconds pendulum calibrated at Helsinki and Sydney, then the time difference in the periods would amount to just 0.00214 seconds. This time difference might not seem to be of any great importance, but we have to remember that the time difference applies to each and every swing taken by the pendulums; after just 468 oscillations, the pendulum in Sydney would already be one second behind in comparison to the same number of swings taken by the pendulum in Helsinki.

Location	g (m/s^2)	$L_{Seconds}$ (m)	$P_{standard}$
(Standard)	9.806650	0.993621	2.00000
Helsinki	9.819	0.994873	1.99874
London	9.812	0.994163	1.99945
Paris	9.809	0.993859	1.99976
New York	9.802	0.993150	2.00047
Sydney	9.797	0.992644	2.00098
Tokyo	9.798	0.992745	2.00088

Table 1.1. Measured values for the acceleration due to gravity (second column) at selected cities (first column). Column three indicates the length of the Seconds Pendulum at the indicated location, while column four indicates the period of swing for a fixed standard length seconds pendulum.

Gravimetry and Kater's Pendulum

The study and measurement of the acceleration due to gravity falls under the label of gravimetry. The primary goal of such measurements is usually to forge a link between any observed gravitational variations and the geological formations responsible for producing the observed anomaly. Indeed, equation (1.4) above tells us that if a pendulum of some fixed length L is set in motion, then its period of swing is determined by the inverse square root of the local value for the acceleration due to gravity. If, for example, the acceleration due to gravity at two different locations are g_1 and g_2, then it follows from (1.4) that $g_2 = g_1 (P_1 / P_2)^2$, where P_1 and P_2 are the measured periods for the pendulum at the two locations. In this manner, a 0.1 percent increase in the acceleration due to gravity at one location compared to another will be revealed as a 0.05 percent decrease in the oscillation period of the pendulum. Field geologists have historically used fixed-length pendulum gravimeters to map out changes in underground rock structure since, for example, a region composed of low density rock will produce a smaller local gravitational acceleration than a region composed of a higher density rock. Many of the Gulf oil fields in the United States of America were first located, in the 1930s, through the use of pendulum gravimeters – rock overlaying an oil-filled cavity having a lower local gravitational acceleration than a monolithic layer of rock.

If a precise measurement of the actual acceleration due to gravity, at a specific location, is required then a simple fixed-length pendulum, it turns out, is not the best experimental instrument to use. This is due to the fact that it is actually rather difficult to measure the length of a pendulum accurately. The problem is that the pendulum length L corresponds to the distance from the support point to the center of mass of the pendulum bob, and if the bob is not perfectly symmetric and/or perfectly homogeneous in composition, then the location of its center of mass is not readily known. A way around this length-measuring problem was first developed in the early nineteenth century by Captain Henry Kater (1777 – 1835).

Henry Kater is one of those highly interesting but sadly under-appreciated figures in the history of science. His early career was spent in the army, and he assisted George Everest in the Great Trigonometric Survey of India[12] (discussed later in Chapter 3) during the early 1800s. His health failing him, however, Kater returned to England in 1808, and thereafter followed a distinguished career at Sandhurst Military College. In 1814 he retired from the army on half-pay to pursue his scientific interests on a full-time basis. He published a number of important memoirs on the standards for length and mass and was awarded the Royal Society of London's highest honor, the Copley Medal, in 1817 for his work relating to the pendulum that bears his name. He was further awarded the Gold Medal of the Royal Astronomical Society, in 1831, for his work relating to the development of a vertical floating collimator that could be used to determine an accurate zenith point; his method, which used a trough of liquid mercury, was generally deemed to be more accurate than the employment of spirit-levels and plumb-lines.

The Kater pendulum is composed of a light bar to which are attached two masses, with one mass being larger than the other. The point of the two masses is that the pendulum is reversible and can be swung from either of two knife-edge pivots attached to the pendulum bar (see figure 1.3). By adjusting the positions of the masses so that the pendulum has the same period of oscillation about each of the pivot points, it can be shown that the period of oscillation is the same as that for a simple pendulum with a length equivalent to the distance between the two pivot points[13]. The advantage over the simple pendulum is therefore revealed since the distance between two well defined knife-edges can be measured to a high degree of accuracy. Making no attempt to derive the equation here (see, however, note 13), it turns out that the gravitational acceleration can be related to the observed periods P_1 and P_2 and the measure of the distances h_1 and h_2 (as illustrated in figure 1.3). The relationship of interest is

$$\frac{4\pi^2}{g} = \frac{P_1^2 + P_2^2}{2(h_1 + h_2)} - \frac{P_1^2 - P_2^2}{2(h_1 - h_2)} \tag{1.5}$$

where h_1 and h_2 are the distances of the two pivot points (knife edges) from the pendulum's center of mass – determined when the bar is balanced horizontally on a triangular fulcrum (see Chapter 4 later). The second term in the equation will be small provided P_1 and P_2 are approximately equal – it will be zero if the periods are exactly equal.

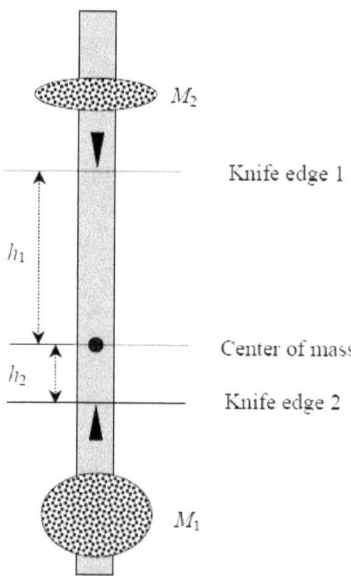

Figure 1.3. Kater's pendulum arrangement. By adjusting the pendulum to have the same period when supported alternately from each of the knife edges, so the period of oscillation will be equivalent to that of a simple pendulum of length $L = h_1 + h_2$.

Kater performed his initial pendulum experiments during the first few months of 1817 at a house located in Portland Place, London. The pendulum was carefully constructed and tested over several months, but eventually a measure of 39.13842 ± 0.0003 inches was derived for the length of the pendulum vibrating in seconds (that is, with a period of 2 seconds). The length of the pendulum was expressed in terms of the British House of Commons standard yard, constructed by John Bird[14] (1709 – 1776) in 1758, when the temperature of the room was determined to be 62 degrees Fahrenheit, a correction also being made for the fact that the house in which the experiment was being conducted was estimated to be 83-feet above sea-level. Converting the inch measure to meters, Kater's experiment gives $L = 0.9941$ m for the two-second period (that is, 1 second vibrating) pendulum, and this in turn provides a value of $g = 9.8114$ m/s^2 for the acceleration due to gravity at Portland Place, London.

Measuring Height by Time

If the mountain won't come to the pendulum, then the pendulum must go to the mountain and having thus been transported the pendulum might just as well measure the mountain's height. While not the best instrument for making precision measurements of the local acceleration due to gravity, a simple pendulum is well suited to measuring a relative change in gravity, and in this manner it can be used to measure a change in altitude.

For all the modern-day fame attributed to Isaac Newton's formula for universal gravitational attraction, he never actually wrote the equation down as we know it – he used words rather than a formulaic description. Indeed, Newton writes in Proposition VII, Theorem VII in Book III of his *Principia Mathematica* (published in 1687), "That there is a power of gravity pertaining to all bodies, proportional to the several quantities of matter which they contain". To this he adds a corollary that "The force of gravity towards the several equal particles of any body is inversely as the square of the distance of places from the particles". This latter corollary is Newton's famous inverse square law statement[15]. What this relationship tells us is that at Earth's surface the acceleration due to gravity is described by a relationship of the form $g_{surface} = K / R^2$, where R is the Earth's radius and K is a constant (related as will be seen later to the mass of the Earth). If we now ask, what is the acceleration due to gravity at a height h above Earth's surface we can use Newton's corollary to write down the expression[16] $g(h) = K / (R + h)^2 \approx g_{surface} (1 - 2h / R)$. To determine the height of a mountain, therefore, we need to record the period of oscillation for the pendulum of any convenient length L, at the bottom and then at the top of the mountain. Usefully for this experiment we don't need to measure the length of the pendulum at all. If the period of oscillation at the base of the mountain is $P_{surface}$, and the period at the top is P_{top}, then equation (1.4) and the approximation given above tell us that $P_{surface}$,

/ P_{top} = 1 - h / R, where remember, h is the height of the mountain and R is the radius of the Earth. Since the acceleration due to gravity $g(h)$ decreases as the height h increase, so the period of the pendulum at the top of a mountain will always be longer than the period at its base, and accordingly $P_{surface}$ / P_{top} will always be less than unity.

The Devil is in the Details

It was just argued that the acceleration due to gravity at a height h above Earth's surface can be written as $g(h) \approx g_{surface} (1 - 2h / R)$. Expanding this expression reveals that for every one meter increase in height above the Earth's surface, the acceleration due to gravity decreases by an amount equal to $\delta_F = 2 g_{surface} / R$, where recall R is the Earth's radius. Taking the Earth's mean radius to be 6371 km and $g_{surface}$ = 9.80665 m/s², then $\delta_F \approx 3.1 \times 10^{-6}$ m/s² per meter. The δ_F correction term for height is generally known as the 'free-air correction', and it is used to reduce a specific gravity measurement (made at some height h) to a sea-level value where h, by definition, is zero. The free air correction for Chomolungma (Mount Everest – see Chapter 3), with its summit located 8.85 km above sea level, is 0.02731 m/s². The free air correction is just one of the many corrections made to actual gravity measurements in order to reduce them to a standard reference system, but we shall delay the discussion of the main correction terms until a little later. It is worth mentioning at this stage, however, one addition correction term that has been ignored so far is the mass of the mountain itself. The free air correction is based purely on the reduction of gravity with height above Earth's surface, and the mass of the mountain, while small compared to the mass of the Earth is nonetheless non-negligible. Indeed, for precise gravitational acceleration determinations not only should the mass of the mountain be corrected for, but so too should the mass of the atmosphere above the measuring station be included in the correction.

With reference to figure 1.4 let us consider the highly simplified (if not quixotic) example of a 1-km radius spherical mountain. A pendulum located at the top of the mountain will require therefore a free air correction of δ_F = 6.2 x 10⁻³ m/s² since the mountaintop is h = 2000 m above Earth's surface. Now, the gravitational acceleration that results from just the mountain[17] (not the Earth) is g_M = 7 x10⁻⁴ m/s², which is about 10% of the free air correction term δ_F. So, while amounting to a small correction, the mass of the mountain does have an effect on the pendulum – we shall have more to say about the gravitational effects of mountains in Chapters 3 and 4, since they can be used to determine the density of the Earth.

The mountain (or mountain range) mass correction term is usually referred to as the Bouguer correction, in honor of the French mathematician and explorer Pierre Bouguer (1698 – 1758) who first considered the problem in detail when trying to correct the pendulum experiments he had made in the Andean mountains, in what is now Ecuador, during the later half of the 1730s. Bouguer actually assumed that the terrain correction could be calculated according to the gravitational influence of an infinitely long, homogeneous slab with a thickness equal to the mountain's height[18]. This approximation is not as bad as it may at first seem since the gravitational force decreases as the inverse distance squared and the influence of the more distant parts of the slab are soon vanishingly small. Why Bouguer was concerned with the affects of the Andean mountains upon his pendulum will be explained in Chapter 3, but for the moment it will suffice to say that he was attempting to determine the shape of the Earth's profile.

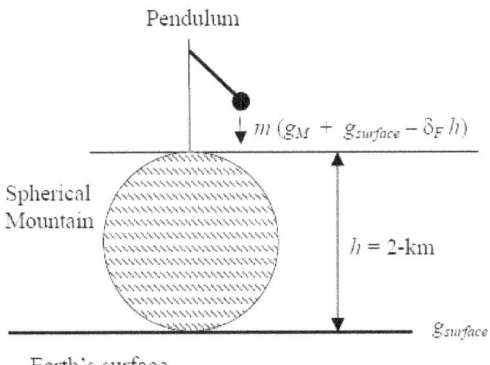

Figure 1.4: A schematic diagram showing the effect of a 'spherical' mountain of height h on a pendulum experiment. While the free air correction amounts to a small reduction in the gravitational attraction due to the Earth, the mass of the mountain results in a small increase in the gravitational acceleration.

More Details and Complications

It is often stated that the period of a pendulum is independent of its starting angle (φ_0 in figure 1.1), and indeed, equations (1.3) and (1.4) appear to vindicate this point. It is, however, an incorrect statement in general and it only follows from the assumption that the angle through which the pendulum swings is small. This was a useful assumption to make (at least initially) since it allowed us to use the small angle dodge which dictates that the expression $\sin(\varphi) / \varphi$ (with, remember, φ being measured in radians – see figure 1.2) becomes closer and closer to unity as φ becomes smaller and smaller. This approximation then allowed us to find the solution for $\varphi(t)$, describing the time variation in the pendulum's angle of swing. If we no longer make this small angle approximation then the solution to the equation of motion is mathematically much more complex, but provided a solution can be found then it should in principle be a much more accurate description of the pendulum's motion.

A complete mathematical derivation for the general solution to the equation of motion will not be given here, but the method proceeds by multiplying both sides of the differential equation of motion by $2(d\varphi / dt)$ and integrating once. In this way the left hand side becomes $(d\varphi / dt)^2$ while the right hand side integrates to $(2 g / L)[\cos(\varphi) - \cos(\varphi_0)]$ – this last term now contains an expression relating to the starting angle φ_0. Ultimately, the equation can be re-arranged so that the pendulum's period is given in terms of the integral

$$P = 4\left(\sqrt{\frac{L}{2g}}\right) \int_0^{\varphi_0} \frac{d\varphi}{\sqrt{\cos\varphi - \cos\varphi_0}} \qquad (1.6)$$

Equation (1.6) is certainly a bit of a beast, and in fact it has no closed analytical solution and it is a so-called elliptic integral of the first kind. Mathematicians have studied such integrals in detail for many years and over this time numerous approximate solutions have been discovered. It turns out, in fact, that the integral in equation (1.6) can be solved for in terms of an infinite series, and that our sought after general solution can be written down as

$$P = 2\pi\sqrt{\frac{L}{g}}\left\{1 \; + \; \left(\frac{1}{2}\right)^2 \sin^2\left(\frac{\varphi_0}{2}\right) \; + \; \left(\frac{1.3}{2.4}\right)^2 \sin^4\left(\frac{\varphi_0}{2}\right) \; + \cdots \right\} \quad (1.7)$$

Only the first three terms of the infinite series approximation have been included in the series (1.7), which at first glance seems rather presumptuous, but it turns out that the successive terms in the equation are getting smaller and smaller very rapidly, and their contribution to the final result is soon entirely negligible.

φ_0 (degrees)	φ_0 (radians)	K_C	ΔT (seconds / day)
0.5	0.0087	4.73 x 10⁻⁶	0.409
1.0	0.0175	1.91 x 10⁻⁵	1.650
2.0	0.0349	7.61 x 10⁻⁵	6.575
4.0	0.0698	3.05 x 10⁻⁴	26.352
8.0	0.1396	1.22 x 19⁻³	105.408
16.0	0.2793	4.88 x 10⁻³	421.632
32.0	0.5585	1.95 x 10⁻²	1684.80

Table 1.2. Circular error terms for various angles of a pendulum's swing. The last column indicates by how much the circular correction term amounts to per day of operation.

Since the first term in (1.7) is unity, we recover equation (1.4) when φ_0 is small, and this is exactly as it should be, since that was the approximation used in the derivation of equation (1.4). The first correction term from (1.7) is $K_C = \frac{1}{4} \sin^2(\varphi_0 / 2)$, which for small angles can be written $K_C \approx \varphi_0^2 / 16$, where φ_0 is expressed in radians. The K_C correction term is usually referred to as the circular error, and table 1.2 provides a few example calculations for the amount by which the initial release angle alters a pendulum's period of oscillation. Clearly the circular error has only a small effect upon each swing of the pendulum, but the correction is cumulative, and the amount of time that a pendulum gains per day of running amounts to an additional 28 minutes when φ_0 is 32 degrees. By gain it is meant that if the pendulum was assumed to be a Seconds Pendulum regulating, say, a clock mechanism then the clock would run 28 minutes fast per day if the pendulum swing was 32 degrees away from the vertical. Even a six-degree angular swing of the pendulum will result in the clock running one minute faster every day.

Conservation of Momentum and Newton's Cradle

In deriving equation (1.4), the mathematical relationship between the period of swing, the gravitational acceleration and length of a simple pendulum, we made use of Newton's second law of motion. This law is the second of three laws that Newton, building upon ideas first expressed by Galileo, introduced in his *Principia Mathematica* in 1687. Written here in their modern day form, Newton's three fundamental statements relating to dynamics are:

N1: A body in a state of rest or uniform motion remains in that state unless acted upon by some external net force.

N2: An applied force F acting upon an object of mass m produces an acceleration $a = F / m$.

N3: To every action there is an equal and opposite, collinear reaction.

We will see all three of these laws applied, at various times, throughout the remainder of this text, but for the present we shall concentrate on **N2,** within which we find another important quantity of motion – the momentum $p = mV$, where m and V are the mass and velocity of a specific object. Recalling that acceleration is simply the rate of change of the velocity over time $[a = d V / dt\,]$, so, for an object of constant mass, **N2** can be re-written in terms of the rate of change in momentum - the net force F acting upon a particle is equal to the rate of change of its momentum with time: $F = dp / dt$. Embedded within this expression for **N2** is one of the most cherished conservation laws of all physics. What we have is the statement that when the net force acting upon a body is zero, that is $F = 0$, so the momentum of the body must be constant: $p = mV = $ constant. Where this particular conservation law becomes important is in understanding the behavior of so-called closed systems. In such systems, objects can move around, they can collide and bounce off one another and change their speeds; they can even change their mass as the result of a collision, but the key point is that the total momentum of all the objects within the system must remain exactly the same at all times. In its simplest form the conservation law says, for example, that if there are two masses m_1 and m_2 that have initial velocities V_1 and V_2 before they collide and velocities U_1 and U_2 after, then the total momentum before the collision must be equal to the total momentum after the collision, and accordingly: $m_1V_1 + m_2V_2 = m_1U_1 + m_2U_2$. One of the most elegant practical demonstrations of the conservation of momentum law is that provided by Newton's Cradle, which at its heart is a set of interacting, slightly modified simple pendulums.

Newton's cradle was not invented by Isaac Newton; indeed it was invented in 1967 by Simon Prebble, and while today it is most often sold as an 'executive toy', its original purpose was to demonstrate the conservation of momentum. Once set in motion, Newton's cradle is hypnotic, and the combinations of possible motion are eye catching. The cradle is typically composed of five metal spheres, each suspended by two wires in a U-shaped frame. Two suspension wires are used so that when set in motion the spheres can only move in one direction – the wires are also of equal length so that when all the spheres are at rest, adjacent pairs just touch each other at a point on their equator – the contact points, in fact, all fall on the straight line that passes through the center of each sphere when they are all at rest. Each sphere has the same mass m, and various combinations of spheres (one, two, and three) are taken to one side and released so as to set the system in motion. The question at this stage, is, what are the before and after collision motions of the various spheres? To illustrate what happens when the pendulum is set in motion we shall use the capital O symbol to represent a sphere. When all five spheres are at rest then the state of the pendulum is written as OOOOO. If the left most sphere is taken aside and released we can describe this as O⇒OOOO, with the ⇒ indicating the direction of motion after the left most sphere is released from its offset position. The four possible starting configurations for a five sphere Newton's cradle are described in the first column of the list shown below (labeled *Start*), and the response of the remaining (stationary) spheres immediately after the collision is shown to the left (labeled *Response*).

Start	*Response*
O⇒OOOO	OOOO⇒O
OO⇒OOO	OOO⇒OO
OOO⇒OO	OO⇒OOO
OOOO⇒O	O⇒OOOO

In the first row, for example, it is seen that when the leftmost sphere hits the chain of four stationary spheres, the response is that it comes to rest immediately after the collision and the rightmost sphere is set in motion. Likewise, when the two leftmost spheres are drawn aside and simultaneously released, they come to a stop after the impact and the two rightmost spheres are set in simultaneous motion. The motion becomes particularly intriguing when the three leftmost spheres are drawn aside and then simultaneously released. In this case the response is that the two leftmost of the starting spheres come to rest, but the innermost, impacting, sphere carries on moving, but now in union with the two, rightmost, initially stationary spheres. Clearly, the spheres are moving in a very specific, symmetrical fashion, and the number of spheres in motion after the collision is the same as the number of spheres in motion before the collision. Since the spheres all have the same mass, we can reasonably conclude that the mass in motion at any one instant is the same, and therefore, the conservation of momentum rule immediately tells us that the speed with which the spheres start moving immediately after the collision must be exactly the same as the speed of the starting spheres at the time of impact. And, indeed, this is exactly what is observed, the angle of swing of the sphere(s) set in motion after impact is exactly the same as the angle through which the starting sphere(s) moved prior to the impact.

Conservation of Energy and the Ballistic Pendulum

Not only does Newton's cradle offer an effective demonstration of the conservation of momentum, it also illustrates (up to a point) another highly important conservation law – the law of the conservation of energy. Looking at the first row of our *Start* and *Response* table (above), the energy that the leftmost sphere has at the moment of impact with the initially stationary spheres is that of its kinetic energy K – literally, the energy of its motion. The kinetic energy is expressed in terms of the mass and velocity, just as in the case of momentum, but now the dependency is upon the velocity squared, and accordingly the kinetic energy at the moment of impact is $K_{imp} = \frac{1}{2} m V^2$. Once again, the conservation of energy law requires that the total amount of energy (in all of its many possible forms) within a closed system must remain constant at all times. Hence, the energy imparted to the rightmost sphere following the impact of the left most sphere must be $\frac{1}{2} m V^2$, and since it has the same mass m as the starting sphere, so its initial velocity must be the same as that of leftmost sphere at the moment of impact.

The conservation of energy law is a very powerful principle, but it is also a very general statement that can sometimes make it difficult to apply. In the case of Newton's cradle, the motion and system behavior is explained simply by considering the conservation of kinetic energy imparted by the drawn apart, leftmost spheres the instant before they collide with the stationary spheres. This same kinetic energy is then imparted to the responding spheres at the instant that they start moving. In reality, however, things are a little more complicated. There is, after all no such thing as a perfect demonstration experiment – and/or free lunch. For instance, at the moment of collision between spheres, the cradle will produce a solid clack sound. This sound wave, of course, carries away a small amount of energy, so the kinetic energy after the collision is slightly less than that before the collision. In addition, at the moment of impact there will be a slight mechanical deformation of the surfaces of the spheres in contact, and this too will rob away a little more of the initial kinetic energy. Furthermore the kinetic energy of the impacting sphere is passed through the initially stationary spheres via a shock wave, and as this shock propagates through each sphere it will heat their interiors by some small amount. Throughout this whole process the total amount of energy is constant, as the conservation law says it must always be, but it is gradually converted into additional less easily tracked forms of energy (sound, mechanical deformation, and heat); slowly, therefore, the initial kinetic energy is robed away from the pendulum and the cradle comes to a stop, the initial energy having all being converted into sound waves and heat.

In spite of the inherent difficulty of applying the conservation of energy at every level and at every instant within a closed interacting system, there are times when just a few very specific forms of energy are dominant and fully describe the system behavior to a high order of accuracy. The simple pendulum, it need hardly be said, is one such system, and in this case the kinetic energy of the bob is cyclically changed into gravitational potential energy, and then back into kinetic energy, and so on. Provided there is no air resistance, and no energy is lost due to frictional heating at the string's support point, then the pendulum will swing in perpetuity – such a state, of course, can never be fully realized since this would contravene another important law of physics – the second law of thermodynamics, which forbids the existence, in spite of many historical claims to the counter, of perpetual motion machines.

The gravitational potential energy U is defined as the energy that an object, of mass m, will have because of its location in a specific (constant) gravitational field. For example, if an object is raised a height h above the Earth's surface, where the acceleration due to gravity is g, then its potential energy will be the product $U = m\,g\,h$. The motion of a pendulum can, in fact, be entirely described according to the conservation of energy. Initially, the stationary bob is drawn to one side of its rest position, and since it is constrained by its support wire, it must move upwards through some vertical height h. The initial energy of the pendulum, therefore, is entirely that of its gravitational potential energy; $U_{initial} = m\,g\,h$, where m is the mass of the bob – this energy, of course, must be put into the pendulum by the experimenters hand; else the pendulum would be a plumb-line at rest. Once released from rest, however, the bob starts moving thereby acquiring a measure of kinetic energy. At the base of its circular arc, the bob is once again at its starting point, $h = 0$, and accordingly, at this mid-point all of the initial gravitational potential energy has been converted into kinetic energy. At this mid-point the bob will be traveling with its maximum velocity V_{max}, and from the conservation of energy, we have $K_{max} = U_{initial}$. Accordingly we derive the result $V_{max} = \sqrt{2gh}$. The process now begins to reverse. At the mid-point the energy of the pendulum is all kinetic, but as the bob moves past the center point it begins to gain gravitational potential energy, and accordingly its velocity begins to fall; the bob coming to rest, with zero velocity, once it has swung through a vertical height h, and all of the energy is once again in the form of gravitational potential energy.

The interchange of kinetic and potential energy that underscores a pendulum's motion makes it a useful tool for measuring velocities. Indeed, the pendulum is an experimental device that can determine the speed of a projectile through the measurement of a height displacement. The first description of such a, so-called, ballistic pendulum appears to be that presented by military engineer Benjamin Robins in his text, *New Principles in Gunnery*, published in 1742. The ballistic pendulum is a wonderfully simple device that, when set in motion, literally puts the swing into a plumb-line. The idea of the ballistic pendulum is to exploit the laws for the conservation of momentum and energy, relating the momentum of the projectile before it hits the bob to the maximum height through which the pendulum moves after being struck – which, as we saw, above, is related to the gravitational potential energy. The renowned, if not eccentric, experimental physicist Charles Vernon Boys (1855 – 1944) used an especially modified ballistic pendulum with a rebound stopping cloth bag to measure the elastic properties of golf balls; a lightweight scribe running over a plate of smoked glass was used to determine the bob's recoil.

The theory behind the ballistic pendulum begins by applying the conservation of momentum before and after the projectile hits the bob. The initial momentum is $P_i = m V_i$, where m and V_i are the mass and velocity (to be found) of the projectile. After impact, with the projectile now embedded within the bob (of mass M), the momentum is $P_f = (m + M) V$, where V is the velocity of the bob, now set into motion. Conservation of momentum requires that $P_i = P_f$ and accordingly, $V_i = (m + M) V / m$. As we saw earlier, however, the velocity V can be related, via the conservation of energy law, to the vertical height through which the pendulum moves, with $V \equiv V_{max} = \sqrt{2gh}$. By experimentally measuring the height h through which the ballistic pendulum moves (and there are various ways in which this can be done), the velocity of the projectile can be determined: $V_i = ((m + M)/m)\sqrt{2gh}$. The highly useful characteristic of the ballistic pendulum is that the velocity of a rapidly moving projectile (e.g., a bullet) can be determined without having to make an accurate time measurement, and this, at least historically, was a major experimental advantage.

While the ballistic pendulum is a rather specialized device, the Charpy pendulum, which is similar in its operation, is used to determine material properties. Invented by French engineer Georges Charpy in 1905, this pendulum is used to measure material strength and failure characteristics. Working like a pendulum axe, the bob is allowed to fall through a preset angle and strike a test sample at the lowest point in its trajectory. A record is then kept of the angle through which the pendulum swings after slicing through the sample. As with the ballistic pendulum, the height of the bob (as determined by the angle of swing) can be used to evaluate the before and after impact energies - any difference between the two is then interpreted as a measure of how much energy is required to fragment the test sample. The American Society for Testing Materials (ASTM) maintains an internationally recognized database of Charpy pendulum impact tests for many different types of materials, providing data on yield strength (at which point the material begins to deform in a non-reversible fashion), material strain and stress response, crack propagation and the mode of fracture according to temperature – the latter being measured according to a brittle moving to ductile scale; brittle fractures produce clean and smooth surfaces, while ductile fractures produce rough and jagged surface breaks. With the Charpy test we find the pendulum being put to a new and fundamental use, indeed, it is being employed to set international standards for material behavior and properties – information, of course, that is vital to mechanical engineers and architects.

The Tautochrone

So far we have only considered pendulum motion constrained to move along circular arcs – the arc swept out by the constant length string supporting the bob. Under this constraint we have seen that the period of oscillation is only independent of the starting angle under the assumption of small displacements. In general, however, the period of a pendulum is a complex function of its starting angle – as shown by equation (1.7). If we relax the circular arc constraint, however, then the question arises, is there a true isochronal pendulum – a pendulum that has the same period of swing no matter what its initial offset angle is? The answer to this question is, in fact, yes, but to produce such a pendulum a special and very particular mechanical adjustment to the simple pendulum described in figure 1.1 must be made. First of all, take another look at table 1.2. This table indicates that the greater the initial offset angle φ_0, so the slower the pendulum runs relative to the period predicted by equation (1.4) for a given length L of the support wire. To compensate for this offset increase one might try to shorten the pendulum wire – that is, by adopting a smaller value of L, as equation (1.4) indicates, the period of oscillation is reduced. The problem, of course, is that the length of the wire must be adjusted continuously so as to correct for each successive angle through which the pendulum swings. This, in fact, is where the idea of the tautochrone comes into play.

The word tautochrone is derived from the Greek *tauto*, meaning *the same* and *chronos*, meaning *time*. The tautochrone, therefore, is the path that a bob must follow in order to produce a truly isochronal pendulum no matter what the initial offset angle happens to be – mathematically we can say that the period of the tautochrone is independent of the starting boundary conditions[19]. One of the key historical figures to investigate the tautochrone was Dutch mathematician Christian Huygens (1629 - 1695). Indeed, it was Huygens who first showed in his classic book *Horologium*, published in 1658, that the period of the simple pendulum is proportional to the square root of the length of the support wire. Huygens also noticed that the period of a finite amplitude pendulum was dependent upon the initial offset angle, and it was this issue, along with his interest in mechanical clocks that brought him to investigate the tautochrone, or isochronal pendulum. Well, to cut a very long mathematical story short what Huygens (in parallel with several other prominent mathematicians, including Isaac Newton) found is that the required shape for an isochronal pendulum arc is a cycloid.

The cycloid had been studied for many years prior to the discovery of its isochronal properties. Galileo studied this curve, for example, and reasoned that it is the strongest arch profile for a bridge. Interestingly, while Galileo actually coined the name cycloid in 1599, he failed to deduce its latent isochronal property. Rather, he mistakenly reasoned in his 1638 *Dialogues Concerning Two New Sciences* that the circular arc satisfied the isochronal condition.

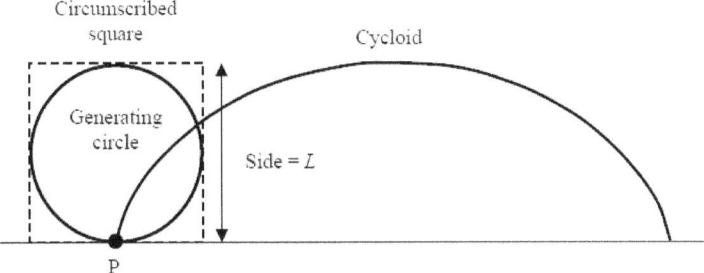

Figure 1.5. The cycloid, its generating circle and circumscribed square. The cycloid is traced out by point P as the generating circle roles without slipping to the right. The length of the cycloid arc is 4L, where L = *2a* is the length of the side of the circumscribed square to the generating circle.

Medieval philosopher, theologian and mathematician Nicholas of Cusa (1401 – 1464) is usually credited as being the first person to study the cycloid in detail. Specifically, Cusa was interested in finding the area under the curve – a problem, in fact, that was only solved some 170 years after his death by the French mathematician Gilles de Roberval in 1634 – the answer is that the area is three times that of the generating circle (figure 1.5). Renowned British architect, mathematician, and Oxford University Savilian Professor of Astronomy Sir Christopher Wren (1632 – 1723) demonstrated in 1658, the same year that Huygens published his *Horologium* that the length of one complete arc of a cycloid is equal to the perimeter of the square circumscribed about the cycloid generating circle (which has a radius a).

The tautochrone is actually an inverted cycloid in the sense that the oscillation period of a frictionless bead constrained so as to move along the curve of the cycloid will be the same no matter what the initial position of the bead happens to be. In addition, it turns out that the cycloidal arc is also the solution to the so-called brachistochrone problem (to be described shortly), which asks, what is the curve of least travel time between any two points? While the cycloid has many interesting mathematical properties, the main reason for Huygens interest in it was the potential to solve a very practical problem: Huygens was literally trying to control time and specifically the going of pendulum regulated clocks - as we shall see in the next chapter.

The tautochrone problem can be solved in a number of different ways – some more mathematically elegant than others. Here, let us take a brute force approach and assume the solution and then work backward, as it were, to show that the answer fits the question. The Cartesian (x, y) equations for a cycloid are given in parametric form through the expressions $x = a(\theta + \sin\theta)$ and $y = a(1 - \cos\theta)$, where $0 \le \theta \le 2\pi$, and where a is a constant. For a small bead of mass m moving along the curve, the distance s traveled, from its starting point s_0, can be written as the integral of the line element $s = \int_s^{s_0} ds = \int_y^{y_0} \sqrt{1 + (dx/dy)^2}\ dy$. Using the parametric equations for the x and y coordinates of a cycloid, the arc length s can be expressed in terms of y as $s = 2\sqrt{2ay}$. At this stage, we apply the conservation of energy to the moving bead. Accordingly, therefore, the constancy of the sum of the kinetic and potential energy gives us the identity $\frac{1}{2}mV^2 + mgy = mgy_0$, where V is the velocity of the bead at height y, and g is the acceleration due to gravity. Substituting now for y and y_0, the velocity $V(s)$ after traveling a distance s along the bead is $V^2 = (s_0^2 - s^2)(g/4a)$. If we now differentiate this expression for the velocity, where $V = ds/dt$, with respect to time, we recover the following result that, $2(ds/dt)(d^2s/dt^2) = -(g/4a)(ds/dt)2s$. Canceling common terms from each side of this expression, we recover, as seen before, the equation for simple pendulum motion $d^2s/dt^2 + (g/4a)s = 0$ - in contrast to our earlier derivation this is an exact formula, rather than an approximate one that only applies in the small angle limit. The equation for the tautochrone has the general solution $s = s_0 \cos(t\sqrt{g/4a})$, which satisfies, as it must, the boundary conditions for the system at the starting point: the displacement $s = s_0$ and the velocity $ds/dt = 0$ at the start time of $t = 0$. The equation for $s(t)$ tells us that the time to travel from any arbitrary starting point s_0 to the origin $s = 0$, is constant – that is, no matter where the bead starts from on the cycloidal hoop the travel time to the origin (the center, lower most point of the hoop) is the same – and this is exactly what is wanted of the tautochrone. The total bead travel time T, taking it from s_0 to $s = 0$, is determined by the condition that $s(t) = 0$, and this requires that the argument of the cosine term $t\sqrt{g/4a} = \pi/2$, which reveals that $T = \pi\sqrt{a/g}$. The period of oscillation is further given by the expression $P = 4T = 4\pi\sqrt{a/g}$. From this relationship, we see that the period of the tautochrone is equivalent to that of a simple pendulum of length $L = 4a$ undergoing small oscillations.

Both Isochronal and Fastest

The brachistochrone problem was first formulated by mathematician Johann Bernoulli. The word brachistochrone is derived from the Greek *Brachistos*, meaning *shortest* and *chronos*, meaning *time*, and it therefore has a similarity to the tautochrone (the isochronal curve problem) just described. Writing in the June 1696 issue of the journal *Acta Eruditorum* Bernoulli simply asked, what is the curve of fastest descent under gravity between any two arbitrary points P and Q? There is an immediate, all be it somewhat trivial, answer to this question in that if Q is located vertically above P then the curve of descent must be the straight line joining Q to P. In general, however, Q will not be located vertically above P and the brachistochrone curve won't be a straight line.

The equation of motion for a small bead moving without friction along a fastest descent curve is conveniently derived by considering the conservation of energy. In this way the initial and final values of the kinetic plus potential energy must be the same. Given that the bead starts from the location (x_0, y_0) and that it is initially stationary, then its speed V at any given point of height $y < y_0$, will be, $\frac{1}{2}mV^2 + U(y) = U(y_0)$ where $U(y)$ is the gravitational potential at height y. The problem to be solved then is, what curve minimizes the travel time T where

$$T = \int_P^Q dt = \int_P^Q ds/V = -\sqrt{\frac{2}{m}}\int_Q^P \sqrt{U(y_0)-U(y)}\frac{ds}{dy}dy$$

where $ds = \sqrt{dx^2 + dy^2} = \sqrt{1+(dx/dy)^2}\,dy$ is a small increment measured along the curve (to be found) connecting Q to P. In a constant gravitational field we have $U(y) = mgy$, where g is the constant acceleration due to gravity, it turns out (without derivation here) that the curve providing the minimum descent time is that of the cycloid. In a constant gravitational field, therefore, the tautochrone and brachistochrone are satisfied by the same curve (the cycloid), and the isochronal oscillation period $P = 2T = 2\pi\sqrt{y_0/g}$, the period of a simple pendulum of length y_0, is also the shortest possible oscillation period.

Within a year of Bernoulli proposing the brachistochrone problem, solutions showing that the required curve must be a cycloid were found by Isaac

Newton, mathematicians Jakob Bernoulli and Guillaume de l'Hopital, and by Newton's arch enemy Gottfried Leibniz[7]. That the solution curves for the brachistochrone and tautochrone are identical is only satisfied under a restrictive set of circumstances. In general, when the force potential varies according to the rule $U(r) = k\, r^n$, where r is the radial displacement, k is a constant and n is any positive or negative number, the solution curves for the tautochrone and the brachistochrone must be found via numerical computation and the resultant loci are not identical. The constant gravity condition, as just described, is one situation where the solution curves are the same and both cycloids, and (rather remarkably) the only other circumstance where they are the same is that under the radius squared potential law: $U(r) = k\, r^2$, where k is a constant. The latter potential corresponds to that experienced within a constant density sphere and the period is identical to that of the linear harmonic oscillator – to be described in Chapter 3. Furthermore, under this specific potential law, the tautochrone and brachistochrone curves are both straight, radial lines.

A Pendulum that Can Never Be: I

The Guinness Book of Records informs us that the world's longest pendulum is housed in the Oregon Convention Center, USA. The pendulum consists of a 40- kg spherical bronze bob suspended on a 21.34-m wire with a 4.57-m diameter swing. The period of the pendulum is a sedate 9.25 seconds. As equation (1.4) would predict, the period of a pendulum will increase as the length of the support wire increases. What, we might therefore ask, is the period of an infinite length pendulum when set in motion close to the Earth's surface? Clearly, there is no such entity as the infinite length pendulum, but our question is nonetheless a physically sensible one. How can this be? If we blindly use equation (1.4) then since the period of swing depends upon the length of the support wire (still assumed to be of negligible weight), so an infinitely long wire will produce a pendulum with an infinitely long period – in other words the bob won't move at all. From this deduction we might imagine that we have shown that an infinitely long pendulum is the same as an infinitely long plumb-line. This deduction, however, would be wrong. The reason for this is that after a while the length of the pendulum becomes irrelevant, and what determines the motion of the pendulum is the proximity to its bob to the Earth. Figure 1.6 illustrates the circumstances, and we see now that when we resolve the forces, Fx will always be equal to the tension in the wire, but $Fy = mg \sin(\varphi)$, where φ is now the angle subtended at the center of the Earth, and where we have assumed that since the bob is close to the Earth's surface the gravitational force acting upon it (g) is constant. We can also see from figure 1.6 that $\sin(\varphi) = y\, /\, R$, where R is Earth's radius. The equation of motion for the bob can now be written down as: $m\, d^2y\, /\, dt^2 = Fy = mg\, (y\, /\, R)$, and this has a solution similar to that for the simple pendulum, but now the period of oscillation is given by the expression $P = 2\pi\sqrt{R/g}$. So, provided the constant gravity condition holds, with $g = 9.81$ m/s², the greatest period of oscillation for any pendulum passing close to Earth's surface is 5063.5 seconds (or 1.4 hours). We shall encounter this special period of oscillation again in Chapter 4, and it corresponds to the minimum possible orbital period for an Earth orbiting satellite – assuming that Earth has no atmosphere to slow the motion of the satellite.

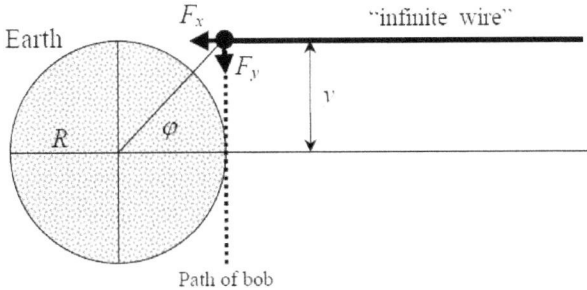

Figure 1.6. The infinite length pendulum. In this *thought experiment* the support wire is assumed to be of negligible mass, and since it is infinitely long its radius of curvature, or the path along which the bob will move, is a straight line.

Circular Orbits

The simplest possible orbit that an object of mass m can have about another, assumed more massive object M is that of a circle of radius R. Isaac Newton realized this fact, and was able to work out the conditions required of the centripetal, that is centrally acting, force required to maintain such an orbit. The force, Newton realized, had to be centrally acting since if the orbit was to be closed (that is a circle, or more generally an ellipse) so the instantaneous motion of the orbiting mass must be continuously 'pulled' away from a straight line path. This latter condition is actually an application of Newton's first and second laws of motion – **N1** and **N2** as described earlier. The fact that our orbiting mass m is moving along a circular path immediately tells us that some force must be continually acting upon it so as to cause a continuous change in its path. The continuously acting force, Newton argued, was that due to the gravitational interaction between the two objects of mass m and M. We shall have much more to say about Newton's gravitational formula in Chapter 4, and for the moment we simple state it as being $F_{grav} = GMm/R^2$, where G (often called "big-G") is the universal gravitational constant. Newton was further able to show in his *Principia* that the centripetal acceleration required to produce a closed orbital path must be $a_{cent} = V^2/R$, where V is the orbital velocity and R, as before, is the orbital radius. An application of Newton's 2nd law, $F_{grav} = m\, a_{cent}$ now gives us the result that: $F_{grav} = GM\, m\, /\, R^2 = m\, a_{cent} = m\, V^2\, /\, R$, which indicates that $V^2 = GM\, /\, R$. Now, given that we are dealing with a circular orbit we can introduce the orbital period P through the expression that in one orbital period the distance moved by mass m is the circumference of its orbit, which is simply $2\pi\, R$. Accordingly, $V = 2\pi\, R\, /\, P$, and hence by substituting for the V^2 term in our earlier formula we have: $P^2 = \left(4\pi^2/GM\right)R^3$. From this relationship we can recover the expression for the period of the infinite length pendulum. Indeed, if m is imagined to skim around a perfectly smooth planet of radius R, which is also the radius of its orbit, then the surface gravity term will be $g = GM\, /\, R^2$, and our expression for the orbital period reduces to $P = 2\pi\sqrt{R/g}$.

In addition to providing an expression for the minimum orbital period of a satellite, the relationship between the period and orbital radius – effectively $P^2 \sim R^3$ – reveals a deeper and much more important result. Indeed, it is Johann Kepler's famous third law of planetary motion, and it was upon finding this result that Newton knew that his hunch, or more specifically Robert Hooke's hunch – as we shall see - about gravity being an inverse square law (that is varying as $1\, /\, R^2$) must be correct.

The Newtonian Domain

The success of Newton's gravitational theory, in spite of its underlying causative mystery (to be discussed later in Chapter 4), was immediate and long lasting. For over two hundred years its application with respect to the dynamics of moving objects, whether on the very small or the very large scale, was never seriously questioned or brought into experimental doubt. But, how sure are we of Newton's formulation? Certainly, within the laboratory and within most of the planetary realm of the solar system Newtonian dynamics appears to hold true to a very high order of accuracy. Indeed, in an outstanding *tour de force* of mathematical manipulation both Urban Le Verrier and John Couch Adams applied Newtonian theory to successfully predict the existence and location of a whole new planet (the planet Neptune – discovered by Johann Gale at the Berlin Observatory in 1846) in order to explain the observed discrepancies in the motion of Uranus. Neptune orbits the Sun at a distance of about 4.5 billion kilometers (equivalent to 30 astronomical units), and Newtonian dynamics seems to work just fine at this range. Dwarf planet Pluto, discovered in 1930, orbits the Sun at a distance of about 6 billion kilometers, and we now know of a multitude of icy worlds with orbits at far greater distances than Pluto, and, again, Newton's gravitational formula still appears to be hold true. At distances of order one hundred astronomical units from the Sun (equivalent to about 15 billion kilometers) we begin to enter a more shadowy realm, skirting the edge of technologies current limits, where objects are so remote that they are not only extremely difficult to detect, but their exact dynamics are almost impossible to infer. Looking deeper into space, however, we find binary star systems, composed of two suns in orbit about a common center. First investigated in detail by astronomer William Herschel in the early 19th century, these great, rolling systems can have separations of many thousands of astronomical units, and yet their orbits, when they are known, are well described by Newtonian gravity. It would appear, therefore, that in the planetary realm and the domain of multiple star systems Newtonian theory reigns supreme.

On the very large scales that encompass structures such as galaxies and clusters of galaxies, where the gravitational accelerations are extremely small, we move into a dynamical domain that can never be experienced in terrestrial laboratories. Indeed, all laboratory experiments, by the necessity of their being made on Earth, fall into a high acceleration domain ($a = g = 9.8$ m/s^2 at Earth's surface) when compared to those that operate on cosmological scales ($a < 10^{-10}$ m/s^2), and this, in recent decades, has resulted in a whole series of possible re-interpretations of Newton's gravitational formula as well as his laws of motion. Interestingly, in the domain where pendulum experiments can no longer guide us, it would appear that confusion, or at least uncertainty, holds court.

Before moving on to consider the low acceleration domain, let us first see how the inverse square law of gravity is confirmed to us, at least on the nearby scale corresponding to the solar system. Indeed, the test is to determine the form that Kepler's 3rd law of planetary motion takes when the gravitational force is assumed to vary as some arbitrary inverse power of the distance: $F_{grav} = GM m / R^n$, where $n > 0$ is now simply some number. With this expression the relationship between the orbital period and the radius of an object moving along a circular orbit, as derived earlier, becomes $P^2 = K R^{(n+1)}$, where $K = 4\pi^2/GM$ = a constant. From this expression it is clear that if we want to recover Kepler's empirically determined third law then the value of n must be 2 – if we adopt any other number for n then there is a disagreement between the theory and the observations. This is not a lock-tight argument that gravity must always vary according to an inverse square law, since it might vary, for example, on scales that are much larger than planetary orbits. Indeed, one of the most important discoveries of mid-twentieth astronomy was that the rotational speeds of stars within spiral galaxies do not obey the expected Keplerian rule (figure 1.7). Rather than the rotation velocity of stars decreasing according to the inverse square root of their distance from a galaxy's center, where most of the visible matter resides, the observed velocities have been found to be almost constant with increasing distance. This result does not actually indicate that the inverse square law of gravity is wrong (see below, however); rather it tells us that there must be more matter, having a gravitational influence, within and surrounding a galaxy than can be accounted for in terms of visible stars and interstellar gas. The push to discover what this dark matter is actually composed of has now pre-occupied astronomers, cosmologists and particle physicists for well over eighty years – with little sight of any resolution to the problem being apparent even at the present time. That dark matter must be an entirely new form of stable matter (that is, matter that has not, as of the present time, been unambiguously detected in any laboratory experiment), entirely different from the baryonic matter that makes stars, planets and us, is a remarkable state of affairs, but it is by no means the strangest known characteristic of the observable universe.

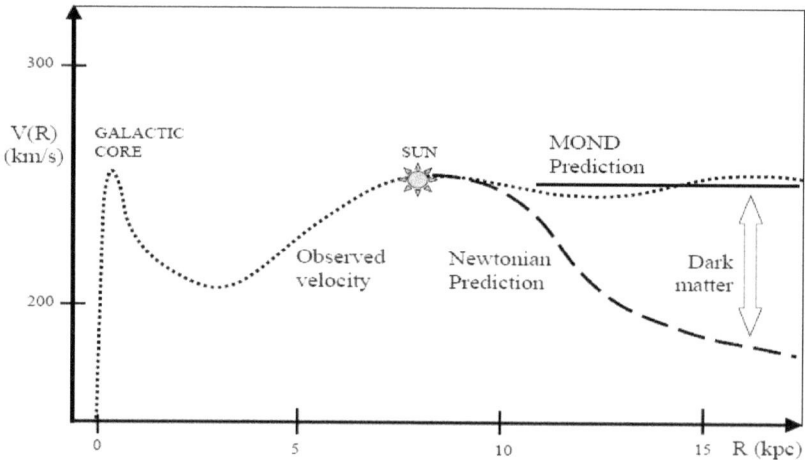

Figure 1.7. Schematic rotation velocity curve for the Milky Way galaxy. The fact that the rotation velocity does not decrease with increasing distance away from the galactic center indicates that there must be very much more mass, dark matter in fact, than can be accounted for in terms of observed stars and interstellar gas clouds.

Must the dark matter problem be solved in terms of new particles? The answer to this, according to whom you ask, is either yes or no. The general assumption is that dark matter[20] has a gravitational influence but does not interact with ordinary matter or electromagnetic radiation – thus making it invisible to our telescopes. Not everyone agrees with the new kind of matter idea, and it has been posited that the apparent existence of dark matter is really a consequence of the way in which gravity 'works' on very large scales when the accelerations produced by ordinary matter are extremely small. This idea can be illustrated if we return to our $P^2 = K R^{(n+1)}$ generalized form of the period - orbital radius relation, where we can see that if n is set equal to unity the period will vary directly in step with the radius: $P = \sqrt{K} R$. Under these condition the rotational velocity for a circular orbit becomes $V = 2\pi R / P = 2\pi/\sqrt{K} = $ constant, that is the velocity is constant irrespective of the size of the orbit. This result is exactly what is observed in the rotation curves of spiral galaxies - the same observations that resulted in the postulate of dark matter. In this fashion, the rotation curves observed for spiral galaxies can be explained if on very large scales the gravitational force is modified to vary as a simple inverse distance law: $F_{grav}^{mod} \sim 1/R$.

Not only can one introduce a modified gravitational law to explain the rotation curves of spiral galaxies, but, as Mordehai Milgrom (Weizmann Institute, Israel) did in the early 1980s, one can posit that, perhaps, Newton's 2nd law of motion no longer holds true in the domain of very small accelerations. In this limiting case of small accelerations, Milgrom suggests, that a force F produces an acceleration a, not according to Newton's 'normal' second law, $F = m\,a$, but rather such that $F = m\,(a^2/a_0)$, where a_0 is a new fundamental constant with dimensions of acceleration (m/s²). The conditions for a circular orbit now become $F_{grav} = GM\,m\,/\,R^2 = m\left(a_{cent}^2/a_0\right) = (m/a_0)\,V^4\,/\,R^2$, and from this we find that $V^4 = GM\,a_0 = $ constant, once again providing a constant velocity, independent of distance, result (see figure 1.7). The current best estimate for the a_0 constant is $1.2 \pm 0.27 \times 10^{-10}$ m/s², and in the solar system, for example, this implies that MOND effects, should they be real, will only become noticeable at distances beyond some 7500 AU from the Sun. At these distances we are well into the Oort cloud of cometary nuclei where it will be very difficult to perform any experiments, spacecraft based or otherwise. This being said the change-over from ordinary Newtonian dynamics to MOND is expected to be a gradual one, rather than an abrupt on-off switch. This dictates, according to several present-day researchers, that MOND effects should be detectable in the orbital motion of the outer most planets, Uranus and Neptune – the level at which any MOND alterations to orbital speed and motion, however, are tantalizingly just smaller than the accuracy to which planetary position in the outer solar system can be measured at the present time. The debate amongst researchers continues apace.

Newtonian mechanics is formulated on the principle that gravity is a force that is instantaneously propagated over arbitrary large distances. This notion of instantaneous interaction, however, was first brought into question when Albert Einstein published his first epoch making paper on special relativity in 1905 (a topic to be discussed in more detail in Chapter 2). At the heart of relativity is the idea that the speed of light is a limiting speed for any form of communication – there can be no form of instant propagation, Einstein argued, be it via light, particles, spacecraft or gravitation. After grappling with the problem for a further decade, Einstein eventually published, in 1915, the first of his research papers dealing with general relativity. Within this more general theory, Einstein made the tremendous intellectual shift of thinking about gravity not as a direct force, but as a manifestation of the curvature of spacetime. In this manner, gravity is transformed into a pseudo-force appearing purely as a manifestation of the changing geometry of space itself. Einstein introduced 10 equations to describe the relationship between the geometry, that is curvature, of space and the dynamics of the matter that is embedded within it, with the result, so aptly summarized by John Wheeler, that, "curvature tells matter how to move, and matter tells space-time how to curve".

While modified gravitational and MOND effects may potentially come in to play in the domain of extremely low accelerations, general relativistic effects most definitely operate in the domain of very large accelerations. The turn-over condition beyond which general relativity takes over from standard Newtonian gravity can be determined from the so-called metric tensor, which in the weak limit can be written as $g_{00} = 1 + 2\phi/c^2$ where $\phi = GM/R$ is the gravitational potential energy, and c is the speed of light. Once the second term in the expression for g_{00} becomes appreciable then general relativistic considerations should be taken into account. More simply put, one can also, somewhat loosely, say that provided the gravitational potential energy governing the dynamics of a particular situation is not large enough to produce velocities comparable to speed of light then Newtonian theory can be safely used. In the solar system the differences between planetary motions predicted by Newtonian mechanics and general relativity are very small, and only in the case of Mercury do they become discernable.

With a sidereal period of just 87.97 days, Mercury orbits the Sun at an average distance of 58 million kilometers (being 0.4 times closer to the Sun than the Earth). Its orbit is the most eccentric, and the most highly inclined (to the ecliptic) of all of the planets[21]. Furthermore, Mercury's rate of orbital rotation, described according to a quantity known as the advancement of perihelion, is the largest of all the planets[21]. Indeed, the rate of Mercury's perihelion advancement is too high – too high that is by standard Newtonian dynamics. That a real discrepancy existed between the observed and the predicted rates for Mercury's orbital rotation became undeniable during the latter part of the 19th century, and astronomers were forced to concede that the hitherto infallible laws of Newtonian dynamics were struggling to explain what was going on. The problem was not that Newtonian dynamics did not predict that Mercury's orbit should slowly rotate about the Sun, but rather that the rate it predicted was too small – too small by just 43 arc seconds per century (that is about 1/84th of a degree over 100 years). At first the discrepancy could simply be dismissed as a combination of observational errors and uncertainty with respect to Mercury's physical properties, but try as they might, astronomers and mathematicians could not account for the observations. Enthused by the success of predicting the existence of Neptune, via a detailed examination of the anomalous motion of Uranus, Le Verrier suggested that perhaps an additional small planet existed between Mercury and the Sun, and that the gravitational perturbations from this extra planet were the cause of Mercury's anomalous motion. The putative planet, in keeping with classical mythology, was even given the name: Vulcan. Indeed, systematic searches for Vulcan were made by amateur astronomers, during the late 19th Century - some observers even reported finding it, only to loose it again – but ultimately the idea of a new planet fell into disfavor simply because it just could not be found[22]. If an additional planet wasn't the cause of Mercury's anomalous advancement of perihelion, then perhaps, it was argued towards the close of the 19th Century, the inverse square law of gravitational attraction itself was modified in the Sun's vicinity. This was a radical suggestion, but a solution to Mercury's anomalous advancement of perihelion problem simply had to be found – no matter how it might upset established (and long cherished) physical theory.

If the inverse square law changed in the near-Sun environment, then what kind of change was required to account for Mercury's anomalous motion? This question was taken-up by American astronomer Asaph Hall in 1894, who showed that the change was quite miniscule[23], requiring the power in R^n to change from $n = -2$, to $n = -2.00000016$. This is a remarkable result in that such a very small change in n can effectively argue out of existence an entire intermercurial planet. The price, however, for changing n away from being exactly equal to -2 is still a very steep one, and by invoking such changes whenever the standard gravitational equations fail to provide a correct prediction brings in to question the very foundations of Newtonian theory as a whole. The story of the inverse square law will be picked up again towards the end of Chapter 4.

With the turn of the 20th Century the problem of Mercury's advancement of perihelion rate was still unsolved. In late 1914, however, a remarkably new interpretation of how gravity 'works' was nearing its completion, and the reasons why Newtonian theory could not explain Mercury's anomalous motion were finally made clear. The person who solved the problem was Albert Einstein, and the answer was provided for by his radically new theory of general relativity. Indeed, when he found that his new geometrical theory could fully account for Mercury's observed advancement of perihelion rate, Einstein had not completely finished and/or published his first paper on general relativity. Einstein even confided to his friend and research collaborator Wander de Hass that when he found that his new model could account for the unexplained motion of Mercury he had a feeling that something actually snapped inside of him – for several days he literally worked in a daze, and was beside himself with "joyous excitement".

The reason why Mercury's orbit should rotate can be thought of in terms of the interplay of two harmonic oscillators: one, OS(radius), describing an oscillation in the radial distance between Mercury and the Sun, and the other, OS(rotation), describing the rate of angular rotation of the radial line (figure 1.8A). Let the radial oscillator have a period P_{rad}, and the rotational oscillator a period P_{rot}. With these oscillators so described we can construct an elliptical orbit E such that, figuratively, we have E = OS(radius) + OS(rotation). Now, for example, if $P_{rad} = 0$, we have a circular orbit: $E = OS(\text{rotation}) = \text{circle}$. If $P_{rot} = 0$, then we have a line, with some fixed orientation, that systematically varies in length with period P_{rad} and $E = OS(\text{radius}) = \text{line}$. In the case that $P_{rad} = P_{rot} = P \neq 0$, we obtain, in general, a continuously repeating elliptical orbit (figure 1.8B). The orientation of the ellipse will be fixed (that is, there will be no apparent displacement of the perihelion point), since after every time interval $t = t_0 + nP$, $n = 1, 2, 3, 4, \dots$, both $OS(\text{radius})$ and $OS(\text{rotation})$ will take-on the same values that they had at time t_0. If, in contrast, $P_{rad} \neq P_{rot}$ then we generate quasi-elliptical orbits that do not quite close, tracing-out a trefoil-like appearance (figure 1.8C). This situation comes about since the oscillators are no longer synchronized and the location of the perihelion point, corresponding to the minimum value of $OS(\text{radius})$, will effectively trace out a circular path with a period corresponding to the absolute difference between P_{rad} and P_{rot}. In general we can think of P_{rad} as being the sum of two terms, one P_{NG} due to the Newtonian gravitational interaction between all the various planets within the solar system, and the other P_{GR} due to a general relativistic correction cause by the great mass of the Sun strongly warping the geometry of space-time in its close proximity. Accordingly, $P_{rad} = P_{NG} + P_{GR}$, and what Einstein found in 1914 was that the additional P_{GR} term could exactly account for the anomalous 43 arc seconds per century motion in Mercury's advancement of perihelion. As the distance from the Sun increases, however, so the P_{GR} correction becomes smaller and smaller, and we recover the Newtonian limit for which $P_{rad} = P_{NG}$. For all the planets beyond Mercury the variously observed advancements of perihelion behave exactly as predicted with respect to the Newtonian determination.

To conclude and summarize this rather long technical section, what we have learned is that the domain of Newtonian dynamics, and therefore the realm of the pendulum, stretches far and wide. Provided that the accelerations are not too high (where general relativistic conditions begin to apply), and not too fast (where special relativistic corrections begin to appear – but see Chapter 2), then Newton's inverse square formula applies and his laws of motion hold true. As to whether Newton's laws and/or gravitational formula require revision under the conditions of very small accelerations remains, at the present time, unresolved.

Our Strange 3-D world

"Why is the world we perceive three-dimensional?" This at first glance seems like a simple enough question, but it is a question that is far removed from having a trivial answer - as philosopher Immanuel Kant realized when he considered the problem in 1746. Writing in his, *Thoughts on the true estimation of living forces*, Kant speculated that the three-dimensionality of space was a consequence of the fact that gravity varied as an inverse square law. Present-day

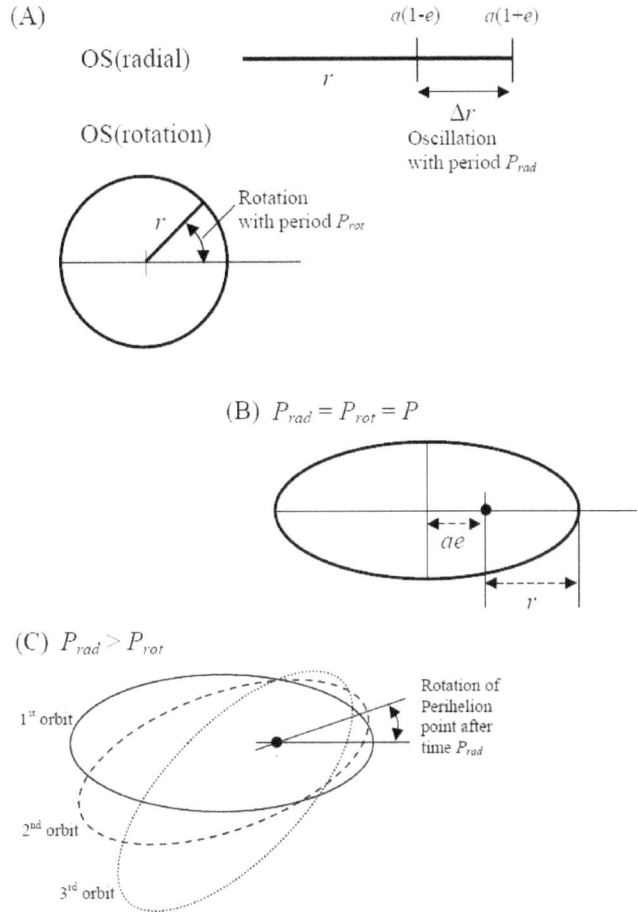

Figure 1.8. (**A**) Schematic form of OS(radial) and OS(rotation). The radial line has a minimum length corresponding to $r = a(1-e)$, where a and e correspond to the semi-major axis and eccentricity of the generated elliptical orbit. The length of the radial line varies from $a(1-e)$ to $a(1+e)$, and the time for the line length variation to run from a minimum, to a maximum and back to a minimum again is P_{rad}. The rotational oscillator, OS(rotation), centered on the origin of the radial line, sweeps through 360 degrees in time P_{rot}. (**B**) When $P_{rad} = P_{rot} = P$ an elliptical orbit, having a semi-major axis a and eccentricity e, is formed through the combination OS(radial) + OS(rotation). (**C**) When $P_{rad} \neq P_{rot}$, so the orientation of the elliptical path rotates and the point of perihelion begins to sweep out a circle.

cosmologist John Barrow (University of Cambridge) re-asserts Kant's point by turning the dependency around, noting that it makes more sense to say that if space has n-dimensions, then gravity will vary as an inverse $(n - 1)$ power law. The question still stands, however, why does the space that we perceive correspond to $n = 3$.

There are a number of intriguing hints as to why it is likely we should find ourselves living in a low-dimensional universe rather than one having, say, 5 or 8 or 18 dimensions. In principle there are no specific reasons why the spacetime geometry of the universe cannot be composed of n-spatial dimensions and m-time dimensions, and that of all the possibilities, the ob-servable universe just happens to have an $(n, m) = (3, 1)$ geometry. This, of course, may be the case, but it seems an unsatisfying answer. John Barrow has considered this topic in some detail[24], and makes an interesting point with respect to the behavior of n-spheres. We can image building-up an n-sphere in a step-by-step process, each time adding in a new dimension. To begin with we take a circle in 2-dimensional space, by rotating this through a third dimension we generate a sphere, we then sweep the sphere through a 4-th dimension and so on until we have swept our original circle through n-dimensions. The volume of an n-sphere S_n is not especially easy to calculate, but we are familiar with the formula for a circle and sphere, for which, $S_2 = \pi R^2$, and $S_3 = 4/3 \ \pi R^3$. As we might expect, the volume of the n-sphere is related to its radius raised to a power corresponding to the number of dimen-sions n; the two-dimensional circle goes as R^2, the three dimensional sphere goes as R^3 and so on. The important point about the n-sphere, however, re-lates to the constant term that multiplies the power of R. Remarkably a de-tailed calculation reveals that the constant term for an n-sphere first increase, going from π to $4\pi/3$, $\pi^2/2$, and $8 \ \pi^2/15$ as n goes from 2 to 5, but then decrease for all n greater than 5. In fact for any given radius R the n-sphere with the greatest possible volume is that achieved in 5-dimensions. If one is more interested in the surface area of the n-sphere, then for any given radius this quantity has a maximum value in 7-dimensional space. This volume and surface area behavior of n-spheres, while not absolutely convincing, is sugges-tive of the idea that we should expect to live in a universe with at least $n < 5$ or 7. The behavior of n-spheres also resonates with Einstein's 1911 statement concerning the appearance of low multiplicity terms in dimensional analysis. In this case, provided $n < 4$ we would only expect multiplicative factors of order π rather than π^2 or higher.

String theory, a currently popular mathematical model for describing the properties of atomic particles and their associated nuclear forces (as well as gravity), perhaps indicates a deeper complexity to the universe since it oper-ates in a (11, 1) geometrical space. We do not 'see' all 11-spatial dimensions, however, because 8 of them are curled-up so tightly that we cannot directly experience them in our everyday lives and/or invoke them in laboratory experiments. Of course, string theory may or may not be true, but the dimensional issue still remains. While string theory requires that the number of spatial dimensions is large, it only requires one time dimension: $m = 1$. Indeed, it is difficult to conceive of there being more than one time dimension, since for any other number we either have no time ($m = 0$), or indeterminate time ($m \geq 2$), and accordingly the concept of stable orbits, or periodic pendulum motion, have no clear meaning. More fundamentally, perhaps, if we want the future to be determined by actions taking place in the past or present, then the number of time dimensions cannot be anything other than one.

With respect to pendulum motion, something quite remarkable happens as we move to consider oscillations in one then two, and finally three dimensions. In the (1, 1) situation we have the simple spring pendulum and the motion is purely up and down along a straight line (see Chapter 3 later). Moving to (2, 1) space we can construct the simple pendulum (figure 1.1) that swings from side to side along a very specific arc (set by the string length L) contained within the two-dimensional spatial plane. Continuing upwards into (3, 1) space, we encounter the possibility of a rotating simple pendulum; the plain of the pendulum's oscillation now being able to rotate and move into the third, extra dimension – an important variant of the 3-D pendulum, the Foucault pendulum, will be described in Chapter 6. We may also construct in (3, 1) space a conical pendulum (to be described below) in which the pendulum bob moves in a circle of some specific radius and within a fixed (2, 1) plane, with the suspension point of the pendulum being located in the additional dimension above the plain of the bobs rotation. We further encounter in (3, 1) space the (theoretical) problem of the spherical pendulum in which a bead is constrained to move on and around the surface of a sphere. For this latter pendulum the trajectories can become very complex, and in principle, for a frictionless idealized pendulum, the bob can visit every point on the surface of the sphere. The unconstrained (3, 1) pendulum problem, in which the bob no longer moves upon the surface of a sphere, is identical to that of the orbital trajectory problem (as discussed earlier). One, two, infinity – this appears to be the range of possible motions available to the pendulum in 1, 2 and 3 dimensions. The remarkable transition of allowed pendulum motions that opens-up beyond two dimensions is further suggestive of the fact that if the universe is to exhibit complex behaviors, such as those necessary for interesting physics (and presumably life) to occur, then it must have at least three spatial dimensions.

Carrying forward the idea that space must be at least three-dimensional, we again come back to Immanuel Kant's observations that there is a correlation between the number of spatial dimensions and the workings of gravity – specifically, Newton's formula. In what is now a classic research paper[25] published in 1918, Paul Ehrenfest showed that by being restricted to three spatial dimensions planetary orbits tend to be stable under gravitational interactions – in higher dimensional space ($n > 3$) virtually all orbits become unstable. Since a stable planetary orbit is a prerequisite for life to evolve, we in essence owe our existence to the fact that space is 3-dimensional and that gravity varies as the inverse square law. Perhaps even more remarkably, Ehrenfest also found that the equations describing the ability of atoms to form bound, stable orbits only have solutions if there are no more than $n = 3$ spatial dimensions. Indeed, the fact that we live in a three dimensional spatial universe appears to be rather a profound observation since it allows atoms to exist and chemistry, vital for life to come about, to work. If there are more spatial dimensions beyond the three we actually perceive, and many physicists believe there are, then they must be wrapped-up in special ways that in generally make them inaccessible to experimental exploration - even the mighty pendulum is excluded from operating within these other worldly domains.

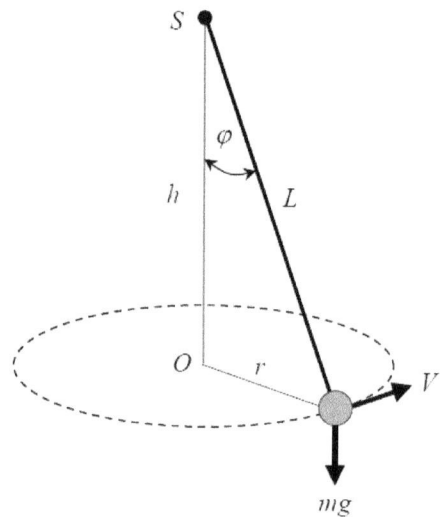

Figure 1.9. The conical pendulum. The bob moves around the circle of radius r at a constant speed V. The angle that the support wire of length L makes with the vertical SO is φ.

The Conical Pendulum

The pendulum systems described thus far produce cyclical motion in the vertical plane only. Moving to the horizontal plane, however, we encounter a new kind of oscillator – the conical pendulum (figure 1.9). In this situation the bob moves in a circle of radius R at a distance h below a suspension point S. By ignoring air resistance, we take the bob to be moving at a constant speed V. In order to determine the period P of the bob, we must first express the speed V in terms of properties of the pendulum. To achieve this we must resolve the horizontal and vertical forces acting on the pendulum bob – the point being, of course, is that for stable circular motion they should be in exact balance.

If the tension in the conical pendulum support wire is T then the horizontal forces must be such that $T \sin \varphi =$ the centripetal acceleration $= m V^2 / r$, where m is the mass of the bob. The vertical forces must also balance, and accordingly $T \cos \varphi = m g$, where g is the local gravitational acceleration. If we divide these two equations to eliminate T we obtain the result that: $\tan \varphi = V^2 / r g = r / h$. Now, we can determine the period of oscillation by noting that in time P, the bob will have traversed one orbit around the circumference of the circle of radius r. So, using the straightforward and well-known velocity = distance / time relationship, we have $V = 2\pi r / P$. After eliminating the velocity term between our two equations we finally derive the period formula for the conical pendulum as

$$P = 2\pi\sqrt{h/g} \qquad (1.8)$$

If we compare equation (1.8) with equation (1.4) we see that the formulae are very similar, and that the period of oscillation of the conical pendulum is the same as that of a simple pendulum with a length equal to the vertical height of the conical pendulum – we note, of course, that $h = L \cos \varphi$, which does technically bring into the expression the length of the support wire and the cone angle φ.

Perhaps the most startling result that follows from equation (1.8) is that all conical pendulums of the same height h have the same period of rotation. Robert Hooke first noted this property in the mid-17th century and he used the result to develop an isochronal conical pendulum. The trick, as Hooke realized was to keep the height (h) constant even as the radius of motion of the pendulum varied. This condition can be achieved by making the flexible pendulum support wire move around the surface of a paraboloid of rotation (figure 1.10).

Hooke was particularly interested in the properties of the conical pendulum, and he adapted it to many purposes. Indeed, on 13 June 1666 he demonstrated to the assembled Fellows of the Royal Society in London a "new contrivance of a circular pendulum applicable to a watch". Since the pendulum's motion was circular, rather than from side to side, the movement of the watch was both smooth and soundless, and these attributes were highly advantageous to its regular running and time keeping.

Not only did Hooke realize that the conical pendulum offered distinct advantages with respect to the regulation of timepieces, he also saw that it afforded a model with which to study planetary motion. Indeed, just a month prior to his clock demonstration Hooke had addressed the Fellows of the Royal

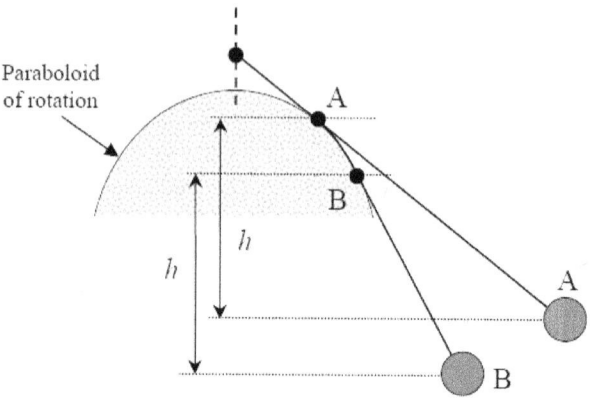

Figure 1.10. Robert Hooke's design for an isochronal conical pendulum. The height *h* remains the same for the pendulum even if the point of contact with the parabolic surface changes. The period of the pendulum is the same irrespective of where the surface contact point is between A and B.

Society on the topic of *Motion in a Curve*. In this demonstration and its eventual write-up Hooke argued that the problem of planetary motion could be addressed through a mechanical explanation – an explanation provided by the conical pendulum. He began his demonstration with the comment, "I have often wondered why the planets should move about the Sun according to Copernicus's supposition, being not included in any solid orbs ... nor tied to it, as their centre, by any visible strings". To this he adds that it is also remarkable that they do not, "depart from it beyond such a degree, nor yet move in a straight line, as all bodies that have but one single impulse ought to do". In this latter statement Hooke has hit upon the very idea that Newton eventually turned into his first and second laws of motion. Likewise, through his conical pendulum demonstrations, Hooke further records, "that circular motion is compounded of an endeavour by a direct motion by a tangent, and of another endeavour tending to the center". Hooke's pendulum experiments also indicated that it was the strength of the two "endeavours" that dictated the shape of orbit adopted by the bob. If there was a difference between the "tangential endeavour" and the "endeavour to the centre" so an elliptical orbit was produced; if the "endeavours" were equal then a circular orbit was obtained. Hooke later developed his ideas on planetary motion during his 1670 Cutlerian Lecture[26], arguing then that, "all Coelestial Bodies whatsoever, have an attraction or gravitating power towards their own centers, whereby they attract not only their own parts, and keep them from flying from them, ... , but that they also attract all the other Coelestial Bodies that are within the sphere of their activity". The idea of a continuously acting, far-reaching, central force was exactly what Newton later turned into his famous equation for gravitational attraction.

Perhaps one of the most interesting points relating to Hooke's conical pendulum experiments is that he mentions Copernicus as an inspiration, but not Johann Kepler. Indeed, while his conical pendulum experiments did allow Hooke to develop the idea of a continuously acting central force, the elliptical trajectories demonstrated were centered on the axis directly below the pendulum's suspension point, rather than on one of the focal points of the elliptical trajectory as is the case of real planetary orbits and as described in Kepler's first law. This was not a critical problem in Hooke's heuristic argument, but it was a critical point in Newton's later mathematical derivation of the laws of planetary motion.

The Governor

Perhaps one of the most useful engineering applications of the conical pendulum is that seen in the speed governor. Indeed, governors are remarkably simple self-regulating devices that use a negative feedback principle to control rotational motion.

It seems appropriate that the first governor was invented to control the drive rate of machinery in windmills. Specifically, Thomas Mead, circa 1787, invented the lift-tenter, which was used to control the spacing between the millstones (a process known as tentering) under gusty wind conditions. Without the tentering device (that is, what we would now call a governor) the gap between the millstones varied as the rotation rate (governed by the prevailing wind speed) changed, and this would affect, not necessarily for the better good, the grain-size of the flour being produced.

Scotsman James Watt further refined the centrifugal governor in 1788, engineering it so as to control the flow rate of steam within steam-powered engines. The centrifugal governor introduced by Watt operated in such a way that a constant drive rotation rate was maintained – indeed, it was an early application of a feedback control mechanism. Fellow Scotsman and physicist James Clerk Maxwell, however, ultimately wrote one of the first detailed mathematical treaties on the governor in 1868, and he showed that they have properties similar to those of the conical pendulum[27] - there are important dynamical differences between the two devices, however. Specifically, by allowing the pendulum masses on a governor to pivot about an attachment collar affixed to the rotation axis, when the rotation rate of the machine increases the masses will swing upwards and outwards, and this motion can be used to close-off a valve mechanism that will slow the rotational increase, thereby slowing and stabilizing the rotation rate. The device is simply marvelous, and marvelously simple.

Maxwell described the basic operating principles of a governor – more correctly a moderator – in his 1868 paper in terms of an applied driving power P and a resistance R relating to some fixed rotational axis within a machine. In this manner, Maxwell writes the equation of motion for the rotating system as

$$\frac{d}{dt}\left(M\frac{dx}{dt}\right) = P - R - f\left(\frac{dx}{dt} - V\right) \tag{1.9}$$

where M is the moment of inertia (which is essentially a measure, in this specific case, of the machine's inherent resistance to changes being made to its rotational state), V is the nominal starting velocity, dx/dt is the actual velocity and f is some constant. The right most term in equation (1.9) is a resistance term that varies according to the value of the initial and final velocity. When the system has attained its final, spun-up, velocity so the left hand side of (1.9) must become zero, and this indicates that $dx/dt = V + (P - R)/f$. The final velocity dx/dt will be permanently increased over V therefore whenever P increases or R decreases – the size of the increase is further moderated by the value adopted for the constant f. A simple example of this kind of moderator at work is that of a conical pendulum rotating within a hollow cylinder. In this set-up, as the velocity of the pendulum increases so the bob will swing outwards, only stopping when it begins to press against the inside wall of the cylinder. The friction established between the bob and inner cylinder wall will thereafter check, that is moderate, any further increase in the pendulum's velocity.

While a governor is similar to the conical pendulum the equation that describes its motion must allow for the vertex angle φ (see figure 1.9) to vary in step with the rotation speed. Indeed, if the angular velocity of the governor is Ω, then the equation of variation for φ is:

$$\ddot{\varphi} = \Omega^2(t)\sin(\varphi)\cos(\varphi) - \frac{g}{L}\sin(\varphi) - k\dot{\varphi} \tag{1.10}$$

In equation (1.10), Newton's dot notation[7] for time differentials has been used, and the equation itself can be solved to determine the vertex angle φ associated with a specific value of Ω - which in the case of the governor is a time varying quantity. The last term on the right-hand side of equation (1.10) is a damping term that is introduced to 'help' the solution for φ settle down to its equilibrium value without excessive undershoot / overshoot oscillations (see Chapter 6 later). We can see that if the damping term is set to zero ($k = 0$) and if the circular rotation rate Ω falls to zero then we recover, as we would indeed expect, the equation for a simple pendulum.

The essential idea of the governor is to use the expansion of the calipers – described by angle 2φ – to adjust the rotation rate Ω to maintain some fixed, pre-set, value Ω_0. The governor, described through equation (1.10), will therefore reduce the rotation rate Ω if it becomes larger than Ω_0 and increase it if drops below Ω_0. In this way the rotation rate of some specific drive shaft is controlled, that is self adjusted, to maintain a constant value. Remarkably, not only do we find the pendulum at the very core of physics and horology, we also find, through its application as a governor and moderator, the modified conical pendulum regulating the great machines of the industrial revolution. Indeed, there is a wonderful resonance between the political and societal use of the terms governor and moderator – the former runs and maintains a system while the latter seeks to limit system excesses and maintain stability.

The Paraconical Pendulum
As a hybrid design between a simple and a conical pendulum, the paraconical pendulum was specifically constructed to look for and respond to asymmetric gravity fields. The name paraconical is derived from the Greek *para*, meaning *about*, and *conical*, meaning cone-like – it is in essence a nearly conical pendulum. Designed by Nobel Prize winning economist and experimental physicist Maurice Allais during the early 1950s, this specialized pendulum is composed of a short rigid staff, an aerodynamic lens-shaped bob (so figured as to minimize air resistance), and a support system that enables the pendulum to swing through three-degrees of rotational freedom (figure 1.11). Indeed, the paraconical pendulum, once set in motion, traces out an elliptical track, in the XY plane, that slowly rotates around the Z-axis. The restricted rotation of the pendulum about the vertical Z-axis is made possible by the large ring situated at the upper support end of the staff. The paraconical pendulum has a number of characteristics in common with the pendulum (to be described in Chapter 5) used by Leon Foucault to experimentally demonstrate the spin of the Earth in 1851. The pendulum also describes what is called the Airy effect, named after the British Astronomer Royal George Biddell Airy. We shall have more to say about Airy's pendulum experiments in Chapter 4, but the effect that is named in his honor relates to the steady rotation of the elliptical track about the vertical axis. In a classical research paper published in the *Memoirs of the Royal Astronomical Society* for 1840, Airy considered the effects due to asymmetric starting conditions as well as the back-reaction of a clock's escapement on the swing of a pendulum (to be discussed in Chapter 2). What Airy found was that if the pendulum traces out an elliptical track with semi-major axis a and semi-minor axis b, with $b << a$, then the period of oscillation for a pendulum of length L is $P = 2\pi\left(1 + b^2/16a^2\right)\sqrt{L/g}$. As we would expect the period P is longer than that of a simple pendulum of length L (oscillating, of course, in just one plane), and that as $b \to 0$, so the period approaches, as it must, that of the simple pendulum. The angular rate at which the major axis turns about the Z-axis was further shown by Airy to be described according to the relationship $\Omega_{Airy} = P\left(3ab/8L^2\right)$.

"Facts alone must guide us, rather than mummified principles, even though they may be most useful for a first approximation. We learn only through experiment, and any thought which permanently withdraws into a set of abstract principles thus sentences itself automatically to a form of sclerosis". So wrote Allais in the second of three remarkable articles[28] published in *Aero/Space Engineering* for September, October and November 1959. His statement is the cry of a scientist who has found a strange result that other researchers find very hard to believe.

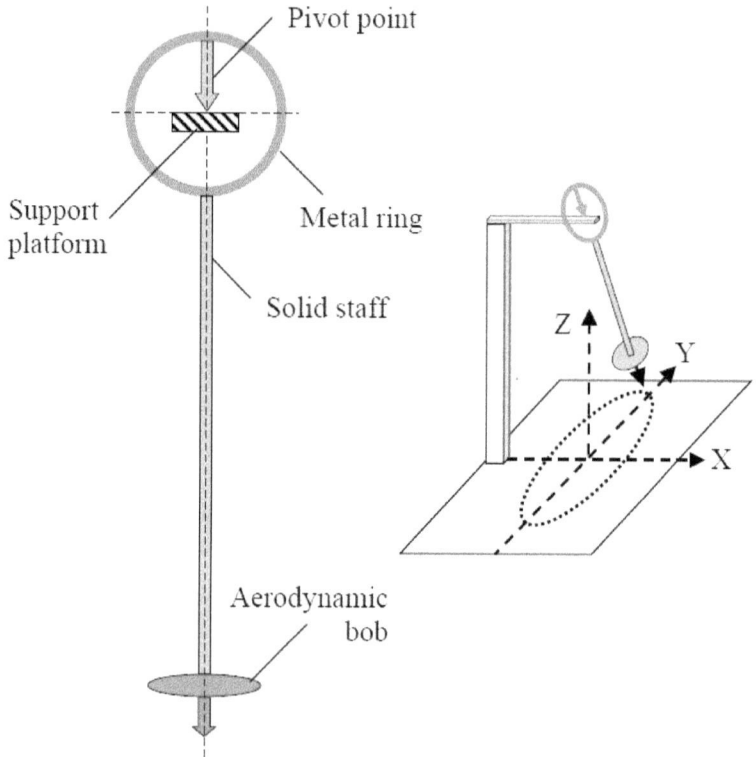

Figure 1.11. The paraconical pendulum.

What was this strange result? Allais and his assistants carefully measured the rate at which the plane of his pendulum rotated, and claimed to find a series of variations with periods corresponding to 12, 24 and 25 hours. Allais accordingly hypothesized that these newly found periods of variation must be due to previously unidentified forces – indeed, forces that, "must be considered as produced by the direct action of a new field"[28]. Working from his experimental data Allais argued that the rotation rate ω of a paraconical pendulum could be described by an empirical equation of the form:

$$\omega = \Omega_{Foucault} + \Omega_{Airy}\left(\tfrac{1}{2}\sin(2\omega t) + \chi\right) \qquad (1.11)$$

where the parameter χ accounts for what is now known as the Allais effect. The first term in this equation is due to the rotation of the Earth itself, while the second term is that of the Airy effect modified to include an additional term χ that accounts for the anomalous Allais effect. For Allais, his pendulum studies had revealed a new fundamental force – in reality, however, it is now clear that he had pushed his deductions far beyond the precision that was allowable in a less than well designed experiment. With the Allais experiment we have the situation in which an unexpected result is interpreted on the wrong side of Occam's razor. The cut from this philosophical knife, of course, is a subtle one, and while it is true that unexpected results can indeed indicate the discovery of new physics, they may equally indicate an unexpected experimental defect – the question, of course, is which answer is more likely. It is now appreciated that while the paraconical pendulum didn't lie, Allais's interpretation of what it was telling him became lost in philosophical translation.

The Mysterious Eclipse Effect

Lodged within the beating heart of any mechanical clock is a pendulum, and the driving force behind the pendulum is gravity. And yet, there is much about gravity, as discussed earlier, that scientists still do not understand. Indeed, ever since the time of its introduction by Isaac Newton in 1678, scientists have struggled to understand exactly how a gravitational influence is propagated. Is gravitational information carried by a wave, a particle, or a magical daemon? On this issue, Newton famously commented "*hypothesis non fingo*". While Newton's equations and Einstein's general relativity tell us the precise consequences of gravitational interactions, how an object will move and so on, they do not tell us what gravity is. Just as time is a great mystery to science so too is gravity. This problem is exacerbated in the realm of the very small, where scientists try to patch together quantum physics and general relativity – no matter how many mathematical twists and turns are made, however, at least to date, no generally accepted theory of quantum gravity has been elucidated. The problem is profound, complex, infuriating and many scientists are employed in the search for this hidden grail.

While many issues have yet to be understood about how gravity works, and how gravitational influences are transmitted, experimental results gathered over the past fifty years with humble gravimeters and simple pendulum experiments may further add to our discomfiture. The experimental results are controversial, and not everyone agrees that a new phenomenon has even been identified. At issue is the possibility of gravitational shielding – literally the weakening of gravity as a result of its interaction with an object. The story of this gravitational anomaly begins with a solar eclipse, when the Moon passed in front of the Sun's disk as seen from Earth, on 1954 June 30.

The eclipse in question started in the heartland of the United States and moved steadily eastward along a thin totality track that took it through the northeastern corner of Canada, across the Atlantic, down over the Scandinavian coast and the Baltic Sea to a final hurrah in Iran. While all this was happening Maurice Allais was conducting his series of precisely timed and tracked paraconical pendulum experiments in an underground laboratory situated in Paris. Allais's experiment ran continuously for a remarkable 30-day period, and during that time only one anomaly in the pendulum's motion was reportedly observed – and the anomaly coincided with the beginning and end times of the solar eclipse that took place on June 30th. A second set of pendulum experiments was performed by Allais during the eclipse of 1959 October 22, and the same anomalous effect was apparently recorded. Since the time of Allais's initial observations, a number of additional research teams have reported the detection of anomalous pendulum and gravimetric phenomena during the times of solar eclipses. Other research groups, in stark contrast, report no anomalies – hence the scientific contention over the phenomenon. On the entirely cynically side one may safely assert that had Allais not been a Nobel Laureate then the anomalous eclipse readings would not have been given any credence by other experimenters, but, as recently as the 22 July 2009 eclipse, experiments were conducted in China to look for transient gravitational anomalies – none were reported. The situation concerning the eclipse effect remains unclear, but the anomalous Allais effect probably sits more reasonably on the pseudo-science side of the boundary separating scientific reason from stubborn romantic quackery.

The standard theory of gravity does not predict, and indeed cannot explain the eclipse timing anomalies described by Allais and his supporters. One possible explanation that has, however, been suggested for the anomalous pendulum behavior is that of gravitational shielding. By this hypothesis gravity is weakened upon passing through an object of density δ, and after crossing a distance L, the gravitational acceleration is modified according to the equation: $g = \mathbf{g}_0(1 - \lambda \int_L \delta \, dl)$. The integral term $\int_L \delta \, dl$ accounts for the total shielding effect, and the λ term is the so-called shielding coefficient. Conventional gravitational theory requires that the shielding term should be exactly zero ($\lambda \equiv 0$), while the most recent experimental results indicate $0 < \lambda < 10^{-15}$. Clearly the shielding coefficient is very, very small, but remarkably, perhaps, it is not zero and possibly, therefore, if the shielding effect really does exist it could be measurable. The jury is still out with respect to the possibility of gravitational shielding being a real phenomenon; certainly the eclipse effect, in its original form as proposed by Allais, is no longer tenable, but if future experimentation proves that $\lambda > 0$, then, once again, it may well be the mighty pendulum that pointed the way to a new physical theory.

Galileo – Again

We began this chapter with reference to Galileo's great work *Dialogues Concerning Two New Sciences*, published in 1638. It was indeed, this measured and reasoned tome that ushered in a new era of natural philosophy, and it literally established Galileo as the founding father of modern dynamics. What sets the *Dialogues* apart from Galileo's other works, however, is its uncharacteristic tempered tone; rather than simply heaping ridicule upon alternative viewpoints (and especially the then predominant Aristotelian ones), Galileo employs a more sympathetic voice in the *Dialogues* and backs-up his reasoning with experimental demonstrations. He wins his argument by the gentle art of dialectic and logical persuasion. These same positive attributes can only rarely be applied against Galileo's earlier works where his underlying arrogance and the vented bile of youth are often very much to the fore. Indeed, by letting such unchecked attributes dominate his writings, Galileo may well have performed brilliantly in a rhetorical sense, but failed abysmally in winning the day and in gaining converts or new supporters. One such text where Galileo's polemic voice rises to magnificence, but fails totally in its reasoning is that of *The Assayer* (*Il Saggiatore*) published in 1623. The book is an unnecessary and misguided vitriolic attack on a treatise written by the Jesuit mathematician Orazio Grassi concerning the appearance of three bright comets in 1618. The dialog specifically concerns the last of the three comets [designated in the modern era as C/1618 W1]. Both Grassi in Italy and Swiss Jesuit astronomer Johannes Cysat observed the comet and attempted to deduce its parallax (and hence distance). By this means they concluded that the comet must be located at a distance beyond that of the Moon. At the same time that Grassi and Cysat were observing the comet English scholar John Bainbridge was using a telescope to trace its path across the heavens, eventually concluding that it was ten times further from the Earth than the Moon. These were pioneering observations, and yet Galileo chose to ridicule them, arguing incorrectly that the comets were merely optical illusions generated within the Earth's atmosphere. It appears that Galileo made no determined effort to observe the comets, and rather than reason that the observations of Grassi, Cysat and Bainbridge provided evidence against the Aristotelian universe, composed of numerous crystalline spheres, he chose to adopt and even endorse an ancient idea for cometary genesis based upon the concept of rising dry vapors.

In spite of Galileo's reasoned failing in *The Assayer*, he nonetheless, in an historic sense, pulls victory from the jaws of defeat. For indeed, within *The Assayer* he passionately argues that mathematics is the language of nature, and it is only through the development of mathematical descriptions and mathematical proofs that any lasting physical theory of the universe, along with its many complex inner behaviors, might be established. It was a powerful piece of writing, and Galileo explains,

It [the universe] cannot be understood unless one first learns to comprehend the language and interpret the characters in which it is written. It is written in the language of mathematics, and its characters are triangles, circles, and other geometrical figures, without which it is humanly impossible to understand a single word of it; without these, one is wandering around in a dark labyrinth.

These very sentiments underscore the workings of modern science, and they underscore the often raised sentiment which questions the unreasonable effectiveness of mathematics to explain the world around us. For all of this, however, we would also argue that similar sentiments to those expressed by Galileo can be extended to the pendulum. We have hardly scratched the surface of the venerable tome that constitutes the known mathematics of the pendulum in this chapter, and yet it is already revealed to us that the pendulum is a device of subtle sensitivity. Indeed, the pendulum, with its oscillations to and fro, is very much the underlying metaphor, even the *genius loci*, of all dynamical systems.

CHAPTER 2
MAKING TIME

The Moving finger writes; and, having writ,
Moves on; nor all thy piety nor wit
Shall lure it back to cancel half a line,
Nor all thy tears wash out a word of it.

The Rubáiyát of Omar Khayyám
Edward Fitzgerald

Time consumes us, it runs our day and it keeps us on track, and yet we have absolutely no idea what time is. We imagine it to flow like a fluid, and yet it has no specific form. We cut it into segments and yet it has no specific substance. We experience only the instantaneous moments of a continuous 'now' throughout our lives. We have no sense or notion where time will take us in the future, and we cannot return to the past. Our hindsight is twenty-twenty clear, but the course of events just moments away is completely impenetrable to our gaze. Time is strange, mysterious and mercurial. Indeed, recently deceased physicist John Wheeler (Princeton University) has observed that "time cannot be an ultimate category in the description of nature. 'Before' and 'after' don't rule everywhere." It is certainly true that we see physical change in terms of the passage of time, but are we absolutely sure that time exists independently of physical change? Indeed, Philosophers have traditionally approached time in terms of two contrasting viewpoints. One viewpoint is that time is a fundamental part, or structure of the universe; a dimension along which perceived events can be sequenced. While the other viewpoint is that time is an intellectual structure, developed by the human consciousness in order to help sequence and make sense of the perceived world. Isaac Newton, as we shall see below, favored the first approach while his arch enemy and intellectual rival Gottfried Leibniz favored the second. Albert Einstein adopted a modified version of the first viewpoint by introducing the idea of spacetime in which space and time are inextricably linked. For all this, however, spacetime is changeable and fluid-like with its geometry changing according to the amount of matter embedded within it.

Question concerning the beginning of time are fraught with philosophical pit-falls and problems. Modern astronomical observations of the cosmos clearly, indeed, undeniably indicate that the universe must be of a finite age and that in the past it was much smaller and in a massively higher density state[1]. The present general consensus is that the observable universe began, that is it came into existence, about 13.75 billion years ago. From a spacetime perspective this means that both space and time, the two are inseparably linked recall, came into existence at the moment of the Big Bang – there was no space and there was no time prior to the occurrence of this singular event. There are, however, loopholes (perhaps, 'fundamental gaps in our knowledge' is a better expression) within the current physical models that enable one to get around this absolutist 'instant of creation' viewpoint. In terms of the physical theories that presently exist, and within which we have great confidence, to describe the big and the small of the cosmos (ostensibly general relativity and quantum mechanics) it is known that at the very moment of the universe coming into existence that they must fail – the physics that underlies the creation event itself is not known to us, and accordingly this allows one to think of time before the Big Bang, and to even contemplate universes that undergo cyclical behavior, regenerating themselves time and time again – indeed, over time immeasurable. Stephen Hawking (Cambridge University) has argued in recent years that time, as we know it, began at the moment of the Big Bang, but that time as we can never know it might have existed before this creative instant – this seemingly contradictory viewpoint is, of course, perfectly sound in that what Hawking is addressing is the notion that information cannot be transmitted across a Big Bang singularity. Under this viewpoint the memory of the universe is reset every time that a Big Bang occurs and nothing that happened prior to the most recent Big Bang can influence present space and time. In contrast to this viewpoint, Roger Penrose (Oxford University), who along with Hawking showed that a singularity at the Big Bang is unavoidable, has recently suggested that the universe might actually be cyclical, forming and re-forming as it were, with events underpinning the end phase of one universe potentially being propagated, in an observable manner, into its regenerated sibling. Remarkably, modern cosmology, for all of its incredible depth of understanding, still leaves room for polar opposite views upon its possible origin to exist.

The smallest time unit that physicists presently recognize is that of the Planck time, which amounts to a miniscule measure of some $t_P = 5.39 \times 10^{-44}$ seconds. According to the presently known laws of physics we can neither measure nor detect, nor make predictions about any system on timescales smaller than t_P. Modern day physical theories, as we presently understand them, enable the properties of the universe to be traced back to about 10^{-35} seconds after the Big Bang, which is a truly incredible achievement, but this time is also some 200 million Planck times removed from the beginnings of the universe itself, and remarkably it is also known (or speculated) that many, many significant events can and indeed must have occurred in this incredibly small fraction of the opening moments of the very first second. Cut it long or thin, time is a great mystery, and this, of course, is exactly why it fascinates us.

Absolute Time and Relativity

The indomitable Isaac Newton thought that there was an absolute time that governed the running of the universe, and he stated this idea at the very beginning of his *Principia*, "Absolute, true and mathematical time, of itself, and from its own nature flows equably without regard to anything external, and by another name is called duration". For Newton and his followers there was an underlying, sequential order to the development of events in the ageing universe, and in principle they could all be ordered and mapped out precisely. Even though there was no clear method by which a universal or absolute time could be measured, a time variable could none-the-less be placed within the equations of physical motion. We are all familiar with the idea of speed and velocity[2], and indeed, the speed V of an object is expressed in terms of the time T it takes to travel a distance S. The time entering into this expression is not, however, an absolute time as such, recorded as it were from some specific historical starting point, the original $T = 0$ (possibly located at the beginning of the universe itself [1]) but it is a time interval – the interval of time experienced between a very definite beginning point and a very definite end point in the objects motion. Likewise, the distance S traveled by an object experiencing a constant acceleration a over a time interval T can be expressed, as Newton revealed, by the relationship $S = UT + (\frac{1}{2})\, a\, T^2$, where U is the initial velocity. Time is embedded within this equation and the predictions certainly agree with the experimental observations, and yet, remarkably, if the initial velocity $U = 0$, then it matters not whether time is running backwards or forwards in the expression since time enters as a squared quantity. The distance traveled in time T is exactly the same as the distance traveled in the time interval $-T$ even though we have no concept for the meaning of negative time. This effect is embodied within what modern science calls the T-Symmetry Principle, which dictates that physical laws of nature are symmetric under time reversal transformations when $T \Rightarrow -T$.

Classical physics is packed-full of equations that allow time to run backwards or forwards in time, and yet it is well known that there are powerful physical concepts, linked to the so-called arrow of time, which 'point' towards just one direction of change. In the field of thermodynamics, for example, hot objects naturally become cold objects, radiating their heat into the cooler surroundings, and never *vise versa* [3]. In the world of the atomic nucleus and the deep realm of quantum mechanics, however, time can run backwards and forwards and it can even stop – apparent movement occurring without the passage of time. Indeed, our very existence depends upon the instantaneous spatial-quivers of numerous protons taking place again and again at the center of the Sun. This remarkable process, known as quantum tunneling, is at the heart of the all-important physics that allows the Sun to convert hydrogen into helium within its core and to thereby generate internal energy [4].

The notion of absolute time, for all of its physical and philosophical convenience, tuned out to be just a mathematical pipe dream. After ten years of mentally toiling with the issue, Albert Einstein was eventually able to do away with the problem of absolute time when he introduced his special theory of relativity in 1905. One of the key underlying concepts of special relativity is the principle of the invariance of the speed of light. No matter whom you are, where you are, or when you are in the universe, the speed of light (in *vacuo*) must always be 299,792,459.08 meters per second. There are no exceptions to this rule and it is indeed a defining characteristic of our cosmos[5]. What this impenetrable cap on allowed speeds implies is that the time at which two spatially separated events A and B occur can be different (that is relative) for different observers. For some observers' event A might follow event B, while for other observers the same two events might be seen simultaneous – it all depends upon the viewing circumstances. Not only do we lose the absolute sequential nature of events in Einstein's universe, clocks can appear to run at different rates depending upon the location and circumstances of the observer. Every observer moving with a specific clock will read normal time; a second is still a second in duration locally. If a clock is moving close to the speed of light, however, an external observer (that is an observer not moving with the clock) will observer that it is running slow; a second can become an hour, a day, a month or an eternity[6]. The slowing down or time dilation of the relativistic clock is expressed through the transformation $1/\sqrt{(1-V^2/c^2)}$, first described by Dutch physicist Hendrik Lorentz in the final year of the 19th Century, where V is the relative velocity between the observer (in the so-called rest frame) and the moving reference frame containing the clock, and c is the speed of light. When $V \ll c$ then the Lorentz transform takes on a value of order 1, but as $V \to c$ the transform takes on bigger and bigger values, becoming infinitely large when $V = c$. Accordingly, the closer the velocity V is to the speed of light c, so the slower will the moving clock appear to run to the external observer. This seemingly bazaar result has been proven true, time and time again, in many different experiments. Perhaps one of the most intriguing demonstrations of the time dilation effect is that witnessed in the decay of cosmic ray produced muons[7]. Cosmic rays careen into Earth's upper atmosphere all the time and they are charged particles, typically protons, that have been accelerated to relativistic speeds by the shockwaves associated with supernova explosions and magnetic field interactions occurring deep within our Milky Way galaxy (see Chapter 6 later on). When a cosmic ray chances to hit an atomic nucleus in Earth's upper atmosphere, however, a debris shower of elementary particles is produced, and some of the particles created are ephemeral elementary particles called muons[8]. Laboratory experiments have revealed that the muon is unstable and that it decays into an electron and a neutrino after about 2.2 microseconds – literally in the blink of an eye the muon has gone. In a classical experiment performed by Bruno Rossi and David Hall in the early 1940s, however, a detailed comparison of the number of muons detected at two laboratories (one located in Denver, Colorado and the other at a research station some 48 kilometers away in Echo Lake) revealed that something strange was going on. Critical to the success of the experiment was not that the two test stations were 48 kilometers apart, but that the detector situated at Echo Lake was some 1624 meters higher in altitude than the detector located in the Denver lab. Remarkably, the two sets of muon detector counts were found to be almost identical, even though the number at the Denver detector should have been lower than that of the Echo Lake detector because of the height difference between the two stations, and the additional decays that should have taken place as the muons moved through the height differential. The results of the experiment only made sense if the muon decay time was dilated (that is, ran more slowly) as a result of the high muon speed relative to the experimental equipment. Indeed, the muons were apparently decaying at a rate ten times slower than that which would have been recorded had they been at rest with respect to the test equipment.

If Einstein cast-out the Newtonian concept of absolute time, then some present-day researchers are trying to develop physical theories that do away with time all together; yet others are considering the possibility of there being two, or even more, time dimensions. Time, of course, that great leveler, will tell how these new ideas will play out.

A Pendulum that Can Never Be: II

How fast can a pendulum swing? The answer to this question is provided, at first thought, in Chapter 1 where it was shown that $T \propto \sqrt{L}$ - so, the answer appears to be the shorter the string length of the pendulum L the faster the pendulum will 'tick'. There is a snag with this explanation, however, in that rapid ticking does not necessarily imply a high maximum speed. The maximum velocity of a bob is attained as it passes through the central part of the pendulums arc, and as shown in Chapter 1 this varies as the square root of the string length and the initial off-set angle (assumed to be small). Indeed, from the conservation of energy $V_{max} = \sqrt{2gL(1 - \cos\theta_0)}$, where θ_0 is the initial offset angle of the bob. While the period becomes shorter, as the string length is reduced, the vertical distance through which the bob descends while traveling along its arc becomes smaller and smaller, and the maximum velocity, accordingly, remains small and finite. To increase the maximum speed attained by a pendulum bob, therefore, one, in fact, must increase the length of the support wire – which brings us back to the infinite length pendulum – as also described earlier in Chapter 1. There is, it would appear, no such thing as a realizable high-speed simple pendulum.

All of the above being given, the question concerning relativistic oscillations is still of potential physical interest. When Niels Bohr first introduced his quantized model of the atom in 1915, the model atom was pictured in accordance with Ernst Rutherford's solar system model, with the Sun being analogous to the atomic nucleus and the planets being analogous to the electrons[9]. What Bohr included in his theory was the quantum mechanical constraint that the electrons could only reside in very specific allowed orbitals –

unlike the planets, which can in principle be at any orbital distance from the Sun. Under these conditions it was realized that the electrons would be moving relativistically, and accordingly the electron mass would be modified according to the Lorentz transform, with its mass becoming increasingly large as $V \to c$. Under relativistic conditions the equation of motion for an electron oscillating under the influence of an elastic, linear restoring force (a condition that we shall explore more fully towards the end of the chapter) is

$$\frac{d}{dt}(mV) = -kx, \qquad \text{where } m = \frac{m_0}{\sqrt{1 - \left(\dfrac{V}{c}\right)^2}} \qquad (2.1)$$

where m_0 is the electron rest mass, k is the elasticity constant, x is the displacement away from the equilibrium position, and V is the electron velocity. The solution to equation (2.1) can, in fact, be described in terms of two elliptic integrals – in a manner similar to that for the finite amplitude pendulum problem described Chapter 1 [see equation (1.6)]. The solution to (2.1), as derived[10] by Robert Penfield and Henry Zatzkis in 1956, gives the period of oscillation as:

$$T = T_0 \left(1 + \tfrac{3}{16} gL \left[\frac{\theta_0}{c} \right]^2 + \cdots \right) \qquad (2.2)$$

where $T_0 = 2\pi\sqrt{L/g}$, the period of oscillation for a simple pendulum of length L, $\theta_0 = a/L$ is the initial offset angle and $x = a$ corresponds to the initial horizontal displacement of the bob. As required, in the non-relativistic limit (approached by letting $c \to \infty$), we recover, as indeed we must, the period of oscillation for a simple pendulum. Furthermore, however, it can also be seen that in the limit that the amplitude of swing $a \to 0$, so the relativistic contribution becomes vanishingly small, and once again the period reverts to that of the non-relativistic simple pendulum. More interestingly, however, equation (2.2) reveals that a relativistic pendulum will have a period that is, in fact, longer than that of the same length non-relativistic finite amplitude pendulum. This seemingly strange result comes about, of course, because the effective mass of the relativistic pendulum bob changes with its speed according to the Lorentz transform [as given in equation (2.1)] and this introduces an additional velocity dependent term into the pendulum equation.

The Movement of Time

In spite of what the physicists and philosophers might tell us, we all know (deep down) that time flows – common sense, that overpowering human arrogance, says that this must be so. We see the movement of time all around us: the shifting shadows cast by the Sun, the seasons of the year, and the ever repeating phases of the Moon. Time and motion, they go together like a hand and a glove. Well, we shouldn't really be quite so glib about such deep philosophical topics, but since there is no clear physical definition of what time actually is we might just as well proceed with our common sense notions.

Perhaps the first time measuring device developed by human beings was the sundial. It is among the simplest of instruments to construct (just like the pendulum), and yet it also allows for a highly refined architecture. At the most basic level the sundial gnomon [derived from the Greek for 'one that knows'] casts a shadow that moves through 15 degrees per hour in perfect synchronization with the Sun's motion across the sky. To go beyond this, however, we need to know a little more about how the Sun moves through the heavens.

The Sun's path when projected onto the celestial sphere maps out the great circle of the ecliptic (figure 2.1), and within a ten degree swath, centered on the ecliptic, runs the zodiacal band, which in turn contains the paths of the Moon and all of the planets. The ecliptic is inclined at an angle ε (called the obliquity of the ecliptic) of 23.5 degrees to the celestial equator, the second great circle that runs around the sky, dividing the celestial sphere into its northern and southern hemisphere halves. It is the variation in the height of the Sun above or below the celestial equator that produces the geometric heating effect responsible for our winter and summer seasons. Since time unrecorded it appears that all civilizations have been aware of the fact that the Sun's altitude at noon, when it crosses the meridian (the imaginary arc running from the north, through the observer's zenith and on to the south), can be used to gauge the time of the year. On the day of the summer solstice, for example, when the northern hemisphere has the greatest number of daylight hours, the Sun reaches its greatest yearly altitude above the horizon at noon, and the gnomon will cast its shortest shadow. On the day of the winter solstice, in contrast, we have the greatest number of nighttime hours, and the Sun attains its least high altitude above the horizon at noon, when the gnomon will cast its longest noontime shadow (figure 2.2). The angle between the Sun's maximum and minimum noontime altitudes is twice the obliquity of the ecliptic: $2\varepsilon = 47$ degree.

To the untrained eye the Sun seems to travel at a constant rate of motion through the sky and around the ecliptic, but in reality, as the Babylonian astronomers some 4000 years ago well knew, its rate of motion varies systematically throughout the course of the year, traveling more rapidly between the winter solstice and the spring equinox than during the following interval from the spring equinox to the summer solstice; the former interval taking 89 days to complete and the latter 93 days at the present time. This, we now know is a result due to Earth's orbit being elliptical. The variation in the Sun's apparent speed as it moves along the ecliptic is only slight, but on a six-month basis it steadily oscillates between a minimum of 28.2 and a maximum of 30.0 degrees per month. These small differences throughout most of history have not been important in any utilitarian sense, and with respect to judging the times of seasonal change they are essentially irrelevant. While the daylight hours, the seasons and the years can be measured by the Sun and its companion shadow, the accurate measurement of shorter time intervals, of say fifteen minutes or a few hours requires us to look below the heavens, and indeed they require the full application of human imagination and ingenuity to be realize.

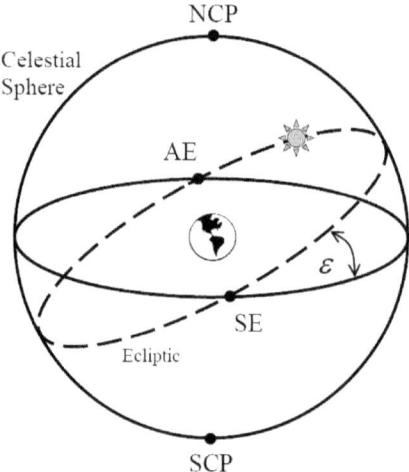

Figure 2.1. The celestial sphere and the path of the Sun along the ecliptic. The points of intersection between the ecliptic and celestial sphere indicate the times of the equinoxes, when there are equal hours of day and nighttime. The summer solstice occurs when the Sun is at its greatest height above the celestial equator, and this corresponds to the day with the longest daylight hours in the northern hemisphere. Six months after the summer solstice the Sun is at its winter solstice location when it is at its greatest angle below the celestial equator.

In perfect union with our minds-eye image both time and water are flowing entities; they approach from somewhere distant and in an unstoppable rush they surge past us, only to disappear again beyond the unreachable, far-off horizon. We only experience the ever changing 'instant' of the flowing river and the passage of time. It seems only appropriate therefore that this commonality of motion should find its unity in the water clock, or clepsydra of the ancient civilizations. The word clepsydra translates directly as *water thief*, and just as time steals away our all too few mortal seconds on Earth, so the steady drip of the water clock counts out the passage of time. Evidence for the design and use of clepsydra can be found as far back as the 16th century B.C., making them the oldest known time-measuring devices after the sundial. The essential operational principle of the clepsydra is simplicity itself and entails the capture or entrainment of a steady flow of water. The flow of time is then marked out either by the raising of a float placed within a water capture cylinder, or via wheelwork and gearing turned by the weight of an entrained water column. The Roman writer and architect Marcus Vitruvius Pollio (born c. 80 B.C.) describes the construction of both sundials and water clocks in his only surviving work *De architectra*. One of the clepsydra he describes is an anaphoric clock (from the Greek 'carrying back') that was specifically designed to drive, through one rotation per day, a circular disk upon which the stars and constellations were portrayed. Through the anaphoric clock, a steady drip of water is miraculously transformed into celestial time and motion. Indeed, the anaphoric clock is a dynamical representation of the very cosmos. Perhaps the greatest clepsydra ever built was that designed by Chinese astronomer and statesman Su-Song (1020 – 1101). Constructed in the town of Kaifeng (in Henan Province) circa 1090, the astronomical clock tower housed a water-driven armillary sphere[11] and a series of 113 striking bells that sounded out the hours.

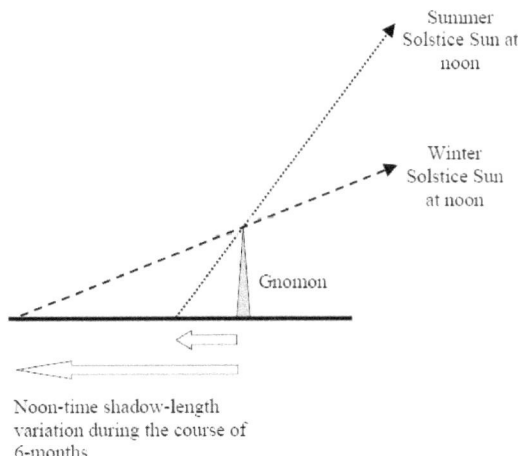

Figure 2.2. The variation in the shadow length cast at noon on the days of the winter and summer solstice. The angular variation in the Sun's altitude amounts to twice the angle of the obliquity of the ecliptic.

Just as the steady meandering of a river reminds us of the stately passage of time, so too does the motion of shifting sand and these images are encapsulated in the hourglass. More often appearing as the humble 3-minute egg timer in modern-day kitchens, the hourglass first became popular in the 14th century, although its origins stretch back much further into history. Constructed of two glass bulbs connected by a narrow tube, the hourglass is a perfect instrument for measuring-out short time intervals. The hourglass does not display time in the sense of a clock, but it counts out the passage of time at a constant rate – remarkably, the flow rate of the sand grains between the glass bulbs is independent of the depth of sand in the upper reservoir - a result that is in contrast to the situation in a water reservoir, where the flow rate varies according to the pressure set by the weight of overlying layers. That the mass flow rate of sand in an hourglass is constant can be demonstrated through dimensional analysis. Following the procedure outlined in Chapter 1; let us assume that the mass flow rate of sand W is expressed in terms of the density of sand grains ρ, the acceleration due to gravity g, and the diameter D of the opening between the two glass bulbs. We now write

$$[W] = k[D]^A [g]^B [\rho]^C = l^0 m^1 s^{-1} \qquad (2.3)$$

inserting the appropriate units for D, g and ρ, equation (2.3) yields:

$$l^0 m^1 s^{-1} = [l^1 m^0 s^0]^A [l^1 m^0 s^{-2}]^B [l^{-3} m^1 s^0]^C \qquad (2.4)$$

Equating powers of the base units for length l, mass m and time s, we find from equation (2.4) the result that $C = 1$, $B = \frac{1}{2}$ and $A = 5/2$, and this tells us that

$$W = \frac{dm}{dt} = k \rho \sqrt{g}\, D^{5/2} \qquad (2.5)$$

Equation (2.5) gives us the expected result that the flow rate is related to the acceleration due to gravity, and that the wider the opening between the two glass bulbs so the greater the flow rate of sand grains. The density term in equation (2.5) further indicates that the more grains there are per unit volume in the sand reservoir, so the more grains will flow through the opening per unit time – note the density of sand grains is independent of the height of the sand in the reservoir (assuming no mass compaction effects), and hence we have our result that the flow of sand from one glass bulb to the other in an hourglass runs at a constant rate[12] – it is an isochronal device.

Just as sand, in our minds eye at least, goes with the sea, so the hourglass was the first time instrument used by mariners. Ferdinand Magellan, who we shall encounter again in the next chapter, equipped each of the ships that set out in his flotilla to circumnavigate the globe with 18 one-hour hourglasses apiece. Throughout the entire voyage an hourglass would be turned, each hour, by the ships page, and the progress and weather conditions would be duly noted in each ship's log.

Celestial Time

Although it hardly enters into the ken of our everyday lives now, the heavens used to be the one clock that everyone could read. Whether a highly schooled and noble king, or a poor illiterate peasant, the passing of time and the measure of the seasons was available for all to read in the sky. While the motion of the Sun through the heavens and the shift in its slave-driven shadow ruled the hours of the working day (figure 2.3), it was the stately motion of the celestial sphere that marked-out the slower time of the seasons. Indeed, the indomitable rolling of the be-caged stars of the zodiacal band tick-off the hours each and every night, and they mark the passage of the seasons over the months. The constellations drift into and then out of our gaze during the course of the year, and the helical rising and setting of bright stars and prominent constellations decry the ebb and flow of the yearly calendar. Indeed, as Hesiod (circa 650 B.C.) so well observed in his pastoral poem *Works and Days*, "*When Zeus has finished sixty wintry days after the turning of the Sun, then the star Arcturus leaves the holy stream of Ocean and first rises brilliantly in the twilight .... when spring is just beginning*"[1] To the ancient Egyptians the helical rising of our brightest nighttime star Sirius was a heaven sent sign of forthcoming rebirth and renewal, since the great, life-providing Nile would soon thereafter undergo its inundation, renewing the fertility of the soils out of which the following year's crop would be grown. Sirius was the yearly celestial clock, and everyone irrespective of learning, social station or power could see it and know the time.

The Greek philosopher Plato, as one might suspect had a rather more complex notion of time. The true universe for Plato was governed by rules that were patterned on an ideal; an intellectual realm of perfect geometrical shapes and forms that were unchanging and timeless. Unlike the ideal realm, however, the real world in which we live-out our daily lives is forever changing and corruptible. Time according to Plato is a manifestation of the never-ceasing progression of change, or earthly struggle that we perceive as the physical world strives (albeit fruitlessly) to attain its ideal form; time for Plato is a kind of bridging mechanism; a perception arising from the attempts of the real world to attain some form of idealized perfection. In his classic dia-

---

[1] The essential meaning of this verse is that 60 days after the time of the Winter Solstice (the shortest day), the bright star Arcturus (in the constellation of Boötes) will rise above the horizon (in the East) just before the Sun. This indicates the time at which Arcturus has its so-called helical rising.

logue *Timaeus*, written circa 360 BC, Plato writes that time was created as the moving image of eternity. In this manner, Plato saw the flow of time as a consequence of the everlasting 24 hour rotation rate that had been imposed upon the first moved and outer most sphere of the stars (the *premium mobile*), and the never ending 'desire' of ordinary (that is earthly) matter to find its final ideal form and resting place. Perhaps, ultimately, Plato's is a rather sad philosophy, since it denies the possibility of ever achieving the perfection that one is incessantly driven to find.

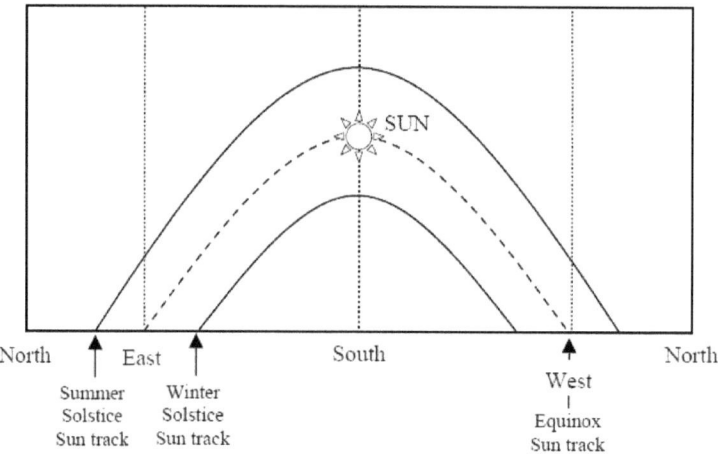

North   East   South   West   North

Summer   Winter

Solstice   Solstice   Equinox

Sun track   Sun track   Sun track

Figure 2.3. The motion of the Sun through the sky. At the time of the equinoxes, when the daytime hours exactly equal the nighttime hours, the Sun rises due East and sets due West. Each and every day of the year the Sun reaches its greatest altitude above the horizon, for any observer, in the due south position at local noon. The greatest northerly rising of the Sun occurs on the day of the Summer Solstice, while the greatest southerly rising takes place on the day of the Winter solstice.

While the stars and constellations tick-off the slow course of the year, the ever-changing Moon rhythmically dances around the zodiac and marks out the days of the month. Full Moon to full Moon, a span of 29.531 days was an interval of time that all could measure - the time being displayed by the Moon's cast of silvery light. To all civilizations that have followed the Moon's motion in detail it has been clear that it is our closest celestial companion, and that it orbits around the Earth. Even the great Tycho Brahe (1546 - 1601), who set the planets in orbit about the Sun and the Sun in orbit about the Earth, had the Moon orbit a central Earth.

Shakespeare, as always, seems to have said it best when he wrote, "the moon she is an errant thief and her pale fire she snatches from the Sun".

Such basic astronomy was obvious to all who had studied the Moon, and certainly as far back as circa 200 B.C. we find Aristarchus of Samos in the one text of his that has come down to us, *On the sizes and distances of the Sun and Moon*, providing a complete and correct account for the lunar phases (figure 2.4). But the Moon has cast us a cruel blow. Its phase cycle, the time interval from, say, new Moon to new Moon is at odds with Earth's yearly track around the Sun. Mathematically we can write T(1-year) / T(1-lunation) = 365.25 days / 29.531 days = 12.368 – the division is none-harmonious and within this result lies the centuries old problem of making the solar and lunar cycle calendars fit into one coherent scheme.

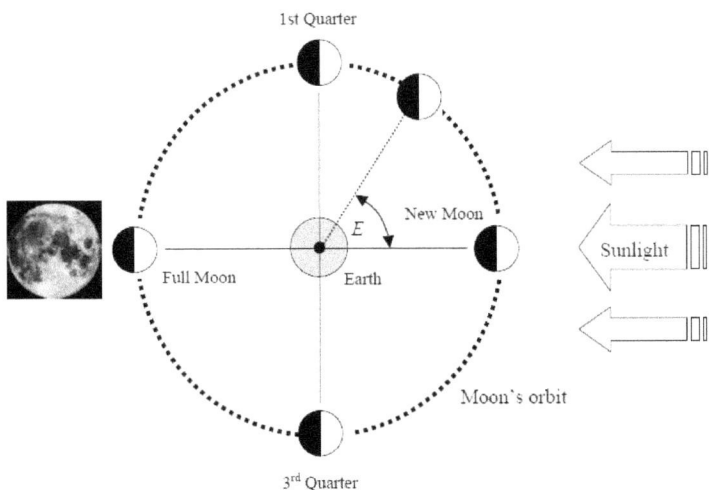

Figure 2.4: The phases of the Moon. At any one instant half of the Moon's surface is illuminated by the Sun. The phase of the Moon is dictated by how much of the illuminated half-fraction a terrestrial observer can see. The phase angle E varies from $E = 0°$ at the time of new Moon, to $E = 180°$ at full Moon. The 1st and 3rd quarter phases correspond to $E = 90°$ and $E = 270°$ respectively.

While the reconciliation of the solar and lunar calendars is a far from trivial task, the continuous and stately phase change of the Moon is reassuringly periodic. Indeed, there is a pendulum-like heartbeat to the Moon's phase, with the illuminated fraction of the Moon's disk varying according to the equation $A = [1 - \cos(E)] / 2$, where $E = T(2\pi / P)$, and where $P = 29.531$ days is the Moon's synodic period with T being the time since New Moon. The pendulum-like analogy is revealed through the time varying cosine term. In the motion of the Moon about the Earth, however, there is an even more surprising pendulum at work. If we assume that the Moon has a perfectly circular orbit of radius R_{Moon}, and that it moves with constant speed such that it completes one orbit in P_{orb} = 27.32 days (the rotation period with respect to the stars), then the orbital period is equivalent to that of a pendulum of length $L = R_{Moon}$. This result follows from equation (1.4) of Chapter 1, when the gravitational attraction term is set to $g = G\,M_\oplus\,/\,R^2_{Moon}$ (at its core this identity is actually equivalent to Kepler's third law of planetary motion[13]).

By trying to be more Greek than the ancient Greeks, and especially more Greek than Claudius Ptolemy, Nicolaus Copernicus (1473 - 1543) cast out the ancient assumption of a central and stationary Earth, replacing it instead by the "glorious lamp" of the Sun. Preserving, however, the ancient idea that all planetary motion must be constrained to move along perfectly circular orbits with a uniform speed, Copernicus set the newly released Earth in motion, third planet out from the central light. The Earth, not the celestial sphere, was now in motion, spinning on its axis once every 24 hours, and orbiting the Sun once every 365.25 days. Time had finally caught up with the great body of Earth, its spin rate, and its spin rate alone would henceforth determine the measure of change. As founding Cambridge Platonist and mathematics lecture John Smith so eloquently wrote in his *Select Discourses* of 1660, "this world indeed is a great horologe to itself, and is continually numbering out its own age".

Mechanical Time

The exact moment when the first mechanical clock was constructed is unknown. The very earliest geared device that might reasonably be considered a clock, or at least a calendar counting machine, is that of the Antikythera mechanism which was discovered in an ancient shipwreck in early 1900. Having resided below the western Aegean seas for some twenty centuries the Antikythera mechanism is hardly in any condition for us to be sure of its intended use, but this hasn't stopped modern scholars from rampant speculation[14]. It seems reasonably clear that the mechanism was a geared machine built circa 100 B.C. with the intention of describing the motion of the Sun and Moon, and possibly the planets, along the zodiac, and it could also have been used to predict the times of eclipses. Interestingly, an Olympic dial is also found on one of the surviving fragments, and this would have traced out a sociological calendar event rather than an astronomical one. We will never know for certain who made the Antikythera device, other than it was clearly made by the hands of a very skilled craftsman, and we will never known for whom it was constructed, but it certainly demonstrates the early origins of the human desire to mechanize the workings of the heavens and to trap the flow of time within the movement of spinning gears.

The origins of the mechanical clock are, to say the very least, obscure. A dark and impenetrable vale obstructs the backward gaze of the historian, and while it is impossible to believe that there were no advanced devices made after the Antikythera mechanism, the fact is, there is a long silence and complete lack of extant prototypes between the time of its construction circa 100

B.C. and the appearance in the late 13th Century of the first *bona fide* mechanical clocks.

The word clock is derived from the Latin *clocca* meaning 'bell', and this word origin betrays the underlying reason for the design of such machines in the first place; the chime of a bell still being the potent sound that reminds us of duties to be performed. In medieval monastic life there was no more important duty than the recital of the daily prayer cycle, with each calling to prayer being performed at the correct time, whether during the day or in the depths of night. And, it is within the ecclesiastical world that we find the origins of the mechanical clock – machines designed to dutifully chime out the hours, rain or shine, day or night, in order that respectful praise could be offered to the omnipotent maker. Even in modern times our everyday language contains linguistic fossils from the medieval period. The word noon, indicating mid-day, is derived from *Nones* the prayers (initially at least) offered at 9 o'clock in the morning. And, indeed, the word o'clock (the concatenated form of 'of the clock') itself was introduced in medieval times when it became possible to expressed time in equal length, mechanical clock hours, rather than canonical hours which referred to specific moments of the day, such as 'before dawn', when the Matins prayer cycle was to be said. With the invention of the mechanical clock, all of a sudden, and unexpectedly, life changed - time was henceforth something that humanity was entrained to live by, rather than live through.

The first artificer of a mechanical clock that is well described, and for which the original plans still exist, was Richard of Wallingford (1292 – 1336)[15]. The brilliant son of a humble blacksmith, Richard was educated at Oxford University and, in 1327, was elected Abbot of St. Albans. It was at St. Albans that he began the design and construction of his great timepiece. Completed circa 1330, the clock was a mechanical model of the heavens, its faceplate showing the annual motion of the Sun through the zodiac, the rise and set times of the brighter stars and the turning of the seasons. Present day visitors to St. Albans Cathedral can view a working replica of Wallingford's clock and it is indeed an impressive instrument.

As with the running of water clocks, gravity is the energizing force behind the operation of early clocks. The foliate provided the mechanical heartbeat of such machines, and its pendulum-like oscillation, back and forth, controlled the rotational advancement of a saw-toothed crown-gear - via the side-to-side motion of the verge and its twin pallets. Encapsulating the quintessential art of motion, the early mechanical clocks divided the workings of the universe into measurable slices of time – the working day, and the motion of the planets were all reduced to the mesmerizing sway of mechanical oscillation. The great *astrarium* designed and built by Giovanni De Dondi (1318 – 1389)[16], Professor of Medicine at Padua University, showed not only mean and sidereal time, but also the zodiacal locations of the planets. Designed according to the properties of Claudius Ptolemy's planetary hypothesis, the *astrarium* not only incorporated elliptical and oval shaped gears to account for the variable motion of the Moon and Mercury, it also calculated and displayed the dates of fixed and moveable feast days. The entire workings of the cosmos and the religious activates of the Christian world were all counted, displayed and made sensible by this one great machine – and at its heart was the stately oscillation of the foliate bar; a horizontal rigid pendulum.

While an undoubted mechanical marvel of its time, Dondi's *astrarium* was doomed to become lost, and presumably destroyed by the time that the mid-16th century arrived. At about this same time, however, perhaps the greatest of all celestial timekeepers was constructed. Designed by Eberhard Baldewein circa 1565 the Dresden Planetenlaufuhr (Dresden Planet Clock) was built for Duke August of Saxony, and it is a glorious temple of engineering wonder[17]. This highly complex device is both beautifully crafted and still extant – indeed, it rivals the Antikythera Mechanism for its brilliance of design and mind-boggling structure. Like Dondi's *astrarium*, the Dresden Planetenlaufuhr is designed according to Ptolemy's Earth-centered cosmology, which posits that the heavenly bodies move at constant speed along circular orbits. Since, however, the Sun, planets and Moon do not actually move as Ptolemy described, the gearing had to be constructed so as to save the appearance, and accordingly the motion is described in terms of compounded (that is epicyclical) circular motion, and with gears cut with uneven teeth spacing. The Moon dial for the Dresden Planetenlaufuhr, for example, is composed of eccentrically mounted gears and a drive-work system that controls the variable rotation rate of a series of concentrically mounted rings – the net result being a description of the Moon's location around the ecliptic, its ecliptic latitude, and its phase. The Moon dial also contains a smaller sub-dial indicating when and where on the ecliptic partial eclipses might be observed. Other dials on the Planetenlaufuhr show the locations of the Sun, the planets Mercury through to Saturn, an astrolabe and a celestial sphere. The clock further chimes hour and quarter hour times, and has a dial that indicates the year, the appropriate date of Easter for the indicated year, and a whole series of Saint Days, religious festivals and annual holidays.

Three hundred and fifty years further-on from the time of Richard of Wallingford the perceived universe had changed beyond all recognition. Gone is the Earth-centered cosmology of Ptolemy and shattered are the ideals of the ancient Greek philosophers. The mathematical brilliance of Johannes Kepler had shown, in the early 17th Century that planetary motion is constrained to follow elliptical orbits with variable speeds, and the genius of Isaac Newton had further explained all astronomical dynamics in terms of the universal theory of gravitation. The universe according to Newton's thinking was fine-tuned and needed only the occasional nudge, from the great architect, to keep everything running smoothly[18]. The universe after Newton became a rare and precisely made clock, its workings being as predictable as, well, clockwork. The Orrey (figure 2.5) was the great symbol of this new era[19]. The first such device was constructed circa 1710, and named in honor of the patron Charles Boyle, 4th Earl of Cork and Orrery, in Ireland. The Orrery reduced planetary motion to the predictable mathematics of mechanical motion and cosmic time was trapped, once again, in the machinations of intricate gearing. Later improvements to orrery design[20], initially introduced by John Theophilus Desaguliers in 1732, incorporated non-circular formers and gears to mimic the elliptical motion of the planets, as well as that of the comets (figure 2.6), thereby illustrating the consequences of the first two of Kepler's laws of planetary motion[13]. All was in harmony at least in the sky, if not on Earth.

For all of its mechanical elegance, the great harmony of the clockwork heavens was doomed not to last, and as we saw in the opening sections of this chapter time, motion, and predictability all began to unravel as we move towards the modern era. The fractal fingers of chaos, the unpredictable stirrings of strange attractors and the power of orbital resonances now underlie our understanding of solar system dynamics (as we shall see in Chapter 6); the steady driving force of time is still there, but the resultant motions are well beyond the workings of any mechanical clock to fully describe.

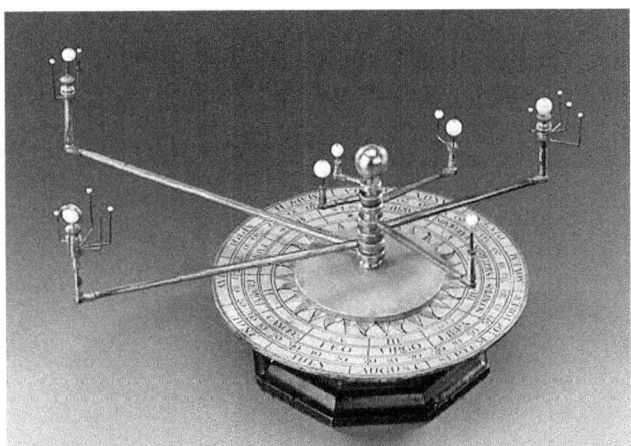

Figure 2.5. An Orrery for demonstrating the relative orbital motions of the planets from Mercury to Uranus. Only two Uranian moons are displayed in the model indicating the machine was produced after 1787, but before 1851 when a third moon was discovered.

The Chaos of Pluto

Ever since its discovery by Clyde Tombaugh on 18 February 1930, Pluto has been both an enigmatic and a problematic object for astronomers to quantify. It is the proverbial object that breaks all the rules and it refuses to be easily classified. Indeed, for 75 years astronomers' listed Pluto as a planet, although it was long known to have relatively few characteristics in common with its other eight companions. All this uncertainty came to a head during the International Astronomical Union meeting held in Prague during August of 2005, at which time Pluto was re-classified as a Dwarf Planet and the archetype of the Plutoids[21]. Even this reclassification didn't please all practitioners[22], and, for example, according to the Minor Planet Center at the Smithsonian Astrophysical Institute, located at Harvard Observatory in Massachusetts, Pluto is officially designated as a Kupier Belt Object[23] number 134340.

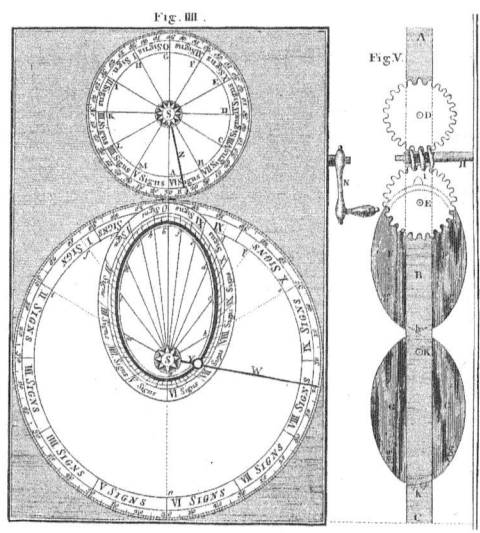

Figure 2.6. The cometarium, as introduced and described by James Ferguson in his *Astronomy Explained upon Sir Isaac Newton's Principles*, first published in 1756. The variable drive speed, elliptical formers are illustrated to the right of the image.

Nothing, for so it would seem, about Pluto fits the standard picture. This misfit trait was further enhanced when in June of 1978 astronomer James Christy, working at the United States Naval Observatory Flagstaff Station in Arizona, discovered that Pluto has a large companion Moon – the dwarf planet Charon. It was later realized that Pluto and Charon actually constitutes a binary dwarf planet system since the center of motion about which they both orbit is located in the space between them, rather than at some point within the interior of Pluto. Not only this, observers using the Hubble Space Telescope discovered in 2006 that two small (about 50 km in size) companion moons, subsequently named Nix and Hydra, orbit about the Pluto-Charon system's center of mass. Additional small moons S/2011 (134340) 1 and S/2012 (134340) 1 (eventually named Kerberos and Styx respectively) were also found in July of 2011 and July of 2012 - the strange world of Pluto just keeps getting stranger and stranger.

While Pluto's discovery, by someone or other, was inevitable, the remarkably fact is that the reasons for initiating the quest to find it in the first place were entirely erroneous. The instigator of the search program was wealthy Bostonian businessman Percival Lowell. Obsessed with the idea that a planet (indeed, Planet X as Lowell called it) existed beyond the orbit of Neptune, he founded in 1894 the Lowell Observatory at Flagstaff in Arizona. The main reason for believing in the existence of Planet X stemmed from the discrepancies that were reported with respect to the observed and the predicted positions of both Neptune and Uranus. The idea that Lowell was exploiting, therefore, invoked the existence of another planet, Planet X, as a means of providing an additional gravitational interaction to act upon Uranus and Neptune. It was this additional gravitational interaction that caused the observed positions to vary with respect to their predicted positions. The problem for Lowell, historically speaking, however, is that the anomalies in the motion of Uranus and Neptune were entirely due to the uncertainties in their orbital parameters and mass. With no anomaly there is no need for an additional planet, and yet, remarkably, one was found.

The orbit of the Pluto-Charon dwarf planet system is markedly eccentric, and indeed for a few brief years it can orbit closer to the Sun than the outermost planet Neptune. That Pluto and Charon have survived gravitational disruption by Neptune is a further remarkable characteristic of its highly inclined orbit, and the fact that for every three orbits about the Sun by Neptune, Pluto and Charon complete two. This 3:2 mean motion orbital resonance ensures that very close encounters with Neptune never occur and accordingly the Pluto-Charon orbit is stable. Even more remarkably, perhaps, is the fact that Pluto and Charon have a whole retinue of companions, the so-called Plutinos, that march to the same lock-step beat of a 3:2 resonance with Neptune, thereby avoiding very close orbit disrupting encounters - the topic of orbital resonances will be picked-up again in Chapter 6.

While the 3:2 mean motion resonance with Neptune has allowed the Pluto-Charon system to remain stable for the past 4.5 billion years, all is not necessarily harmonious with respect to its continued stability in the deep future. This is a truly startling result, but one that cuts to the very core of the complexity of gravitational interactions acting between multiple bodies over extended periods of time. Gerald Sussman and Jack Wisdom, both at MIT, first showed that the orbital motion of Pluto is chaotic in a series of papers[24] published in 1988. To complete their computations a special electronic orrery was constructed to help speed along the complex calculations. Indeed, the need for the electronic orrery was entirely practical since it effectively resulted in the extensive orbital calculation set being produced 50 to 60 times faster than on a stand-alone computer (a VAX 11/780). The orrery cut the computation time for a simulated 100 million year run from about one year to approximately one week. With the electronic orrery over-drive in place Sussman and Wisdom investigated the possible evolution of Pluto's orbit over an un-precedented 845 million years simulated time span. Since there is always some uncertainty in the initial starting conditions a whole series of calculations were performed, with the divergence of initially neighboring model Pluto's being closely followed. Remarkably, they found that the Pluto-Charon orbit is chaotic with initially close trajectories diverging more and more over time. The use of the term chaotic should be qualified at this stage, and it is taken to mean that the specific location of Pluto in its orbit about the Sun becomes unknowable in the deep future, it does not mean that Pluto will be scattered to some new orbit or ejected out of the solar system. Here the chaotic diver-gence of neighboring trajectories means that it is not possible to predict ahead of time exactly where Pluto will be in its orbit about the Sun on time-scales longer than 10 to 20 millions of years. In practical terms, however, this is not an especially disastrous problem since on shorter timescales of say many hundreds or even many hundreds of thousands of years we can still accurately predict exactly where Pluto will be in its orbit. Philosophically speaking, however, the remarkable point we learn from Pluto's chaos is that we cannot, for so it appears, usefully run numerical predictions into the in-definite future: there is limit to the steady and reliable clockwork of the New-tonian heavens.

The Heartbeat of Time

Although the predictable clockwork paradigm of the heavens has long been brought into question, the swing of the pendulum is still, very much, the potent symbol for describing the passage of time. The successful harnessing of the pendulum, as a control mechanism, within clocks and timepieces, however, was an advancement that came along relatively slowly, and relatively late in the history of horology.

The isochronal properties of the simple pendulum (discussed in Chapter 1) were first investigated during the mid to late 1500s. As with many discov-eries attributed to Galileo Galilee we cannot be sure if he really was the first observer to note the constant periodicity of a pendulum's swing, but as an old man he recounted such a story to his less than trustworthy biographer Vincenzeo Viviani. Writing seven years after Galileo's death Viviani ex-plained that in 1584, when still an undergraduate student Galileo had noticed the constant period of swing betrayed by a suspended lantern in the Cathe-dral at Pisa. The story is problematic since the lamp that Viviani refers to, still called Galileo's lamp to this very day, was not actually installed at the Cathe-dral until well after Galileo graduated – such, however, are the vagaries of history and the recollections of old men. The first real hint that Galileo was actually considering the properties of the pendulum do not appear until 1588, and specifically 1602, when Galileo wrote to a friend explaining an experi-ment that he had conducted with two pendulums of equal length, one having a lead bob and the other an equally-sized cork one. The experiment, Galileo explained showed that the period of swing was independent of the pendulum mass – a result that probably required some considerable degree of guess-work on Galileo's part, since such experiments are notoriously difficult to 'make work'. The fact that the period of swing for a pendulum varies system-atically with the length of the support string [as described in Chapter 1] is often attributed to Galileo, but it was first clearly articulated by the French mathematician and Minim monk Marin Mersenne. In 1641 Galileo, then totally blind and infirmed, further explained to Viviani that a pendulum might be used to regulate the going of a mechanical counting device (having charac-teristics more like a metronome than an actual clock). Indeed, in the last few months of his life, Galileo described the workings of an escapement mecha-nism to his son Vincenzio, but nothing practical, in the sense of a working prototype, seems to have developed from these discussions. Indeed, the first description of a true pendulum regulated clock (figure 2.7) is attributed to Christiaan Huygens (1629 - 1695) who presented the new design in his *Horo-logium Oscillatorium*, published in 1673.

As we saw in Chapter 1 the isochronal nature of the pendulum only ap-plies in the situation of small angles of oscillation, and this was a problem since early clocks typically required large angles of swing in their going. Huy-gens solved this problem by making use of the tautochrone property of the cycloid.

Many improvements to the running and accuracy of mechanical time-pieces appeared during the later part of the 17th century. New designs for clock escapements were introduced, and watches driven by spirally wound strips of metal were developed by Huygens, Hooke and many others. It was a time of great experimentation, and inevitably controversies arose. Robert Hooke and Huygens, for example, soon fell-out over who produced the first design for the balance spring clock (although history appears to favor Hooke's priority claim with respect to conception, it was Huygens who man-ufactured the first working example of such a watch). The matter of priority was not just one of intellectual pride; there was money to be made in the production of clocks, and a great financial prize was waiting for the first per-son to build a clock that could keep time to better than a few seconds per month – as we shall see shortly.

The regular going of a clock is not solely due to the isochronal (or near so) properties of the pendulum. Key to the process of constant going is the escapement – a mechanical device that transfers energy to the actual gearing of the clock and essentially allows the number of pendulum oscillations to be counted (this was the basic idea behind Galileo's counting device). Indeed, the definitive *tick-tock* sound of a clock is due to the gear train stopping and starting as the escapement oscillates between its drive and locked positions.

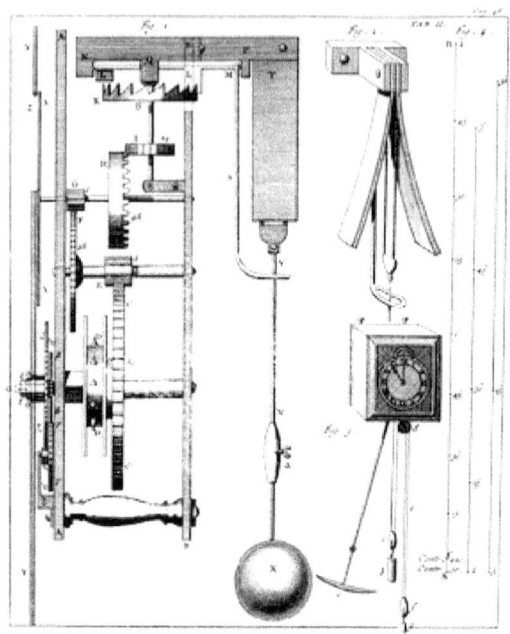

Figure 2.7. The first design for a pendulum regulated clock. The cycloidal checks used to correct for large amplitude swings can be seen to the upper right hand corner of the diagram.

Not only does the escapement control a clock's counting cycle, however, it also imparts a small impulse of energy (from the drive weights or coiled spring) to the pendulum itself. These small energy-jolts will counteract any frictional energy losses suffered by the pendulum mechanism and will, thereby, keep it oscillating. This, indeed, is an important action and throughout the entire history of mechanical time keeping, from the ancient Greek clepsydra, to Richard of Wallingford's tower clock, to the modern day, the workings of the escapement and the minimization of frictional energy losses has been the motivating force behind horological advancement. For all this, however, the design of watches and clocks is all about tribological compromise – a perfectly frictionless clock will not, in fact, keep perfect time. If, for example, the pendulum moved in an entirely frictionless manner, then any small jolt that it might receive (while dusting or during winding, for example) would permanently change its speed. With a small amount of friction and an escarpment mechanism, however, the pendulum can effectively forget its past (that is any small energy gains or losses) and keep regular time. Astronomer Royal, George Biddell Airy, in a famous paper entitled, On the disturbance of pendulums and balances and a theory of escarpments (published in the *Transactions of the Cambridge Philosophical Society* in 1827) considered the effects of small additional accelerating forces upon a pendulum. Accordingly, Airy modified the simple pendulum equation to read

$$\frac{d^2x}{dt^2} + \frac{g}{L}x = f \tag{2.6}$$

where f is a small, constantly applied perturbing acceleration. While equation (2.6) has no general, always applicable, analytic solution, it does have approximate solutions provided that f is small. Indeed, following the methodology described in Chapter 1, solutions to (2.6) can be found by setting $x(t) = a\sin(\omega_0 t + \varphi)$, where $\omega_0^2 = g/L$ and where it is required that the amplitude a and phase φ are time variable quantities (as opposed to the earlier assumption that they are constant). Accordingly, Airy was able to determine the fractional increase in the amplitude Δa and the period ΔT in terms of integrals relating to the variation of the applied additional force f over the pendulums arc While Airy considered many different ways in which the small additional force f might arise (e.g., circular error – as derived from equation 1.7), he was able to show that it was not possible to make Δa and ΔT simultaneously zero. The best escapement mechanism, however, turned out to be the so-called dead-beat escapement – this type of escapement provides the drive impulses symmetrically about the lower point of the pendulum's arc.

The reason why an escapement is required to keep a pendulum from gradually slowing down can be seen by looking at a second modification to the equation for the simple pendulum. In this case a frictional term is introduced leading to the equation

$$\frac{d^2x}{dt^2} + \gamma\frac{dx}{dt} + \frac{g}{L}x = 0 \tag{2.7}$$

where it has been assumed that the frictional term varies according to the speed with which the pendulum is moving – this assumptions relates to what is otherwise known as Stokes law and we will have much more to say about this law, and the solutions to equations such as (2.7) in Chapter 6, but for the moment we simply need to know that it modifies the oscillating sine term for $x(t)$ through the addition of a time decreasing exponential component $\exp(-\gamma t/2)$. The larger the value of γ, that is the greater the amount of friction, so the more rapidly the pendulum comes to a stop. It is through the escapement, and its small energy jolts, that the exponential, amplitude-destroying term is counteracted. This all-important action of the escapement does come at a price, however, and the period of the pendulum's swing is slightly modified over the ideal case, becoming

$$T = \frac{2\pi}{\sqrt{\left(\frac{g}{L} - \frac{1}{4}\gamma^2\right)}}$$

(2.8)

The period of oscillation for an escapement regulated pendulum is slightly longer, therefore, than that for a similar length, friction-free pendulum. While some friction is vital to the steady running of a clock's mechanisms, too much friction is clearly a bad thing; it will cause moving parts, for example, to wear out quickly and then jam. Simply adding a lubricant to reduce friction, however, is not necessarily beneficial either, since lubricants are temperature sensitive (being more viscous and sticky when cold) and they will trap dust and small metallic particles that will compromise gear machination and also advance wearing. The understanding and correction where required of frictional effects within clock mechanism is all part of the clockmaker's life-long learning skill, and indeed such knowledge was of vital importance, as we shall see shortly, in the solution of the navigational problem associated with longitude determination.

In 1820 the Danish physicist Hans Christian Oersted discovered that passing an electric current through a coil of wire could produce a magnetic field – this was a monumental discovery, and one that eventually led James Clerk Maxwell, in 1865, to the understanding that light must be some form of electromagnetic radiation. It was also realized, however, that an appropriately constructed electromagnetic coil might be used to provide an impulse to a pendulum and there-by keep the running of a clock regularized. The first patent for such a device was awarded to Scottish clockmaker Alexander Bain in 1841. A year later Swiss engineer Matthaus Hipp devised a pendulum controlled switch that provided an impulse only when the pendulum arc dropped below a certain set level, rather than proving an impulse during every swing as in the case of Bain's patented system. During the 1851 Great Exhibition in London, British clockmaker Charles Shepard displayed an electrically regulated clock system that caught the eye of then Astronomer Royal, G. B. Airy, who promptly ordered a set of such clocks (a master clock and set of slaves) for the Greenwich Time Service. This government-run service distributed time signals for then burgeoning British railway networks, and was responsible for the famous 1 p.m. daily time-ball drop at Greenwich Observatory – an event orchestrated to enable ship captains to set their chronometers before a sea-voyage.

The Clock of Boys

Charles Vernon Boys (1855 – 1944) was a British experimental physicist who invented a number of wonderful instruments. He was a pioneer of high speed photography, he wrote about the physics and geometry of soap bubbles, and he invented a hyper-sensitive device for measuring radiation. In later life,

Boys wrote a gardening book entirely devoted to weeds, and he performed experiments to determine the influence of sound upon the behavior of garden spiders. Famed writer H. G. Wells actually took classes from Boys, and in his *An Experiment in Autobiography* (Vol. 1, published in 1934), we find the following wonderful (albeit disconcerting from a student's perspective) eccentric description, "I thought him one of the worst teachers who ever turned his back upon a restive audience, messed about with the blackboard, galloped through an hour of talk and bolted back to the apparatus in his private room…. Boys shot across my mind and vanished from my ken with a disconcerting suggestion that there was a whole dazzling universe of ideas, for which I did not posses the key". Most importantly, for our topic, however, Boys invented a method for the manufacture of long thin strands of fused quartz fiber. Indeed, this latter invention made his laboratory a rather dangerous place to work, since he used a crossbow and bolt attachment to pull-out the strands of fiber from a pool of molten quartz. In 1935, politician and 1st Baronet to the Privy Council, Sir Richard Paget wrote a poem to celebrate the occasion of Boys' eightieth birthday, and in the second stanza of his friendly doggerel we find,

> *Why did his bold, untrammeled thoughts*
> *Conceive the scheme of fusing quartz,*
> *Using an arrow, as it fled,*
> *To draw a microscopic thread,*
> *And from the fusion to "unreel"*
> *A gossamer more true than steel,*
> *Which every Physicist enjoys?*
> *The fact is this: Boys will be Boys.*

The thin, strong and highly durable strands of quartz fiber enabled Boys and others researchers to construct high precision torsion balances and pendulums – about which we shall have much more to say in Chapter 4. For the moment, however, it is towards a pendulum clock designed by Boys that we turn. His design, which was described[25] in 1877, is innovative in that it attempts to bring together, and then solve, many of the practical problems associated with the construction of highly accurate, long-running, pendulum regulated clocks. The pendulum and regulating arrangement for Boys' clock is shown in figure 2.8. The first innovation in his design was to encase the pendulum in an evacuated glass case – this would reduce the effects of air resistance on the pendulum's swing. Second, Boys also allowed for variations in the ambient temperature by using a mercury compensation bob. The idea here is to counteract any expansion, or contraction, of the pendulum's length due to changes in the background temperature. If the temperature increases and the pendulum rod expands, this is equivalent to increasing the length term L in equation (1.4), and the pendulum will run more slowly (that is, its period of swing will increase). To counteract this expansion effect a small pool of mercury is placed in a cup whose base has been attached to the end of the pendulum, and as the temperature increases so the mercury pool expands and pushes a metal cap upward. The upward movement of the metal cap acts to compensate for the increased length of the pendulum rod, and accordingly its period remains approximately constant (we shall have more to say about temperature compensation in the next chapter). Having compensated for temperature and air-resistance effects, Boys then adds electro magnets (k k' in figure 2.8) to his clock in order to provide small periodic impulses to the pendulum bob to counteract any mechanical friction that might exist at the pendulum's support. This, the argument goes, should keep the pendulum's period of swing constant over extended periods of time. To make sure that the impulses provided by the electromagnets are constant over time Boys designed a self regulating power supply system (driven by the gear train illustrated in figure 2.8, and two additional electromagnets M M'). In addition the electromagnets are only allowed to operate when the pendulum is at the exact center of its swing, and accordingly a mercury pool at the base of the vacuum glass is used to turn the pendulum into a regulating switch. It is this switch action that in fact provides the output of the clock, since when at the mid-point of its swing the pendulum triggers the activation of another electromagnet (E in figure 2.8), and it is the movement of arm F that provides the 'ticking' action of the clock.

With the clock designed by Boys we begin to enter the realm of the precision electronic regulation and control of time. Indeed, in the modern era, many of the same ideas explored by Boys are actually used to regulate the going of electronic clocks, the only difference being that a quartz crystal oscillator has replaced the mechanical pendulum.

The Pendulum Transformed

It might at first appear in the modern era that the simple pendulum no longer has a place within the running of our daily lives and scientific experimentation. This is only partially true. Certainly the rhythmical swing of the regulating pendulum is only rarely seen on clocks today and pendulum experiments are mostly confined to high schools and undergraduate physics laboratories. The essential essence of the pendulum, however, is still very much with us – only its form and operating frequency have changed. Today the timepiece pendulum has been replaced by the rhythmical oscillation of electrons moving from one allowed atomic orbit to another[9]. The physical principles might have changed but the end result is identical, the steady pulse or clock cycle is essential to the successful running of all the various electronic appliances that pervade our modern lives – from lap top computers, to television sets, cell phones and even washing machines.

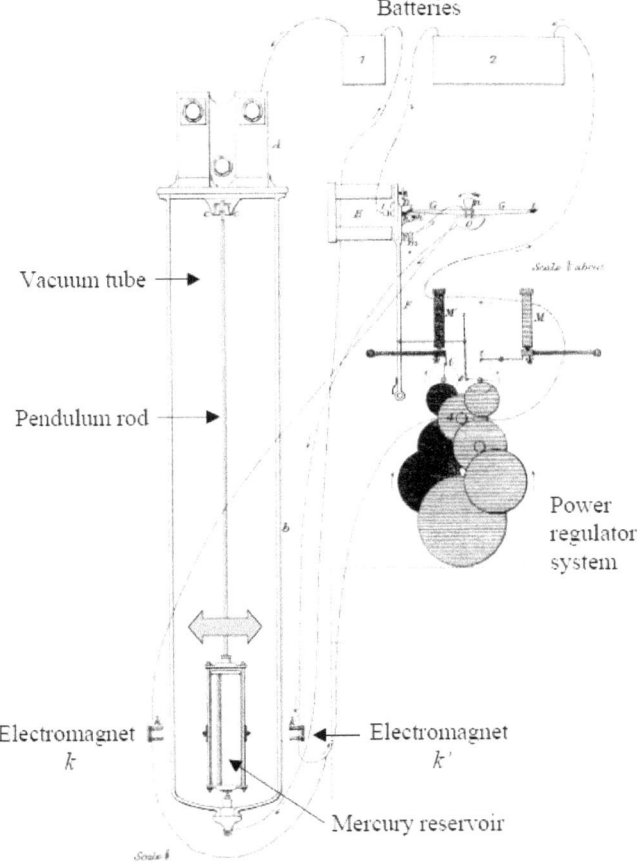

Batteries

Vacuum tube ⟶

Pendulum rod ⟶

Power
regulator
system

Electromagnet
k

Electromagnet
k'

Mercury reservoir

Figure 2.8. Diagram illustrating the main components in Boys' 1877 astronomical clock. Image adapted from ref. 25.

The piezoelectrical properties of quartz (silicon crystals) were first investigated in the 1880s by Jacques and Pierre Curie - the latter eventually sharing the Nobel Prize for Physics in 1903 with his wife Marie Curie and Henri Becquerel for their joint investigations into radiation phenomena. By being piezoelectric a quartz crystal generates an electrical impulse upon being deformed, and it was by harnessing this property that American physicist Walter Cady developed the first quartz crystal oscillator clock in 1921. The great advantage of such clocks was immediately realized since they ran at a rate at least tens times more accurately than the very best mechanical clocks. Indeed, a quartz crystal clock will typically achieve a going accuracy of order several seconds per year. The stability of the quartz crystal oscillator was also soon recognized in the newly burgeoning broadcast industry and by the late 1920s such devices were being used to regulate transmitting frequencies. The frequencies for quartz crystal oscillators range from between 10^5 to 10^7 Hertz, indicating pulse times that vary from a few hundred thousandths to a few ten millionths of a second – the modern regulating pendulum, like the pace of our very lives has speeded up dramatically over the past century.

The ultimate timing device and the standard by which time is determined in the modern era is that given by atomic clocks. Here the pendulum has been transformed into the rhythmical 'quantum jumping' of electrons between allowed orbitals within the caesium-133 atom. Such atomic clocks were first manufactured in the early 1950s and they were soon used to calibrate the standard measure of time. Indeed, in 1967 the second was defined in terms of 9,192,631,770 oscillation cycles between two energy levels of the ground state of the caesium-133 atom. Modern atomic clocks run at rates that conform to an uncertainly of fractions of a nano-second (10^{-9} or billionths of a second) per day, and they are the ultimate source of our proper time standard against which all other events, such as the length of the day and Earth's spin rate, are measured.

Present research in ultra-short time keeping is looking to push the accuracy of atomic clocks to ever higher levels by utilizing the pendulum-like oscillations of neutrons. Indeed, since neutrons are so closely bound to the atomic nucleus their oscillation rate is far less affected by external perturbations (unlike electron oscillations) and this effect alone, at least in principle, would yield a 100 times improvement in 'ticking' accuracy. Presenting their pioneering ideas in the journal *Physical Research Letters*, Victor Flambaum (University of New South Wales, Australia) and co-investigators explain[26] that by using lasers to carefully control the orientation of test atoms they can utilize the neutron oscillations to govern a clock that would be accurate to within 1/20th of a second over 14 billion years – a timescale equivalent to the present age of the universe.

International Atomic Time and the Leap Second

The number of time standards in current usage is quite bewildering – but they are all necessary for a whole host of technical reasons. Since pre-history the Earth's rotation rate has been used to define the length of a day, and since the time of the ancient Egyptians each day has been furthered divided into 12 hours of daytime and 12 hours of nighttime. At first the hours of division were unequal and adjusted so that each night and each day, irrespective of the time of year and season, were 12 hours long. It was only after the introduction of mechanical clocks that the idea of there being 24 hours of equal duration per day became standard. The division of hours into 60 minutes, and then each minute into 60 seconds is derived according to the sexagesimal counting system of the ancient Babylonians[27].

In terms of the Earth's rotation being measured with respect to the repeat alignment of the stars, one (mean) sidereal day is 86,164.0998 seconds long, and one sidereal year corresponds to 365.256363 days (where in the latter case the day is defined as being 86,400 seconds long). Such definitions all sound lock-tight and well defined, but there is a problem – a big and indeed, Earth-sized problem. At issue, of course, is the fact that Earth's spin rate is not constant over time and that it is slowly, but assuredly and measurably, slowing down. The current rate of increase in the length of the day is about 2×10^{-3} seconds per century. This slowdown is related to the tidal interaction between the Moon and Earth's oceans, the redistribution of land-masses on Earth's surface and variations in global sea level. Such gradual changes seem hardly worth worrying about, but over millennia the differences add up and are quite astounding, with, for example, lunar and solar eclipses reported by ancient Babylonian observers occur at local times some 6-hours earlier than would be predicted if the Earth had not been spinning down[28].

For a coordinated time system based upon the Earth's rotation some decision must ultimately be made about the location of the beginning and end point for the day. While the details will be explored below, the beginning and end points for time measurement (and the zero point of terrestrial longitude) were set, by international agreement in 1884, as the Greenwich (London, England) meridian. This established the so-called Greenwich Mean Time (GMT) standard, with one day corresponding to the time interval between successive transits of the Sun across the Greenwich prime meridian. An ambiguity in the use of GMT soon arose, however, in that astronomers considered the day to begin at noon, while the civil authorities, for legal reasons, deemed the day to begin at midnight. The system of Universal Time (UT) was therefore introduced in 1928 with each new day – starting at 0 UT – corresponding to local noon on the Greenwich meridian. Since 1961 Coordinated Universal Time (UTC) has been adopted as the time standard for regulating clocks, commerce, air travel, GPS navigation and electronic communications. UTC itself is regulated according to International Atomic Time (TAI), which in turn is based upon an averaged time standard derived from numerous atomic clocks operated at National Institutions and research centers across the globe. Given the stability of TAI, and the variable length of day as measured by Earth's rotation (solar time), UTC drifts with respect to solar time by as much as a second per year. To circumvent this discrepancy, beginning on 1 January 1972 leap seconds have been periodically added to UTC. By adding leap seconds, typically every 18 months or so, UTC will remain in essential agreement with solar time. Should leap seconds not be added to UTC then, given enough millennia, it will drift more and more with respect to the time standard governed by Earth's rotation, ultimately resulting in absurd (from a human experiential sense, that is) correlations such as sunrise taking place at UTC midnight and so on.

Even though the leap second is vital with respect to keeping UTC synchronized with solar time, there have been recent calls to abolish it. The argument being that for many commercial computing systems it is a problematic issue to deal with. While raised as a topic for discussion at the International Telecommunications Union (ITU) 2012 meeting in Geneva, the decision to abolish the leap second was deferred, so that further studies could be made – the topic will again be reviewed at the World Radio Conference in 2015. If the leap second is abolished, then the time standard used by humanity will (for the first time) be entirely governed by atomic clocks (through TAI) and no longer tied to the spin of the Earth and/or human intervention – atomic time would then indeed rule absolutely.

Time Bounds the Earth

The ancient Greek philosophers first developed the idea that the Earth was a perfect sphere (a history that we shall explore in Chapter 3). Given for the moment, however, that the Earth is spherical we can immediate ask the question, how might the location of any specific point on its surface be specified. To identify the location of a point on Earth's surface, geographers have long used a coordinate system based upon the imaginary circles of latitude (φ) and longitude (λ). Circles of constant latitude run parallel to the Earth's equator, which is defined as having latitude $\varphi = 0°$. The North Pole is located at a latitude of $\varphi = 90°$, and the South Pole is located at a latitude of $\varphi = -90°$, the positive and negative signs being used to indicate a northerly or southerly direction. The circumference of a constant circle of latitude is given by the relation: $C(\varphi) = 2\pi R_\oplus \sin(90 - |\varphi|)$, indicating that the circle of latitude varies from that of Earth's circumference at $\varphi = 0°$, to zero at the poles where $\varphi = \pm 90°$. The arcs of constant longitude, on the other hand, all have the same circumference of $2\pi R_\oplus$ and are also all great circles running through the North and South Poles. The arcs of longitude are spaced east and west around the Earth and all run perpendicular through the equator. At an international conference convened in Washington D.C. in 1884, it was agreed by the attendant representatives that the zero of longitude $\lambda = 0°$ should be taken as the great circle that runs through the central position of the meridian telescope situated at the Greenwich Observatory in London, England. As with many international treaties, however, very little changed for many years; the French Government, for example, continued to use the meridian through the Paris Observatory until 1911. Likewise, the Ordnance Survey that produces topological maps in England only switched to using the 1884 agreed upon meridian (as first established by Astronomer Royal G. B Airy) in 1957.

While latitude runs from -90° (the South Pole) to 90° (the North Pole), longitude runs from 0° to 180° East of Greenwich and from 0° to 180° West of Greenwich. The division of an arc of longitude running from the North to the South Pole is such that $\Delta\lambda(1°) = \pi R_\oplus / 180 = 111.195$ km. It has been assumed in this calculation that Earth is a perfect sphere, but as we shall see in the next chapter this isn't actually the case. Irrespective of reality, however, the unit of the nautical mile, still used to this very day in international treaties concerning territorial waters, is defined as the length of one minute of latitude along any meridian. From the calculation above this corresponds to a distance of 1 nautical mile = 111.195 / 60 = 1853.25 meters, although with the introduction of the *SI* system, the nautical mile was redefined to be exactly 1,852 meters (so much for the consistency of definitions). Also still used by mariners to this very day is the knot, this being defined as a speed corresponding to the number of nautical miles covered per hour.

The two latitude and longitude coordinates (φ, λ) uniquely specify the location of any point on the Earth's surface. The practical problem with this system, however, is that it is not very convenient to use maps that have been traced upon the surface of a globe. It is far easier in terms of engraving, printing, usage and storage to draw maps on flat sheets of paper. This shift from a curved three-dimensional surface (the sphere of the Earth) to the two-dimensional plane represented by a piece of paper introduces many problems, resulting as cartographers throughout the ages have long understood in the distortion of both distances and angles. We will pick-up on the cartographical problems in the next chapter and concentrate for the moment on how a navigator might determine their position on Earth from astronomical observations.

Latitude can be determined, whether isolated at sea or on dry land, by observing the altitude of the visible celestial pole (recall figure 2.1 and see figure 2.9) – a result that has been known since antiquity. The geometric proof of this statement is straightforward and follows from the definitions that $A + b = 90°$ and $\varphi + b = 90°$, revealing after a little algebra that, indeed, $A \equiv \varphi$ as required.

The determination of longitude is much more complicated than that of latitude and its measurement was one of the most important challenges to face astronomers in the 17th and 18th Centuries. At issue was the development of a method for finding longitude that could be applied under sea-going conditions (typically far from ideal), and that did not require excessive calculation and or analysis. Galileo suggested that if tables of the locations of Jupiter's four large moons (Io, Europa, Callisto and Ganymede) could be constructed then these might be used to determine longitude. The problem with this method, however, is that it requires the use of a telescope (which is not easily held steady at sea), a micrometer for measuring the angular separations, and clear skies – not only this, Jupiter is not always visible in the night sky. In order to help observers identify which moons might be visible on a specific date and time, mechanical jovilabe models (literally analog computers) were constructed to track the relative motion of the moons as they would be seen from the Earth (figure 2.10). Although the Galilean moon method found great favor in 17th Century France, when Danish astronomer Ole Rømer set out to measure the transits of the Jovian moons in 1671 he soon found that it was a far from straightforward exercise. Indeed, while Rømer did not solve the longitude problem, he did discover, for the first time, an effect relating to the finite speed of light.

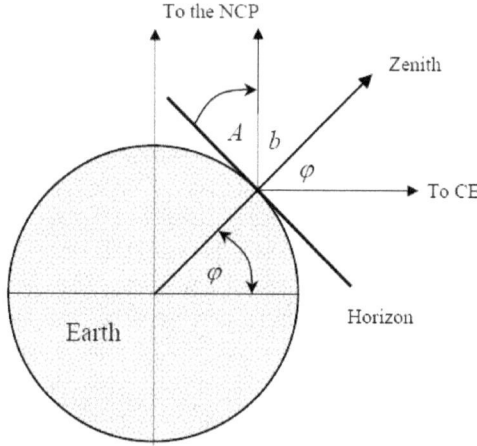

Figure 2.9. The altitude A of the visible celestial pole is equivalent to the observer's angle of latitude φ. CE is the celestial equator.

Extra spice and incentive were added to the longitude problem when on July 8th, 1714 the British Government announced a £ 20,000 prize for its solution via a practical method. A corresponding Board of Longitude was established, headed-up by the Astronomer Royal and selected luminaries from the scientific world; the hunt for a utilitarian method of measuring longitude was on, and a veritable fortune was in the offering for the prize-winner. Appearing before a Parliamentary Committee on June 11th, 1714, no less a luminary than Isaac Newton surmised that there were essentially two practical means of determining longitude; by mechanical clock, or by observation of the Moon. The latter method, naturally favored by the astronomers, entailed the careful mapping of the Moon's journey through the zodiac, such that on any given day, at any given time, its position would be exactly known with respect to a specified set of bright stars. The problem with this method, however, is that the Moon traces out a very complicated motion on the sky, which takes about 18.9 years to repeat. If this method was going to work then very precise observations of the Moon's position, as well as highly accurate maps showing background star positions, would need to be constructed, and then the whole condensed into practical and easily read sets of tables. The essential idea behind the lunar method is that the exact location of the Moon at a specific time would be tabulated for an observer at Greenwich. An observer located east or west of Greenwich would, however, see the Moon at a slightly different location in the sky to that which had been tabulated, the result of a parallax effect, and this difference between the observed and calculated position of the Moon would provide the observer's longitude.

Incredible amounts of time, effort, money and thought were applied to the lunar method, but eventually a humble carpenter's son decided that the way to win the longitude prize was through the production of a highly accurate chronometer – a clock, that is, that could be carried at sea without losing appreciable amounts of time. John Harrison (1693 - 1776) was a master craftsman, and he knew that longitude could be determined through the use of two clocks. If one clock, for example, is set to Greenwich Time, and the other set to local time, then for every one hour difference in the times displayed by the clocks the observer would be 15 degrees of longitude (either East or West) away from Greenwich. Here the basic idea is that the Earth rotates through 15 degrees every hour such that after 24 hours it has spun on its axis once (i.e., accumulated 360 degrees of rotation).

In 1735, following many years of experimentation, Harrison's first great clock, rather unceremoniously known as H1 (figure 2.11), was delivered to the Longitude Board for testing. H1 is a glorious timepiece, made of brass and pear wood (which is self lubricating and won't rust at sea), and has numerous correction mechanisms to compensate for temperature changes and the rolling motion of a ship. Indeed, as Huygens had discovered in trials conducted in November and December of 1662 an ordinary pendulum clock, whether equipped with cycloidal cheeks or not, simply won't work reliably at sea, the to and fro motion of the ship breaking the steady rhythm of the pendulum's swing. Even though it had many compensation mechanisms, the sea trials of H1 were not a resounding success. It certainly kept good time, but not good enough to win the prize outright. The British Government did, however, awarded Harrision several hundred pounds (a veritable fortune at that time) with which to continue his researches and after the construction of two intermediate clocks he eventually produced, in 1764, his prize winning masterpiece – simply called H4. This fourth clock was another marvel of design and construction, and it pioneered the use of jeweled bearings, again with the intent of avoiding rusting problems and ensuring maintenance-free motion. While the pendulum had been at the heart of the first three of Harrison's clocks (indeed, for H1 he invented the remarkable low-friction grasshopper escapement based upon the interplay of three interconnected, restricted-motion pendulums – see figure 2.16 later), H4 was constructed according to the balance spring design pioneered by Robert Hooke and Christian Huygens in the early 18th Century.

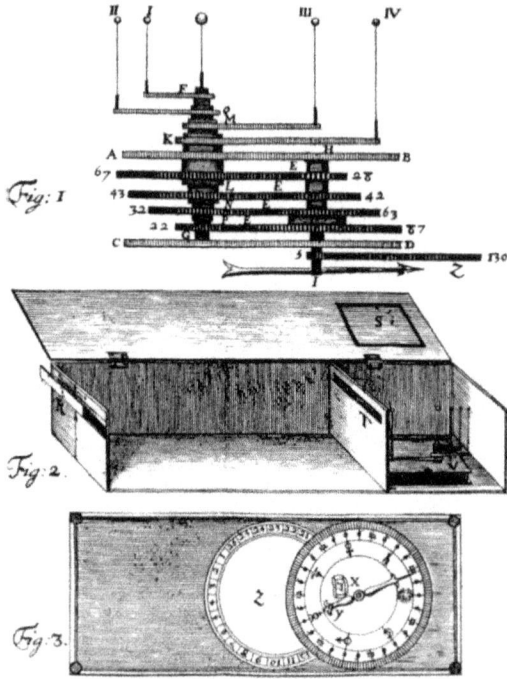

Figure 2.10. The jovilabe – a mechanical device showing the correct relative motion and positions of the Galilean moons. The original such machine, made by Rømer in Copenhagen, was destroyed in a fire in 1728.

The practical problem of determining longitude at sea had been solved. In its wake a whole new industry was established to mass-produce nautical chronometers, and nearly countless numbers of observatories sprang up around the world to deliver accurate local time signals. Time, motion and location, and one might also add profit, were all now inextricably bound together, and humans have lived by this partnership ever since.

The present web of time that thoroughly binds us to the Earth grew out of the Prime Meridian conference convened in Washington D.C. in 1884. Not only was the Prime Meridian fixed to Greenwich at this meeting, but terrestrial time itself was rationalized. The world was divided into 24 standard time zones, each one being 15 degrees wide in longitude – the Earth and its human cargo were thus imprisoned; the boundary lines cutting across the globe like a series of impenetrable prison bars. Each clock within a given time zone would, under the standardized scheme, display the same time irrespective of where in the zone it is located, and accordingly the large number of local times was done away with. Each new day was further deemed to official begin at midnight Greenwich Mean Time (or equivalently 12 UT).

The Prime Meridian Conference was ultimately an astounding success, in spite of there being a number of non-signatory countries and the rather slow implementation of its recommendations. In many ways the Conference is a superb example of the Victorian zeal for organization and standardization. We may well measure time with atomic clocks in the modern-era, the rotation rate of the Earth not being stable enough to synchronize or even regulate our measurements (as will be further described in Chapter 6 later) or our commerce, but we still abide by the standard time zones agreed upon in 1884. Indeed, while modern physicists strive ever harder to banish time from their theories, our every-day lives are constrained by a system of time conventions that are as absolute in their form as the definition first offered by Isaac Newton in the late 17th century.

Figure 2.11. H1 – The first clock developed by John Harrison in his attempt to solve the longitude problem. The timepiece stands 3-feet (about 1-m) high and weighs-in at 72 pounds (32 kg). The two spheres, attached together via a horizontal spring, swing back and forth at a constant rate in spite of any ship movement. The dial faces show the day of the month (lowest dial), the hour of day (middle-right dial) and the appropriate minutes and seconds.

Hooke's Animadversions

Jingle-jangle tunes play in the pit of creation,
which once gave birth to the music of the spheres.
Gears turn; ratchets click; time passes almost
imperceptibly.

Engines of Ingenuity (2001)
Kit Williams

Ask any astronomer and they will tell you that accurate tracking and polar alignment are the two biggest problems encountered when setting up a portable telescope. Polar alignment in the northern hemisphere is aided, at least at the present epoch, by the fact that the north celestial pole (NCP) is situated close to the relatively bright star α *Ursae Minoris* (Polaris). This convenience, however, as we shall see later, hasn't always been available, and within the publications of Robert Hooke, for example, we find the description of an unofficial constellation, the *English Rose*, being used to locate the north celestial pole[29]. This particular group of stars, defined by Hooke circa 1680, due to the precession of the equinoxes, is now some three degrees away from the NCP. Southern hemisphere observers fare no better than their northern companions since there are no particularly bright or conspicuous stars to indicate the location of the SCP. So, why do astronomers go to all the fuss of aligning their telescopes? The reason, of course, is that once the rotation axis of the telescope is aligned with the celestial pole it is automatically aligned with Earth's spin axis. Rotating the body of the telescope about this axis in an East to West direction (i.e., in the opposite direction to that in which the Earth spins) will therefore result in any object (a star, a planet or galaxy) remaining fixed within the observer's field of view. Such equatorial mounts have long been used by astronomers, but the very first clockwork driven equatorial mount was described by Robert Hooke in his 1674 work entitled *Animadversions* – and at the heart of the clockwork described by Hooke was a conical pendulum (recall figure 1.9). Hooke's *Animadversions* was primarily concerned with the observational work of renowned Polish astronomer Johannes Hevelius's and in particular with his 1673 publication *Machina Coelestis*. Hooke was critical of Hevelius firstly for not using telescopic sights on his instruments, they were, after all, readily available at that time, and second Hooke was critical of what he believed to be Hevelius's unrealistically accurate naked-eye determinations of small angles. Indeed, the subsequent row that developed between Hooke and Hevelius on this second criticism was only smoothed-over once the Royal Society sent Edmund Halley to Danzig (now Gdansk) on a diplomatic appeasement and test-of-claims mission in 1678. Hevelius's exceptional eyesight and skill as an observer were accordingly vindicated by Halley, although it was also clear that, in general, Hooke's criticisms were also true; Hevelius was the last of the great naked-eye, in the manner of Tycho Brahe, astronomers. Hooke, in contrast, was the new visionary who saw that the future of astronomy lay in the improved design and construction of evermore-innovative telescopes and the clockwork required to drive them.

Throughout his *Animadversions* Hooke continuously emphasizes the importance of instrumentation, and he describes numerous entirely new devices for astronomical research – devices that, Hooke understood, would put nature to the question. One such innovation is an equatorially mounted quadrant, complete with telescopic eyepiece, mirrors for the superposition of two images, geared micrometer screws for fine-scale adjustments, and the first ever equatorial clockwork drive (figure 2.12).

In figure 2.12, the bob of the conical pendulum is labeled F, while the drive-weight is marked with an x. The great advantage of the conical pendulum, as we saw at the end of Chapter 1, is that it provides a continuous, jerk-free regulation to the clockwork's motion, and this, of course, is vital for accurate astronomical observations. While Hooke was not openly advocating the use of an isochronal conical pendulum (recall figure 1.10) in his new machine, he does provide the clock with a variable speed adjustment. Accordingly, the observer can change the period of the pendulum, and hence fine-tune the turning rate of the equatorial drive shaft by raising or lowering a vertical beam of wood connected to a rope and pulley system. Adjusting the height of the vertical beam effectively changes the height (h) of the conical pendulum and as we saw earlier, through equation (1.8), this will alter the pendulum's period of oscillation.

As a modification to his new equatorial instrument, Hooke also shows in his *Animadversions* how it could be transformed into a device for the direct measurement of altitude and azimuth angles. To perform this modification, however, the development of an entirely new universal joint, indeed Hooke's joint, was required. This wonderful device, still seeing employment in numerous engineering applications to this very day, was first demonstrated to the Royal Society in 1667, and it allowed the drive shaft to provide a constant azimuthal rotation. Necessity, as ever is the mother of invention, and to Hooke's ever fertile mind the future of science lay in the development of new instruments with the express purpose of extending and enhancing our all too feeble and far to easily fooled human senses.

While Hooke was responsible for designing and seeing through production the initial inventory of instruments at the Royal Greenwich Observatory (founded in 1675), he did not have the opportunity to develop his conical pendulum drive to control any of its instruments. This situation was somewhat reversed in 1858 with the installation of a new 12 ¾ -inch equatorial telescope in the observatory grounds. While a much larger instrument eventually replaced this telescope, the drive mechanism for the mounting was a marvel of Victorian engineering. The telescope mount was driven hydraulically, with a cistern on the observatory roof providing a constant head of water to drive a turbine located some 30-feet below in the observatory basement. The turbine was adjusted to rotate at four revolutions per second, with this spin being down-graded, via intermediate gearing, to one rotation in 24-hours – the required drive rate for the telescope and dome. In order to ensure the smooth rotation of the telescope a governor mechanism (recall Chapter 1) was designed to control the turbine's rotation rate. The governor in this case mechanically compared the rotation rate of the turbine against that of a reference 2-second period of rotation conical pendulum. If the turbine was found to be running faster than the pendulum then the governor would automatically throttle back the water supply, so as to slow the turbine down, and likewise, if the turbine was running slow compared to the pendulum, the governor mechanism would increases the water flow rate. Hooke would, with little doubt, have been greatly pleased by the construction and efficiency of the Greenwich equatorial drive mechanism, and doubly pleased by the fact that its smooth running was controlled by his much beloved conical pendulum.

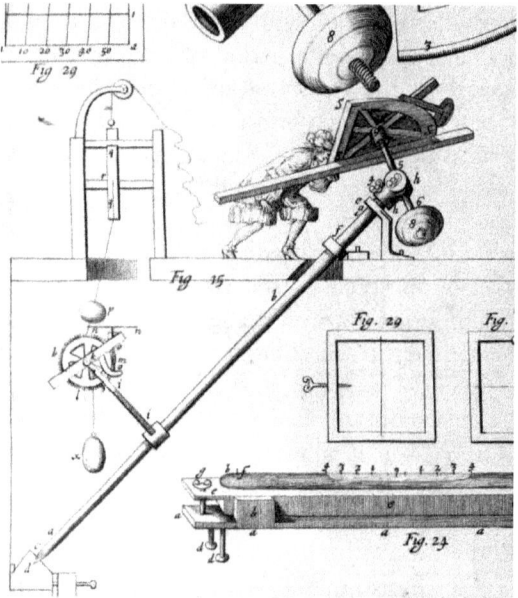

Figure 2.12. Robert Hooke's design for an equatorial mount driven by clockwork. The conical pendulum controlling the clockwork's going can be seen to the middle left in the diagram. The shaft and pulley system behind the observer is the control mechanism for adjusting the rotation rate of the pendulum. And, one might question, is that Hooke himself portrayed as the observer?

The Heavens Declare

The human brain specializes in teasing order and structure out of chaos[30], and the random spacing and brightness distribution of the stars has inspired the identification of countless animals, shapes and entities. The stars are literally a Rorschach test for the human psyche. Indeed, all civilizations throughout recorded history have found their Gods, demons and mythical heroes in the sky. Since 1922 the International Astronomical Union (IAU) has officially sanctioned 88 constellations, but the number has varied significantly throughout history. There were just 48 constellations listed in Claudius Ptolemy's 2nd Century AD *Mathematica Syntaxis* (*Almagest*) and these were largely based upon earlier 6th Century Babylonian designations. Many additional constellations have been proposed over the centuries, but it seems that those proposed by French astronomer Nicolas Louis de Lacaille (1712 – 1762) have had the greatest staying power. Lacaille is especially remembered for his early mapping of the southern hemisphere and in his *Coelum Australe Stelliferum*, published posthumously in 1763, he proposed 14 new constellations – all of which have now become standard. Amongst his new southern hemisphere stellar arrangements was the constellation of *Horologium Oscillitorium* (now shortened to *Horologium*). Proposed in honor of the pendulum clock this small constellation (figure 2.13) is composed of 6 principle stars, the brightest of which, α Horologii, is only just above 4th magnitude brightness, with the remainder being of 5th and 6th magnitude and therefore only just visible to the human eye. By introducing the constellation of *Horologium*, Lacaille chose wisely since it celebrates and symbolizes the great mechanical and astronomical achievements of his time, and indeed, it seems only appropriate (to co-opt the writings of Psalm 19) that the heavens should declare the glory of the mighty pendulum.

Not only is the pendulum clock described in the heavens, but so too is the Pulfrich effect. This phenomenon relates to the physiology of perception and the manner in which the visual cortex deciphers images. In his classic book *Light and Color in the Outdoors* (first published in 1937) Marcel Minnaert describes a heavenly version of this effect as demonstrated through the oscillating double star phenomenon - although the effect was widely discussed much earlier by John Herschel in the 19th Century. Herschel first encountered the oscillating effect with respect to the star Enif (the brightest star in the constellation of Pegasus) and a faint, near-by, optical double[31] companion. The phenomenon is so striking that Enif has unofficially been named the Pendulum Star. The oscillating (Pulfrich) effect is due to the fact that one star is much brighter than other, and that it takes longer for the light from the fainter star to stimulate the eyes retina. Herschel explained that if one centered a low power telescope upon the star Enif and lightly tapped the telescope tube, then the faint companion star appears to lag slightly behind its brighter companion, giving the effect of a pendulum like motion. Minnaert notes that the effect can also be seen when looking at Alcor and Mizar, the binary system located in the handle of the Big Dipper asterism[32], with low-power binoculars. Another example of the oscillating effect is that shown by stars in the open cluster NGC 2362 in the constellation of Canis Major. The brightest star in this cluster far outshines its companions and by exhibiting erratic, pendulum-like oscillations when the telescope tube is tapped it is has become known as the Mexican Jumping Star (named, in this case, after a similarity to the behavior of Mexican jumping beans).

And finally, *Let there be light*: so begins the Book of Genesis and the statement encapsulates just one cultural interpretation of the creative act that oversaw the origin of the universe. The night sky literally demands our attention and it seemingly requires that we must study and measure its function and form, but for all this, it seems only appropriate that the pendulum and the pendulum clock, two of humanities greatest constructs, have been back-projected onto the great vault that is the celestial sphere.

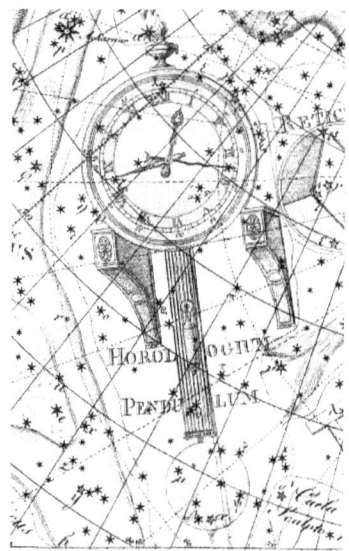

Figure 2.13: *Horologium Oscillitorium* (here shown as *Horologium Pendulum*) in Johann Bodes 1801 *Uranographia* star atlas. In this rendition, the brightest star in the constellation, α Horologii, is located in the bob of a gridiron pendulum.

The Pendulum is Dead. Long Live the Pendulum

While we all think we know what time is, and while we also know that our sense and understanding of time is ephemeral and uncertain, the time that bounds our daily rituals is measured by devices that are far from mundane. Clocks are physical, almost living entities – they adorn our bodies and they rest, centrally dominant, upon mantelpieces and they hang upon walls, and like Big Ben in London, England they can define a nation (figure 2.14). Clocks can be massive and imposing, they can be small and almost hidden from view, they can astound and they can amuse. Clocks can be cheep and throw-a-way and they can be gold-leaf finished and diamond studded; signifiers of rich status and abundant wealth. For all these characteristics, however, they all measure-out 60 seconds to every minute - time is indeed the great leveler.

The imposing tower of Big Ben, completed in 1859 with its 300-kg, 3.9-m pendulum that swings once every 2 seconds, has little to do with the process of conveying the time of day – rather, it is the embodiment of power, stability and governance – the time on its four dials, while accurate to a fraction of a second, is almost an irrelevant by-product of its function. In utter contrast to the sobriety of Big Ben, the British artist Kit Williams has designed and constructed a number of large and imposing public clocks, such as the Wishing Fish Clock (located at the Regent Arcade in Cheltenham, England – figure 2.15), but these devices are intended to delight and fascinate the observer. Once again, the actual time displayed is largely irrelevant; the main point, perhaps, of Williams's clocks being that time is mysterious and something that we may never fully understand in detail.

Figure 2.14. The iconic clock tower of Big Ben. Named after the massive, 13.76-tonne bell that chimes the hours, the 96.3 meter tall Elizabeth Tower is located at the Palace of Westminster and is the symbolic heart of the British Houses of Parliament. The clock dials are 7 meters in diameter. Photograph by David Iliff and courtesy of Wikimedia Commons.

Figure. 2.15. The Wishing Fish Clock designed by British artist Kit Williams and housed in the Regent Arcade in Cheltenham, England. Built by Michael Harding and installed in 1985, the clock stands some 14 meters high, and weighs-in at over 3000 kilograms. Image by Herbythyme and courtesy of Wikimedia Commons

The Corpus Clock (figure 2.16), designed by British engineer and inventor John Taylor, cuts a symbolic path that falls somewhere between those conveyed by Big Ben and the Wishing Fish Clock designed by Kit Williams. It is a beautiful device that has been exquisitely made; it is somehow old and yet it is entirely new, it is whimsical and it is serious, and it extends a bridge from the deep past to the present, and the never to be repeated instantaneous now. The design uses a special, low-friction grasshopper escapement to convert and regulate the swinging motion of the pendulum into the rotational motion of the clock's gearing; most inspiringly, however, the escapement mechanism has been metamorphosed (at least in name) into a Chronophage (from the Greek for *time eater*). The grotesque yet beautifully crafted Chronophage steps over the rotating outer, saw-toothed dial ring and symbolically consumes the time that has just past, reminding us of our mortality and the deeper mysteries of existence, and of how time itself is central to that mystery.

Modern society, just like the Chronophage on the Corpus Clock, thrives on the instantaneous moment – the eponymous now. In the words of Stewart

Brand[33], "civilization is revving itself into a pathological short attention span", and our ever growing expectation of instantaneous satisfaction and thirst for the constantly new only drives our attention away from the past and effectively denies us a future, since a life lived only in the instantaneous now has no future or past. This existence, should it ever really come about, would be a sad state of affairs, but with the establishment of the Long Now Foundation, in 01996, the future may yet have a physical link with the past and *visa versa*. The Foundation was established to foster responsibility over a timeframe encompassing the next ten millennia, and for this reason years are accordingly expressed in a five-digit format. With this noble rational in place, two major projects have been initiated – one, The Rosetta Project, intends to establish a digital library of human language, while the other intends to oversee the construction of a clock that will run for at least 10,000 years. The prototype Long Now Clock is a sublime and highly potent mechanical device (figure 2.17), and while the full-scale version is presently under construction in a vast cavern cut inside of a mountain located on the Texas/New-Mexico border, it is a pure marvel, and a superb symbiotic amalgam of great human intelligence and engineering skill. Behind all the technical innovation and superb construction, however, the Long Now 10,000-year clock has essentially one aim, and that is to remove the phase-locked-now from our lives. Dan Wolf, former Vice President of Communications for the Walt Disney Company, has eloquently written[34], "a traditional clock depicts time in the context of our lives. This clock [the 10,000-year clock] depicts our lives in the context of time. The jump is from prime time to primal time…. In its company there is nothing special about now".

Various versions of the Long Now clock have been developed over the past 15 plus years by American engineer and inventor Danny Hillis, and the final configuration has been appropriately described as the world's slowest (analog) computer. That this is appropriately so is a reflection of the clocks design requirements – it has to run, non-stop for 10,000 years (or equivalently some 315.5 billion seconds). To minimize issues such as gear wear the clock is designed to operate in a digital fashion, via levers, cams and gears, updating its configuration once every 5 minutes. The duty cycle of the clock, however, is governed by a massive pendulum adjusted to have a period of 10 seconds. It is pleasing, if not comforting, to see that the pendulum will carry the 10,000-year Clock project forward, and indeed, this was a deliberate initial design choice. Given that the Clock will require maintenance over many hundreds of human generations, one of the key design requirements was that the working and operation of the clock should be as transparent as possible – properties that the mighty pendulum clearly satisfies.

By looking to measure future time the designers of the 10,000-year clock found themselves looking to the past for inspiration. Not only does the ancient pendulum provide motive regulation, the time displayed on the clock-face itself is modeled upon the anaphoric representations used in the clepsydrae of ancient Greece. The operation of the clock is further synchronized to the Sun, enabling it to self-adjust (if necessary) to solar time and the erratic vagaries of Earth's spin. A heat sensor that is illuminated for a short period of time at local noon determines this latter coupling, and it effectively solves the UTC leap second problem described earlier. By allowing the sun synchronizer to adjust the pendulum-derived time, the displayed clock-time will automatically remain in harmony with apparent solar time as determined by the rotation of the Earth.

Figure 2.16. The Corpus Clock is located at the Taylor Library of Corpus Christi College at the University of Cambridge in England. Officially unveiled by astrophysicist Stephen Hawking on 19 September 2008, the 24-carat gold-plated clock face is some 1.5 meters in diameter. The dial has no hands, and displays the time via a series of illuminated slits arranged in three concentric rings. The clock is entirely mechanical in its working, and yet the pendulum can stop at irregular intervals without affecting the accuracy of the display. The Chronophage is located at the top of the clock face and serves to regulate its going and symbolically *eats* each of the old seconds of time as they pass by. Image courtesy of Wikimedia Commons.

Figure 2.17. Prototype of the Long Now Clock. Housed in the Science Museum in London, England, construction of the prototype was finished in 01999. The time dial is flanked by two gravity-driven weight columns, which drive a torsional pendulum located at the base of the central gear stack. A governor can be seen at the top of the right most weight-drive column. Image courtesy of the Long Now Foundation. http:/www.longnow.org

Time and place are the very cornerstones of our existence; they literally make us what we are. And, for all its simplicity it is the pendulum that has historically enabled humans to chart their quixotic course through space and time – with the Long Now 10,000-year clock the humble, albeit mighty pendulum will continue this tradition as it measures-out with indomitable stride our myopic journey into the future.

Time goes, you say? Ah no!
Alas Time stays, we go;
Or else, were this not so,
What need to chain the hours,
For youth were always ours?
Time goes, you say? – ah no!

The Paradox of Time (1886)
Henry Austin Dobson

CHAPTER 3

THE SHAPE OF THE EARTH

It suddenly struck me that that tiny pea, pretty and blue,
was the Earth. I put up my thumb and shut one eye, and
my thumb blotted out the planet Earth. I didn't feel like
a giant, I felt very, very small.

Neil Armstrong (1930-2012): The first person to walk on the Moon

The view from my office window shows an incredibly flat Earth. Indeed, living in Saskatchewan, set in the middle of the Canadian Prairies it is very hard to believe that the Earth is a sphere – or more correctly a spheroid. While there have been the occasional historical (and invariably misguided) attempts to argue that the Earth is actually flat, the observation that it is most definitely spherical was correctly made over two thousand years ago. The requirements to make such a deduction about Earth's shape are both straight-forward and delightful, and they can be performed by anyone with a mind and inclination to make them[1].

Little Pendulum Island

Little Pendulum Island is located off the northeastern coast of Wollaston Foreland, Greenland – it is a bleak and desolate place. Barely 10-km long, the island is situated 5-km due east of Sabine Island (formerly known as Inner Pendulum Island), and it was named by Arctic explorer and Naval Captain Douglas Clavering. Indeed, the island was so-named since from the 14th to 29th of August 1823, the small outcrop of snow-capped rock was home to Captain Edward Sabine and his observatory for pendulum experiments. Here was cutting edge science at its extreme, and the science being conducted had no less a goal than to map out the topology of the Earth: to literally measure and quantify its lumps and bumps.

Sabine was the quintessential scientist/explorer, devoting his life to travel and study. In the same year, 1818, that he was elected a Fellow of the Royal Society of London, Sabine traveled as the designated astronomer with John Ross on his first Artic expedition. While Ross had been charged to seek out a Northwest Passage, Sabine busied himself by studying Earth's magnetic field and by making measurements of local gravitational variations with a seconds pendulum. While Ross called the expedition home, far too hastily according to some of his officers, after encountering sea ice in Lancaster Sound, Sabine was soon back on his travels. In 1819 he returned to the Artic with Edward Parry, once again making observations of Earth's magnetic field with the aim of identifying the location of the northern magnetic pole. From 1821 to 1823

Sabine traveled extensively, first traversing the African and then the American coastlines, making seconds pendulum experiments as he progressed. In early 1823, Sabine began to travel northward via New York, and Trondhjem (Norway) to once again explore the high Artic, Greenland and Little Pendulum Island. The results of Sabine's extensive pendulum measurements were published in 1825, and they represent the first detailed assessment of the figure of the Earth. In later life Sabine continued his pendulum and geomagnetic field experiments, became briefly, before its abolition in 1828, a member of the Board of Longitude, was elected General Secretary of the British Association (1839), and eventually took office as President of the Royal Society (1861).

Sabine espouses all the great characteristics and drives expected of a Victorian-era explorer and gentleman of science. He willingly suffered great physical hardships and yet rejoiced in the study of the Earth and in the discovery of its ever abundant fauna and flora. Indeed, the measure of the Earth has been a task that has stretched across human history, has pushed many an explorer's endurance to their very limits, and has cost countless lives. The Earth has not yielded-up its secrets easily and the annotation of its surface and profile has been one of humanities greatest achievements. While the story, as we shall see, ends in the modern era with laser ranging and stereographic imaging from Earth orbiting spacecraft, it began much more humbly, but no less brilliantly, with the ideas developed by the ancient Greek Philosophers.

A Classic Result
The suggestion that the Earth must be a near perfect sphere was made for esthetical reasons by the Pythagorean School of philosophers in the 5th century B.C. Their reasoning, as one might well expect, was based upon the ideals of geometry: given, they argued, that the Earth is at the center of the universe, then it should be a perfectly shaped object, and the most perfectly shaped solid is a sphere – unlike the cube, or other polyhedra, it has no edges or corners to distract from its beauty. The Pythagorean esthetic was later developed by the indomitable Aristotle, who bolstered his arguments with the notion of final causes. Cold, rocky objects, Aristotle reasoned moved according to the directive of finding the center of the universe. Since other cold, rocky bodies might impede the movement of a given rock attempting to find the center of the universe, it would 'strive', to get as close as it could to the center[2]. Then, and this is the clever bit, Aristotle argued that the equilibrium shape adopted by a swarm of solid particles all of which were trying to get to the center of the universe must be a sphere - this profile in effect maximizes the closest packing of objects in the sense that all of the particles are at their closest possible distance to the center. This argument also underscores the more subtle point that from Aristotle's perspective the Earth was not *the* center of the universe, but more fundamentally, as a result of its very consti-

tution and under the action of final causes, it simply *resided* at the center of the universe.

That the Moon moves around the Earth, and that its variable degree of illumination (its phases) are due to the changing angle subtended between the Moon, Earth and Sun was well known to the ancient astronomers. That the Moon might be eclipsed by the Earth was also well known to the ancients, and since the Sun is an extended object, it was also clear to the ancients that the Earth casts a conical shadow into space. Accordingly, if, at the time of the Full Moon phase, the Moon happens to pass through this conical shadow, then it will no longer be illuminated by the Sun, and an eclipse will take place. The key point, however, as noted by Hipparchus circa 150 B.C. is that as the Moon passes through the Earth's shadow the shadow boundary is curved. Indeed, the shadow boundary is always curved, no matter what eclipse event is being observed, and the only shape that can produce this effect is a sphere.

Shadow Lands

Next to the pendulum perhaps the simplest and most fundamental of experimental (and utilitarian) devices is that of the sundial – or more simply, a gnomon for casting a thin shadow upon the ground[3]. As discussed earlier in Chapter 2, the sundial was for many centuries the only device by which the sunlight hours could be measured with any accuracy. More than this, however, as the ancient astronomers well knew, the sundial can delineated the cardinal points of the compass, and tell the approximate day of the year[3]. And, as if all its other attributes weren't enough, the sundial (well, actually two sundials) can be used to measure the size of the Earth[1]. The first person to do this, as far as is known, was Eratosthenes (circa 275 B.C. – 194 B.C.) who spent most of his life in the great City of Alexandria.

We do not know exactly how Eratosthenes gathered his data, and nor do we have his original manuscript for study. What we do know of Eratosthenes is provided for us by Cleomedes who was writing several centuries after the events supposedly took place. The key observations attributed to Eratosthenes are that he realized on the day of the summer solstice[4], while the Sun shone directly down a well located in the southern Egyptian town of Seyene (long-ago re-named Aswan - the location of the great Nile dam), it was observed to be some 7 degrees away from the vertical at the City of Alexandria located approximately $D = 5000$ stadia to the north. In terms of shadow lengths, at Syene it would be zero, while in Alexandria the shadow length would be (in modern terminology) L x tan(7°), where L is the length of the sundial gnomon (placed vertically in the ground) and 'tan' is the trigonometric tangent function. On your calculator you will find that tan(7°) = 0.12278, so the shadow length will be about 1/8th the height of the gnomon. The geometry of the situation described by Eratosthenes is shown in figure 3.1. Now, in fact, one need not know what the tangent of an angle is to work out the size of the Earth – the key point, however, is that the 7 degree angle is measured and known. What

Eratosthenes reasoned was this: the 7 degree angular offset to the Sun from the vertical at Alexandria was in fact the same as the angle, subtended at the center of the Earth, between Syene and Alexandria[5]. That this distance across the surface of the Earth was also known to be $D = 5000$ stadia further tells us that the number of stadia contained in one degree along the Earth's circumference corresponds to the ratio $D / 7$ (the units are the odd mix of stadia per degree[6]), and consequently if C is the circumference of the Earth then $C / 360 = D / 7$, and consequently $C = (360 / 7)\ 5000 = 252,000$ stadia. What an incredible piece of thinking this really was, not only does it bravely and correctly extrapolate local measurements to a global result, it is also a remarkably accurate determination of the size of the Earth. Even though there is great uncertainty as to what the stadia measurement actually corresponds to in kilometers, Eratosthenes' result was within 5 to 10 percent of the currently accepted average value for the Earth's radius[7].

About a century after Eratosthenes made his estimate for the size of the Earth, the basic experiment was repeated by Posidonius. Rather than using the Sun, however, Posidonius used altitude measurements of the bright star Canopus at culmination[8], from two different locations a known distance apart. The method of Posidonius is potentially more accurate than that of Eratosthenes since a star is a small, bright point highlighted on a dark sky, whereas a shadow cast by the Sun tends to be fuzzy and difficult to measure accurately. Perhaps needless to say Posidonius deduced a value for the size of the Earth that was only slightly different to the one found by Eratosthenes. By the beginning of first century AD therefore, it seems that the size of the Earth was reasonably well known even though no one person had ever traveled around its girth. The mathematicians and astronomers had literally encircled the Earth with logic and geometry.

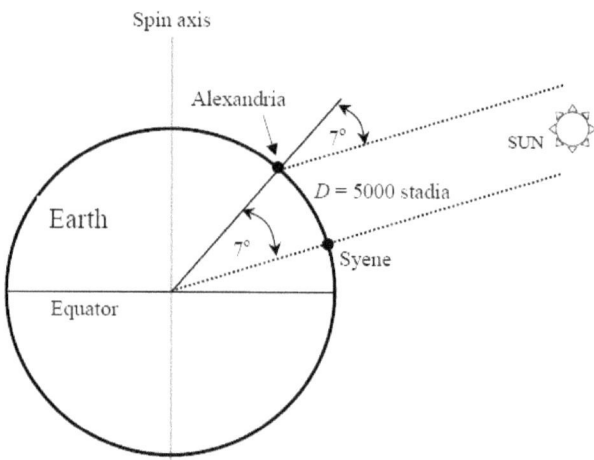

Figure 3.1: Eratosthenes method for measuring the size of the Earth.

There and Back Again

Greed and conquest were the primary reasons for early European exploration. And, perhaps not surprisingly, making more money was the reason behind the first successful circumnavigation of the world – although circling the Earth was not the primary objective of the expedition. On August 10th, 1519 Ferdinand Magellan (1480 – 1521) sailed from the Spanish port of Saville to undertake an important trade mission. Under his command were five ships and a total of 270 able seamen and officers. His objective was to find a westward route to the Spice Islands (now the island of Maluku) in Indonesia. The aim of the mission was in essence to assert Spanish control over the highly profitable clove, nutmeg and mace markets. By finding a westward route, via Cape Horn, the southern most tip of South America, the Spanish intended to side-step the Treaty of Tordesillas, signed with the Portuguese in June of 1494. This particular treaty split the world into two halves, establishing, as it did, an invisible circle of reference that ran along the north-south meridian near 40° West midway between the Cape Verde Islands and Cuba. Engineered by the papal office of Pope Alexander VI, the treaty was established to resolve a festering diplomatic and trade dispute, and it decreed that the lands to the east of the dividing meridian belonged to Portugal, while lands to the west belonged to Spain. The voyage of Magellan was instigated to prove that the Spice Islands fell under the treaty rights of Spain. Well, so much for the historical background. The point is that while Magellan didn't successfully complete the first circumnavigation of the world (he was killed in the Philippians at the Battle of Mactan on April 27, 1521) a mere 18 of his original 270 crew members did – commanded by Juan Sebastian Elcano (1476 – 1526), who sailed in to Seville on the one remaining ship of the expedition (Magellan's ship *Victoria*) on September 8th, 1522. It had taken three years and one month to sail around the world. The voyage didn't prove that the world was a perfect sphere, but it certainly showed that it you headed off westward you would eventually return home along an eastward path (and *vice versa*).

A Map without Wrinkles

Navigators, pilgrims and armies have used maps to guide their progress for countless centuries. And, maps have been produced in all shapes and sizes, and in a multitude of styles. While early maps offered very little useful information about the general topography of the land, they did show the towns to be encountered and directions to be taken as a journey progressed. Scale was, for the most part, inconsistent on early maps, and large swaths of domain would simply be described as *terra incognita*.

Claudius Ptolemy in the first century A.D. produced one of the first maps of the then known world. To the modern-eye it doesn't reveal a great deal of information. Most of the coastlines were only guesses and, of course, there is no hint of the land masses of North or South America, Antarctica or Australia, and only northern Africa is shown. The great innovation that Ptolemy introduced in producing his world map, however, was the coordinate system. The map was projected onto an imagined conical 'hat' situated above the Earth's North Pole, and the cities and towns were located according to a pseudo latitude based upon the length of the longest (or midsummer's) day. At the equator the longest day is 12 hours long, while at the poles it is 24 hours long. The longitude of the cities (which was always given as an easting) was based upon a reference meridian situated in the Canary Islands, since this was the westernmost limit of the known world.

While Ptolemy reasoned and accepted the fact that the Earth was a sphere, he didn't worry too much about the projection used for his world map since only a small segment of the globe was then known. As time passed, however, and with the first successful circumnavigation of the globe by Magellan's expedition, the problematic issue of representing a three-dimensional spherical surface on a two-dimensional page came to a head. The problem being that one cannot simply wrap the surface of a sphere with a piece of paper without producing wrinkles. The first cartographer to usefully solve this thorny mathematical problem was Gerard de Cremer (1512-1594)[9], who is most commonly known through his Latinized name, Gerardus Mercator (literally, Gerard the Merchant).

Just 16 years after the survivors of Magellan's expedition had completed the first circumnavigation of the globe, Mercator published (in 1538) his first world map. It was an odd-looking affair, being a double-cordiform projection – such maps have an appearance reminiscent of a pair of rounded butterfly wings although technically the two lobes (or wings) have a heart-shaped appearance. His first world globe was manufactured in 1541. While these early works provided the observer with the latest in information concerning the location of landmasses, oceans and cities they were not particularly practical. Indeed, Mercator knew full well that the maps he produced were essentially useless to the practicing navigator. What was needed, he reasoned, was a map from which a traveler could easily derive a set of bearings that would get them to their desired destination as quickly as possible - time, after all, is money.

Mercator's breakthrough came about in 1569. It was already clear that no projection of a sphere onto a flat piece of paper could be constructed so that area, direction and scale could all be conserved. That is, distortions of one form or another will always occur on any (ancient or indeed, modern) map. The eventual compromise that Mercator accepted was to allow the scale to vary. In this fashion, the parallels of latitude and longitude were laid out perpendicular to each other in a grid formation, the key twist, however, was that the east-west and north-south stretch applied to each square varied according to latitude (figure 3.2). While this scale variation resulted in the more northerly regions having greatly distorted coastlines, it did make the map conformal – and this, for the navigator, was of great utility. A conformal map is one that conserves angles, and accordingly Mercator's projection[10] ensures that a straight line drawn between any two points on the map provides the correct compass bearing for the navigator to follow. Perhaps oddly (at first thought) a straight line on a Mercator map does not actually correspond to the shortest distance between those two points on the Earth's surface – this latter distance is given by a great circle arc (or orthodrome) which corresponds to a curved line in Mercator's transformation system.

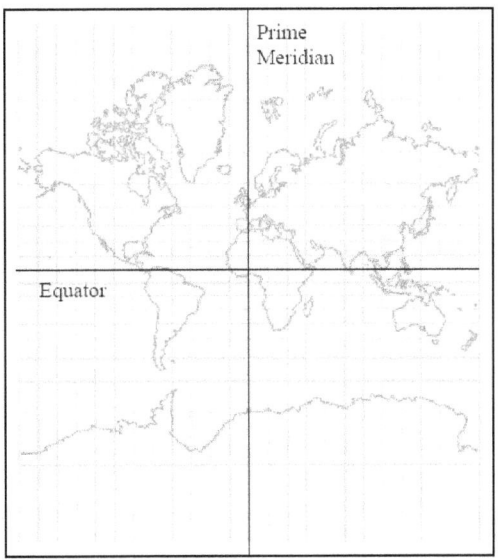

Figure 3.2. A modern-day Mercator world map. The scale distortion increases in the Polar Regions as indicated by the stretched-out coastline of Antarctica. Constant compass bearings (also called rhumb lines and loxodromes) correspond to straight lines on a Mercator map. Background image from *MathWorld*.

Richer's Adventure

It might at first seem that French astronomer Jean Richer (1639 – 1696) drew the shortest straw from the stack. Late in 1671 the French Government under the urgings of the Director of the Paris Observatory, Giovanni Domenico Cassini (1625 – 1712), agreed to send Richer to Cayenne, the capital city of French Guyana. His primary tasks, as outlined by Cassini, were to make positional observations of the planets Mercury, Venus, and Mars, as well as to observe the motion of the Sun and the Moon. Clearly, there was nothing particularly profound, astronomically speaking, that could be obtained by making such observations in Cayenne – surely, one might argue, such measurements could be made more accurately and much more conveniently in

Paris? Well, yes, they could, and indeed, that is exactly what Cassini was going to do - as it ever was and is, the boss got the easier job. The point of all this adventure, along with the expense and danger of ocean travel, however, was to make a series of simultaneous observations of the planets (especially Mars since it was then close to being at opposition and therefore at its closest distance to Earth). By such combined sets of observations it was hoped to determine the fundamental measure of astronomy – the, so called, astronomical unit (AU). The astronomical unit is the average distance of the Earth from the Sun – or more technically, it is the semi-major axis of its elliptical orbit about the Sun. All measurements in astronomy are based upon knowing the astronomical unit; it provides the scale for the solar system, the baseline for measuring the distance to the stars, and ultimately the scale for the entire universe. Ever since the time of Nicolaus Copernicus (1473 - 1543) astronomers have been measuring the astronomical unit in one manner or another, and in the modern era the AU has been determined to a high order of accuracy – indeed, it is the most accurately known number in all of astronomy with 1 AU = $1.4959787069 \times 10^{11}$ meters.

The journey undertaken by Richer was the first purely science focused expedition to be made – it was an all-together new kind of enterprise. There were no requirements of Richer to make maps, chart coastlines, collect plants, animals, birds, rocks, fossils or any of the other typical trappings required of European explorers at that time. His job was solely to do astronomy and to collect data relating to planetary positions. The aim of the observational campaign was to gather simultaneous observations of the position of Mars from Cayenne and from Paris such that its actual distance from Earth could be calculated in kilometers. The essential geometry behind the observations is shown in figure 3.3. By knowing the sky position of Mars from the two observatories, the angle P can be determined and then through an application of the sine rule the distance to Mars, D_{Mars}, can be calculated: $D_{Mars} = R_{\oplus} \sin(P/2)$, where $R_{\oplus}$ is the Earth's radius. Now, it was already known from Kepler's third law[11] that Mars travels about the Sun on an elliptical orbit with a semi-major axis of 1.524 AU. Since the distance to Mars was being measured when it was at opposition (its least separation from Earth), then the distance to Mars must correspond to the difference 1.524 – 1.000 AU, or D_{Mars} (km) = 0.524 AU, and from this the astronomical unit in standard measures (i.e., meters or kilometers) can be determined.

Using the combined observational data from Paris and Cayenne, Cassini was able to determine a horizontal parallax of 25 arc seconds for Mars, and a horizontal parallax of 9 ½ arc seconds for the Sun. This latter parallax provides an estimate of 21,712 $R_{\oplus}$ for the astronomical unit, which is about 8% smaller than the modern day value – this is a remarkable result none the less, and one that fully testifies to the success of Richer's mission[12].

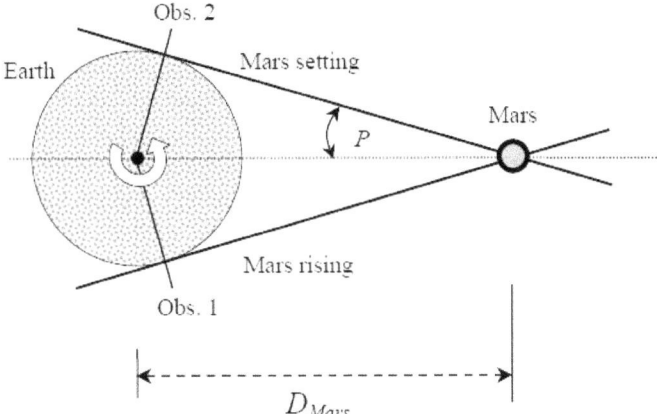

Figure 3.3. Simultaneous measurements of Mars, just rising for observer 1 and just setting for observer 2, can be combined to determine the angle P.

For over a year Richer diligently made observations of the planets from Cayenne. He also observed a lunar eclipse and transits of the satellites of Jupiter enabling a determination of the longitude for his observatory to be made - thus placing it firmly on the world map at latitude 4° 55' North, longitude 52° 18' West. Not only this, Richer also made detailed and repeated observations of a second pendulum that had been carefully calibrated beforehand in Paris. This minor part of Richer's expedition mandate turned out, in fact, to be his most profound contribution to science, and indeed, it is probably the only reason that his name is known today (history can otherwise be, and indeed for most scientist entirely is, a very forgetful task master). What Richer was attempting to establish through his pendulum observations was the constancy of its rate. The aim, of course, being to see if the length of the seconds pendulum could be used as a natural standard. His results, however, showed that the Paris adjusted seconds pendulum ran 2 minutes 28 seconds more slowly per day in Cayenne. The hopes of establishing a natural unit of measure according to the length of the seconds pendulum were dashed – the going of the seconds pendulum is location dependent and not fixed.

Newton's Equatorial Bulge
When Newton was a young schoolboy, in the early seventeenth century, the picture of a perfectly spherical Earth (the odd mountain chain and deep valley wrinkle aside) was firmly established - indeed, the astronomers and ocean going navigators all confirmed its sphericity. In later life, however, Newton put the adopted sphericity of Earth to the question, and found the answer wanting. In addition, and again in characteristic style, Newton's new insights called into doubt the observational results obtained by several of Europe's most respected savants. At the heart of the debate was the determination of

Earth's true profile – was it an oblate spheroid or a prolate one? Figure 3.4 illustrates the basic problem that needed to be solved. As is well known, the distance from the center of a sphere to any point on its surface is a constant – indeed, it is the radius of the sphere. An oblate spheroid, on the other hand, has a greater distance across its equator than it does from pole to poles; a prolate spheroid, in contrast, is longer from pole to pole than it is across its equator. One might say that an oblate spheroid has been squashed at the poles, while a prolate spheroid has been squashed in at the equator.

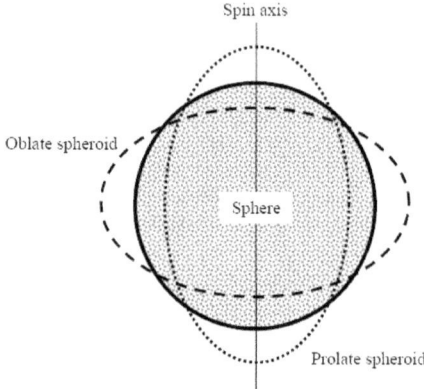

Figure 3.4. Cross-sections through a sphere (shaded), an oblate sphere (dashed line) and a prolate sphere (doted line).

Giovanni Cassini, in Paris, interpreted Richer's pendulum observations to indicate that there must be less mass under Earth's surface at the equator, and that therefore the Earth must be a prolate sphere. This conclusion was further supported by Giovanni's son, Jacques Cassini, when in 1718 he summarized the results derived from an extensive geodetic survey that stretched along a north-south meridian through France. Specifically, Jacques Cassini noted that the surface length corresponding to one degree of latitude decreased systematically as one moved northward, and this is exactly what one would expect if the Earth were a prolate sphere.

When Isaac Newton reviewed Richer's pendulum observations in his *Principia Mathematica* he concurred that they indicated Earth's gravitational attraction diminished towards the equator. Unlike Cassini, however, Newton argued that gravity was 'weaker' towards the equator not because there was less mass underneath the Earth's surface there, but because the Earth's radius was larger at the equator – that is, Newton argued, Richer's pendulum observations showed that the Earth was an oblate spheroid. By using the same data Newton and Cassini had come to two totally different conclusions, and this, of course, begged the question: who was right?

While the Cassini's looked to the geodetic survey results to bolster their claim for a prolate Earth, Newton turned to the heavens in support of his argument. Indeed, in Book III of the *Principia* (third edition, published in 1726), proposition XVIII (theorem XVI), Newton states "that the axes of the planets are less than the diameters drawn perpendicular to the axes". This is Newton's way of saying that the not only the Earth, but in fact all of the planets are oblate spheroids. To support the validity of his proposition Newton actually made reference to some of Cassini's astronomical observations of the planet Jupiter. Specifically, Newton noted that in 1691 Cassini had recorded that the east to west (that is the equatorial) diameter of Jupiter was in the ratio of 16:15 with respect to its polar diameter.

Newton had much more to his argument than just the observed polar flattening of Jupiter, however, and indeed, his argument was based upon a thorough theoretical analysis of the shape that a deformable, rotating but self-gravitating body must adopt. Defining the ellipticity as the ratio $\varepsilon =$ (equatorial radius – polar radius) / polar radius, Newton predicted that $\varepsilon \approx 1/230$ which is equivalent to the Earth's equatorial radius being some 28.5 kilometers longer than its polar radius (modern day observations make the radial difference to be 21.4 km and $\varepsilon = 1/298$).

By the early 1730's the stage was set for a classic show down. Who was right? Newton, the temperamental but brilliant English hero, or the Cassini's the bastions of French scientific honor and pride. Neither side was prepared to budge. Both parties felt that they were justified in their opinions, and neither side accepted that their respective arguments had been shown to be wrong. It was the French savants, however, who eventually made the key move to settle the issue - once and for all. Clearly, they argued, two detailed geodetic surveys must be conducted, one at a high latitude and one at the equator. The ground distances along one-degree north-south arcs measured at two such locations would settle the argument. If, they noted, Earth were prolate then the arc length in the more northerly survey would be distinctly smaller than that of the one degree arc at the equator; the Earth must be oblate if the more northerly arc was the longer of the two.

Discontent Goes to Peru

The French Academy expedition to measure the Earth's equatorial arc left Paris for Peru in 1735 – it was to be nearly ten years before the dispersed and thoroughly fallen-out participants were to return to France[13]. Astronomer Louis Godin was the nominal, but as it turned out completely incompetent, leader of the expedition, and he was accompanied by geographer Charles-Marie de La Condamine and mathematician Pierre Bouguer (the same Bouguer that we encountered earlier in Chapter 1), and their destination was the elongated plane of Quito in what is now Ecuador (but was then located in Spanish ruled Peru). The expedition was charged with the responsibility of making an extensive survey of the region with the express aim of mapping out two arcs along the Earth's surface. The first was to be a three-degree north-south arc, and the second was to be an east-west arc. For a prolate Earth, the degree of arc in the east-west direction must be longer than a degree of arc in the north-south direction. It all sounds simple enough when written down on paper, but in practice the expedition suffered numerous extreme hardships, personality clashes, murder, lack of money, poor weather and almost endless technical problems.

At the heart of geodesy is the triangle, and by triangulation the shape of the Earth can be measured. The process, once again on paper, is straight-forward. Begin with any baseline of known length – say the distance between two distinct geographic features - two hill tops, or perhaps two church steeples. From each end of the baseline the angles between the baseline and a third distinct feature (another hill top, or building, or a specially constructed beacon) are determined. Having measured these two angles, the third angle of the triangle is then known since all three angles must add up to 180 degrees. Now, using one side of the mapped out triangle as a second base line the angles to a fourth distinct feature on the landscape are measured. This process is then repeated for as long as is required. If the Earth were a perfect sphere or radius $R_\oplus$ = 6371.00 km then 1 degree of longitude would correspond to a distance of $\pi R_\oplus$ / 180 = 111.195 km, so to determine the shape of the Earth the triangulation procedure must extend over at least a hundred kilometers (and preferably much more). The triangulation process and how a meridian length of arc is measured are illustrated in figure 3.5. Once the triangulation has been completed the final step is to determine the latitude at the beginning and end points of the triangular web. This can be done by carefully studying the altitudes to a series of stars as they culminate in the sky.

Bouguer arrived back in France in June of 1744. Some eight months later, after an epic journey down the Amazon River, La Condamine also returned – and almost immediately the two academicians fell out again. Squandering both his money and scientific respect, Godin stayed-on in South America, essentially electing himself to a professorship at the University of Lima in 1744; he eventually returned in disgrace to Europe but was finally able to acquire the Directorship of the Naval Academy in Cadiz. It was in many ways a sad ending to what should have been a great voyage of discovery and adventure. Presumably the bitterest blow for Bouguer and La Condamine, however, was that after all their trials and tribulations, their hard-won equatorial arc measurements were too late; that the Earth was an oblate spheroid had been effectively proven some six-years before their return by a rival expedition sent to Lapland in 1736.

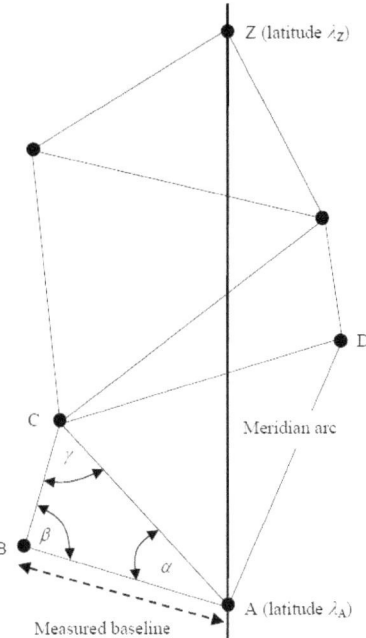

Figure 3.5. A schematic triangulation survey. The original baseline length between points A and B is carefully measured out, and then the angles α and β are determined. The angle γ is then known since $\alpha + \beta + \gamma = 180$ degrees. In addition the length of the other two sides of the triangle AC and BC can be determined through the so-called sine rule: $AB \,/\, \sin(\gamma) = AC \,/\, \sin(\beta) = CB \,/\, \sin(\alpha)$. Once the length of the side CA is known then the angles to point D are measured. The whole procedure of angle measurement and triangle length determination then continues until the final point Z is reached. Astronomical measurements at points A and Z then provides the length of arc in degrees $\Delta\lambda$, while standard trigonometry gives the distance AZ in kilometers. By dividing the two results $AZ \,/\, \Delta\lambda$ the number of kilometers per degree of latitude is determined.

Maupertuis Goes to Lapland

By all accounts the mathematician Pierre Louis Moreau de Maupertuis (1698-1759) was an ambitious, self-centered and conniving man – he also had a strong personal dislike for his slightly younger Paris Academy colleague Pierre Bouguer. The decision to sponsor a second meridian arc survey was made by the Paris Academy after the Peru expedition had all ready set sail. Under the general direction of Maupertuis, however, the second expedition was to head north, their destination the Artic Circle.

The Arctic expedition left Paris on the 20th of April 1736 and started its geodetic survey in July of that same year. The site of their survey was in the region of the Tornea River situated along the Gulf of Bothnia in Lapland. From all accounts the Academicians suffered incredible hardships during their survey; wanting for good food, shelter from freezing rain and relief from the multiple plagues of flies, gnats and midges that swarmed about them. After three months, however, their survey work was complete, and by early November they had obtained the astronomical observations that would determine the difference in latitude of the two extreme points of the surveyed arc. As the deep freeze of winter tightened its firm and frigid grip on the Lapland landscape, however, the expedition headed back to the town of Tornea in late November, and there they hunkered down until March of 1737.

The deep winter months were not spent in ideal retreat, however, and Maupertuis determined that the length of the meridian arc at the Artic Circle was 111.09 km, some 0.6% greater in length than the meridian arc measured in France (a distance of some 110.46 km). This was a staggering result, and clearly indicated that the Earth must be a prolate sphere. Indeed, the increase in the arc length was so great that the expedition members decided to double check their astronomical observations which had established the longitude difference between the extreme ends of their geodetic survey – the geodetic measurements they felt to be beyond any suspicion of error. From late March to early April of 1737 they performed a new series of astronomical observations, and found only very small differences with respect to their earlier results. The outcome was clear, the meridian arc in the Artic far exceeded the size of the arc measured in France, and Newton's equatorial bulge was accordingly elevated to the status of a proven terrestrial feature. Leaving Tornea in June, only to suffer a shipwreck in the Gulf of Bothnia, Maupertuis and his colleagues eventually returned in triumph to Paris in August of 1737.

The final result from the Artic survey indicated that the ellipticity of the Earth was $\varepsilon = 1 \,/\, 178$, and while this was a triumphant result, it was also rather unexpected, being much larger than anyone had previously predicted. From a modern perspective the ellipticity derived from the Lapland survey is clearly in error and much too large. While Maupertuis gloated over the fact that his survey results had stolen away any 'glory of first proof' that might have been attributed to the Peruvian expedition and especially to Pierre Bouguer, Bouguer can at least have the (posthumous) satisfaction of knowing that his measurements are generally deemed to be the more accurate and trustworthy of the two surveys. And, while the 1-degree meridian of arc in the Artic Circle is longer than that measured in central France, the 1-degree meridian arc measured at the equator was smaller still. Writing in his *Figure de la Terre*, published in 1749, Bouguer found that the length of a meridinal degree of arc on the Earth's equator was 109.92 km.

Running Hot and Cold

In addition to the geodetic work conducted during the Lapland survey an important series of pendulum experiments were also conducted. As the team moved further northward they found that their seconds pendulum, which had been carefully calibrated in Paris, ran faster and faster. By the time that the survey had reached its northern most point, close to the town of Pello, it was determined that the seconds pendulum made an additional 59 oscillations per 24 hours time interval.

When Maupertuis and colleagues performed the pendulum measurements they made sure that the apparatus was housed in a near constant temperature environment. Indeed, the pendulum had better and more comfortable quarters than the expedition members. This experimental refinement was not solely for the comfort of the pendulum observer; rather, it was to make sure that any variation in the running of the pendulum was not due to changes in the length of its support wire. This specific problem was a key point picked up upon by Isaac Newton when he discussed Richer's pendulum observations, and it accounts for the fact that a warm wire expands, while a cold wire contracts. Any slight contraction or expansion in a pendulum's support wire with temperature is a problem since the length L will have changed and accordingly so too will the period of swing.

The amount by which a wire expands ΔL upon being heated and undergoing a change in temperature ΔT is determined by the composition of the wire and it is expressed in terms of the coefficient of linear expansion α. Accordingly, the expansion is $\Delta L = \alpha L_0 \Delta T$, where L_0 is the length of the wire at temperature T_0. We obtain the result, therefore, that the expansion is directly proportional to the initial length L_0 and to the temperature change being considered. In this manner, for example, the expansion equation dictates that the greater the initial length L_0, so the greater the expansion ΔL that results for a specified change in temperature ΔT. Table 3.1 shows the expansion and period change effects that come about for a pendulum of initial length $L_0 = 1$-m at a temperature of zero degrees centigrade when it is heated to $40°$ C – the temperature of very hot summer's day.

Table 3.1 indicates that for a specific temperature change a brass wire will expand nearly twice as much (technically 1.667 times as much) as a steel wire. In terms of the change in the period of swing, under the $40°$ C change being investigated, a quartz strand pendulum will loose time at a rate 40 times slower than that of a brass wire pendulum. This result is expressed in the last column of the table, and it is found that a pendulum with a brass wire support at $40°$C will lose 1 second over its $0°$ C comparison pendulum after 2500.5 oscillations; a quartz wire pendulum, in contrast, will take 100,000 swings to lose 1 second.

Material	α (K^{-1})	ΔL (mm)	P_{40} / P_0	$N_{lose + 1}$
Brass	2.0 x 10^{-5}	0.80	1.00040	2500.5
Copper	1.7 x 10^{-5}	0.68	1.00034	2941.7
Iron	1.2 x 10^{-5}	0.48	1.00024	4167.2
Quartz	4.0 x 10^{-7}	0.02	1.00001	100,000.0

Table 3.1: The effect of temperature change upon a pendulum. Column 1 indicates the composition of the pendulum support wire, and column 2 indicates the coefficient of linear expansion. The expansion in a 1-m length wire resulting from a 40° C temperature change is shown in column 3. Column 4 indicates the slight increase in the period at 40° C (P_{40}) compared to the period of swing at 0° C (P_0). The last column indicates the number of swings required for the hotter, slightly elongated pendulum to lose 1-second over its shorter, cooler comparison pendulum.

Give a Little, Take a Little: A Constant Length Pendulum

The electromagnetic pendulum regulated clock developed by Charles Boys was described in the last chapter, and one of his refinements, recall, was to introduce a mercury compensation pendulum (figure 3.6) in which the expansion and contraction of the mercury was used to raise or lower a weighted plate in order to keep the pendulum length constant under variations in ambient temperature. The mercury compensation pendulum was invented not by Boys, however, but by renowned clock maker George Graham (c. 1673 – 1751) in the early 1720's. John Harrison further introduced an alternative device, the gridiron compensation pendulum in 1726 - we encountered Harrison earlier, in Chapter 2, with respect to his longitude prize-winning clock H4. Harrison's gridiron pendulum was made of linked brass and iron rods. Table 3.1 indicates that for a given change in temperature a brass rod will expand more than an iron rod by a factor of 1.67, and accordingly the idea of the gridiron arrangement (figure 3.7) is to trade-off these different expansion effects in order to keep the pendulum length constant. The key requirement is to construct a network of alternating brass and iron bars in such a way that the net expansion is as near to zero as possible[14].

The same temperature effects that change the period of a pendulum's swing are also evident in the running of pocket watches. Rather than using a pendulum, of course, such timekeepers run according to the elasticity of a spirally wound spring. The spring is attached to a weighted balance wheel and it is the constant back and forth motion of this wheel that regulates the ticking of the watch. Robert Hooke studied the use of a balance wheel and spring in 1658, but it was Christian Huygens who developed the first fully functional spiral-spring clock in 1674. It was again John Harrison, however, who developed, in 1753, a temperature compensation spring system for his longitude prizewinning H4 chronometer. The trick this time was to add a bimetallic compensator to the rim of the balance wheel so that any temperature variations could be offset by a mechanical displacement, expansion or contraction, of the wheel diameter. The bimetallic strip, as invented by Harrison, is made up of two thin strips of different types of metal (typically steel and copper, which have different coefficients of expansion), riveted or welded together. The different expansion coefficients of the two metals will cause the strip to bend when it is heated. Perhaps the most common application of bimetallic strips in the modern era is that within the thermostats that control the heating of our homes and work places. In this latter application the unwinding of a bimetallic coil is used to make or break an electrical switch, which in turn controls the heating element.

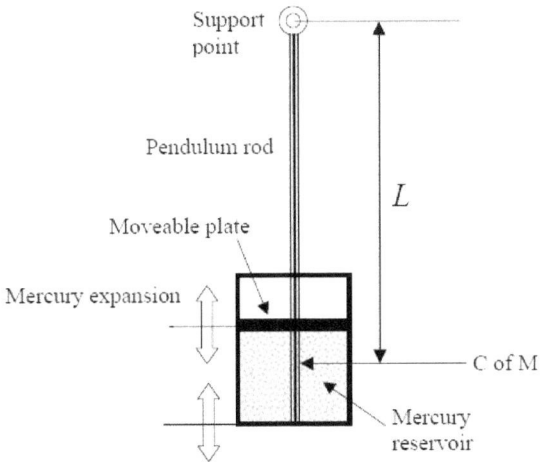

Figure 3.6. The mercury compensation pendulum. As the temperature rises, so the height of the mercury reservoir increases and this changes the effective length of the pendulum.

The Great Arc

While the length of arc surveys conducted in France, Lapland and Peru are all inspiring examples of hard won results, gained under less than ideal circumstances, perhaps our highest admiration should be reserved for the architects of the great trigonometric survey of India. Under the guidance and responsibility of the British army, Lieutenant Colonel William Lambton began the survey of India in 1802. It was a mammoth task that he and his many assistants embarked upon. The aim of the project was to survey the entire arc running from the southern most tip of India to the foothills of the Himalayan Mountains in the far north. The entire mapping project, of which the Great Arc was eventually just a small part, was to take sixty-four years to complete, and cost many lives in the making. When considering just the meridional arc component, however, Sir John Herschel eulogized in 1848 thusly, "The Great Meridional Arc of India is a trophy of which any nation, or any government, of the world [would] have reason to be proud, and it will be one of the most enduring monuments of their power and enlightened regard for the progress of human knowledge"[15] - high and heady praise indeed.

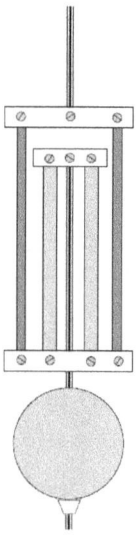

Figure 3.7: The gridiron pendulum. The various metal rods are arranged so that the total length of the pendulum remains constant even when the temperature changes.

The mild-mannered, ever-so precise, but highly likeable Lambton placed the first series of Great Arc survey triangles in Punnae at latitude. 8° N, near Cape Comorin in 1806, and then ran the triangular grid steadily northward. Twenty years later Lambton and the trigonometric survey were closing in on Nagpur, in central India. The survey had by this stage crisscrossed the Indian nation, and moved some 10 degrees north along a longitude of 78° E. Early on in the survey Lambton had been assisted by Henry Kater, who we encountered earlier in Chapter 1, and in 1818, Lambton's eventual successor and future Surveyor General of India, Colonel George Everest (pronounced *eve – rest*) joined the survey team. Character wise, Everest was almost the exact opposite of Lambton – he was brash, dictatorial and had few diplomatic skills. Like Lambton, however, he had a passion for surveying and was an absolute stickler for accuracy in all measurements. With Lambton's death in 1823, Everest took charge of the survey and over the next twenty years pushed the triangulation ever northward, eventually terminating the sequence just beyond Kaliana. The Great Arc covered 21 degrees of latitude and it is the longest directly measured meridian arc on Earth.

The main inspiration behind the trigonometric survey of India was the establishment of a baseline from which a systematic topographic survey could be performed. The scientific aim, on the other hand, was to determine the size and shape of the Earth, and Everest used the data from the various completed sections in 1830 to determine a polar flattening of 1 / 300.8 (a result that is within 1 percent of the modern-day value). Everest retired from military and government service in 1843, and returned to England, where he received great praise for his labors. Work on the trigonometric survey continued after Everest's retirement, however, under the guidance of Andrew Waugh, and it was Waugh who was able to show that Peak XV, located in the northern Himalayan reaches of Nepal, was the highest mountain in the world. Towering to an estimated 8,840-m (just 8-m lower than the currently accepted value) above sea level[16], this record setting peak clearly required a name. Since Waugh was unable to discover a local name for the mountain, he suggested in 1856 that it might be named after Everest, "that illustrious master of geographical research". Although it has since been discovered that to the Tibetans the mountain is known as Chomolungma (meaning *Goddess Mother*), and to the Nepalese as Sagarmatha (meaning *Ocean Head*), the name Mount Everest has stuck, and while Everest himself, in the modern-era, is an obscure historical figure everyone has heard of the mountain named in his honor. Ironically, there is no evidence to indicate that Everest ever saw the mountain that was given his name. As with all records there is still considerable debate concerning the exact height of Mt. Everest. The Nepalese Government, as of 2010, has adopted the official figure of 8,848-m while the Chinese Government, argues that is should be 4-meters smaller, being measured according to the base-rock height rather than that of the overlying snowcap. An American climbing team, sponsored by the National Geographic Society (NGS), in 1999 derived a GPS-based height of 8,850-m, but this figure, while currently used by the NGS, has not been accepted as definitive by the Nepalese Government. Irrespective of which number is used, however, it is known that Mt. Everest is still getting higher as a result of the Indian subcontinent continuing to slide under the Eurasian plate – the growth rate is currently estimated to be about 6-centimeters per year.

The Challenger Deep and Planetary Status
Having reached the roof of the Earth, at the summit of Chomolungma, what are the lowest terrestrial depths to which one might explore? For this, of course, one will be below sea level and indeed enwrapped in the absolute blackness of the Hadalpelagic Ocean. Since the first successful scaling of Mt. Everest, in 1953, by New Zealander Edmund Hillary and Nepali Sherpa Tezing Norgay, more people, indeed, many more people (by a factor in excess

of 2500) have subsequently stood upon its summit than have plumbed the darkness of Earth's greatest undersea trench – the Challenger Deep.

Located at the southern end of the Mariana Trench, an arching 2550 km subduction zone in the western Pacific Ocean, where the Pacific Plate dives underneath the opposing Mariana Plate (a small tectonic region located to the east of the more extensive Philippine Sea Plate), the Challenger Deep drops to a staggering 11.03 km at its deepest recorded point. The first indication that the seafloor along the Mariana Trench dropped away to such an incredible depth was found during the 1872 – 1875 *Challenger Expedition* sponsored by the Royal Society of London. The scientists and crew on this remarkable expedition, which was entirely dedicated to oceanographic exploration, circumnavigated the world aboard HMS Challenger recording details of ocean currents, sea temperature and sounding depths. On March 23, 1875 while located over the Mariana Trench a dragline depth of 8.184 km was recorded. Some 70 years after the *Challenger Expedition* ended, another Royal Naval vessel, once again called HMS Challenger, surveyed the entire length of the Mariana Trench, recording a maximum depth of 10.9 km via an echo sounding technique. The greatest recorded depth of 11.034 km was obtained during a Soviet survey expedition of the Mariana Trench in 1957. Just three people have ever seen firsthand the seafloor of this remote and desolate domain, Swiss oceanographer Jacques Piccard (actually the son of the submersible's designer, Auguste Piccard), American Navy Lieutenant Don Walsh and Canadian James Cameron. The first two of these pioneers descended to a depth of 10.916 km aboard the bathyscaphe *Trieste* on January 23, 1960. Since that time only two other remotely operated submersibles have explored the stygian darkness of the Challenger Deep. Perhaps most remarkably of all, however, given the crushing 1.25 metric tones per square centimeter pressure at an ocean depth of 11 km, abundant life forms have been recorded and sampled from the sedimentary ooze at the base of the Deep.

The quest to continue the human exploration of the Challenger Deep has begun afresh in recent years, and several highly financed research teams are vying for the spoils of first successful descent. As one of the major players in this new challenge, Canadian filmmaker and Hollywood director James Cameron set a record depth for a solo dive of 8.166 km on March 7, 2012. Using a specially designed 7.3-m long submersible, called the *Deepsea Challenger*, Cameron's record-breaking dive was made in the New Britain Trench located off the coast of Papua New Guinea. The next step, and indeed the very reason for constructing his submersible, was to explore the darkness of the Challenger Deep, and this Cameron successfully did, for several hours, on March 25th, 2012.

The vertical difference between the summit height of Mt. Everest and the detritus ooze at the base of the Challenger Deep amounts to a variation equivalent to just 0.3 percent of Earth's radius. By any standards this represents but a small ripple upon the surface of Earth's greater body. This relative smoothness of Earth's surface has been taken as an example by astronomers in their recent attempts to define what criteria an object must satisfied if it is to be called a planet. While being far from universally accepted the International Astronomical Union (IAU), at its 2006 meeting in Prague, introduced a three-point definition for planetary status. The object must orbit the Sun, the object must be spherical due to its own gravity, and the object must have cleared its immediate orbital path of smaller objects. Although definitions play an important role in the sciences in general, the IAU planetary definition serves no genuinely useful function. It was introduced purely on the grounds of excluding Pluto and other Kuiper Belt Objects[17] (especially the newly discovered KBO Eris, which is physically larger than Pluto) from the planetary class. The IAU rules fail in many ways, and there is no sense of any definition consensus developing between dissenting parties any time soon – the introduction of dwarf planet and plutoid categories has only made the situation more of a farce. Be all this as it may, the point here is that within the current IAU definition for planetary status is the notion that the object must be spherical – although this is a non-precise term. Clearly, the Earth is not perfectly spherical but its surface variations, from the loftiest mountaintops to the deepest seafloor trenches are but small wrinkles compared to its overall radius. This observation essentially begs the question therefore as to exactly how high can a mountain be before the general sphericity criterion is broken.

How can we estimate the height of the largest possible mountain that might exist upon a rocky planet? A first approximation can obtained by considering the force that the base of the mountain exerts upon the underlying mantle – let us, for the sake of simplicity, assume that the mountain is a cube with sides h. If the mountain is composed of material of density ρ, then its mass will be $m = \rho h^3$. The pressure exerted at the base of our mountain located on the surface of a planet with surface gravity g will accordingly be $P_{base} = g\, m / h^2$, where the h^2 term is the mountain's base area. Assuming the host planet is a sphere of radius R and composed of the same material as the mountain, then we can re-write the base pressure expression as $P_{base} = \frac{4}{3}\pi G \rho^2 h R$. The mountain will begin to sink into the surface of its host planet once the base pressure exceeds the yield strength Y of the mantle material. With this condition in place, the maximum height of a mountain can now be estimated as $P_{base} = \frac{4}{3}\pi G \rho^2 h_{max} R \geq Y$. Setting R to the Earth's radius, and adopting characteristic values for granite ($\rho = 2750$ kg/m^3 and $Y = 140 \times 10^6$ Pa), we find that $h_{max} \approx 10.4$ km. In light of this approximation, Mt. Everest appears to be about as high as any mountain might possibly be on Earth – although it still has some room to grow. If we turn the problem upon its head, we can further estimate the minimum size for a granite planet to support a mountain of some specified size. For example, if the mountain is to represent no more than, say, a 1% variation in the planet's surface (that is h /

$R = 0.01$) then it turns out that $R_{min} \approx 2575$ km. What this latter result indicates is that any planet composed of granite must be larger than about 5000 km across (nearly half the size of the Earth) if its surface is to have no height variations larger than 1% of its radius. For ice worlds, such as Pluto and Eris, the equivalent minimum size for surface variations to be smaller than 1% is $R_{min} \approx 1340$ km. Indeed, this result indicates that Pluto and Eris are about as small as ice worlds can possibly be and still be spherical at a variation level smaller than a few percent.

The Figure of the Earth
Ultimately, if they are truly going to be of any navigational value, all maps must be reduced to some agreed upon reference ellipsoid or geoid model for the Earth. The reference ellipsoid is a smooth, mathematically constructed model surface that approximates the general shape of the oblate Earth. There are actually many reference ellipsoid models currently in use and this in spite of international agreements and standards, but their basic properties are similar and they are all described according to the equatorial radius R_E, the polar radius R_P and the ellipticity (ε). The International Earth Rotation and Reference System (IERS) 1989 reference ellipsoid adopts the following parameters:

R_E (m)	R_P (m)	ε
6,378,136.000	6,356,751.302	1/ 298.257

The equatorial radius of the reference ellipsoid is correspondingly some 21,384.7 meters longer than the polar radius.

The reference geoid is arguably more complicated to define than the reference ellipsoid, and rather than being perfectly smooth (as in the case of the reference ellipsoid) its shape accommodates more of the Earth's 'bumps' and 'troughs'. A profile of the EIGEN-CG01C geoid model is shown in figure 3.8, where it is clear that the Earth has a truly odd (that is non-regular) oblate form. The geoid is an equipotential surface, and it can be though of as the surface to which a water layer will conform to. It is also the surface for which, at any point, the direction of gravity is acting in the vertical direction. The geoid and reference ellipsoid are technically quite different surfaces, but at their extremes they only differ by several hundreds of meters.

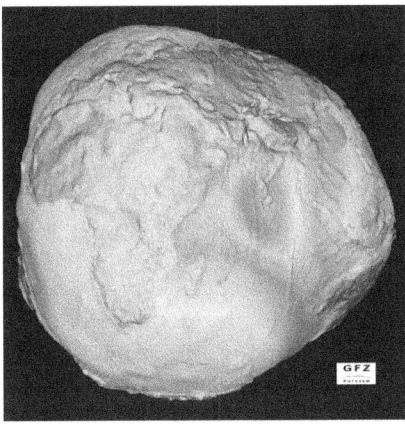

Figure 3.8. The global gravity field (geoid). Image by GZF, Potsdam.

Rather than try to accommodate the variation of gravity at every point over the entire Earth's surface geographers tend to reduce its variation to a formula that varies only in latitude – that is the longitudinal variation in the geoid is ignored. The world geodetic system formula, agreed upon in 1984 (WGS84), for the variation of surface gravity with latitude (λ) is written[18] in all its numerical glory, as:

$$g_0 = 9.7803267714 \left(\frac{1 + 0.00193185138639 \sin^2 \lambda}{\sqrt{1 - 0.00669437999013 \sin^2 \lambda}} \right) \qquad (3.1)$$

and this formula allows for the surface gravity to be slightly higher (by 0.53 percent) at the Earth's poles ($\lambda = \pm 90°$) than at the Earth's equator ($\lambda = 0°$). A seconds pendulum calibrated at the Earth's equator will therefore run more slowly at either of the Earth's poles, making some 228 fewer swings in a 24 hour period.

Earthquakes and Polar Drift
It is the strangest of feelings to experience an earthquake – the whole world seems to change, and all of ones senses are altered. Indeed, it is a surprising affront to normality that the ground, the very essence of solidity, should tremble and shake and rumble and slide. But heave and quiver it surely does when fault plains deep underground shift and buckle. The energy released during an earthquake can be truly colossal and such phenomena represent one of the most deadly of natural forces, leaving in their wake a trail of disaster, death, destruction and mayhem: it is the Earth at is most dangerous; raw and primordial.

Devastating earthquakes have been recorded throughout human history, and their origins have long been debated. To the ancient Greeks, who lived in one of the most earthquake prone and active volcano rich parts of the Mediterranean, the rising and shifting of the ground was a direct consequence of the ignition of dry, combustible vapors that had accumulated in underground caverns – earthquakes literally flowed from the underworld and the gateway to Hell. Learned opinions on the origins of earthquakes hardly changed from those of the ancient Greeks until well into the Renaissance.

In a remarkably modern sounding theory, Robert Hooke suggested in a 1687 lecture that regions of land and sea could change over time; with the sediments built-up on the seafloor being raised by earthquakes to form mountains. This heaving and shifting of surface landmasses, Hooke further suggested, hinted at the possibility that Earth's interior was a dynamic, shifting, heterogeneous, and ever-moving region. Not only this, the varying locations of landmasses, Hooke argued, should result in an observable polar wander; an apparent shift in the direction of the Earth's spin axis as indicated by the north and south celestial poles. By this, Hooke did not mean that Earth's spin axis was actually changing its orientation, but that the surface land masses were shifting with respect to the spin axis: "the axis of its [Earth's] rotation hath and doth continually by a flow of progression vary its position with respect to the parts of the Earth". This was a radical new theory – and none of his contemporaries believed a word of it. In spite of its negative reception, Hooke, ever inventive and determined, suggested that extremely long base-line telescopes might be constructed to accurately monitor the location of the north celestial pole – indeed, be suggested that church spires and tall buildings might be used to house large objective lenses having focal lengths of many hundreds of feet, out of which telescopes could be fashioned. Having made this suggestion, however, Hooke, as ever, was far too busy to see the project through to completion.

While Hooke's ideas relating to Earth's surface and interior were inherently active and dynamic they were not periodic. Rather the upheavals due to earthquakes were random and unpredictable (much the same situation still holds to this very day). In contrast to Hooke's model, his contemporary Edmund Halley developed and described in 1692 a much more ordered, hollow Earth model. Halley built his ideas upon the direct observation of variations within the geomagnetic field, and upon an (incorrect) argument outlined by Isaac Newton. Specifically, Newton had argued in his *Principia* that the Moon must be composed of material having a much higher density than that of the Earth. The latter result is a rare example were Newton got his numbers badly wrong (although, of course, the theory was correct), but none the less, given, as Halley supposed, that the Moon was much denser than the Earth, he reasoned that the problem could be explained if Earth's interior was largely composed of empty space. In this manner, Halley suggested that the Earth and Moon were made of material having the same density, but that Earth's interior was composed of four co-axial spherical shells, with each shell separated by a volume of air. Halley also speculated, much to the delight of later science fiction writers[19], that each shell could be inhabited. Halley additionally speculated in 1716, that the region between Earth's surface and the first interior shell (a zone deemed to be about 500 miles thick) might also contained a form of luminous matter that was responsible for the production of aurora – the Northern Lights. Not only did a hollow Earth 'save' Newton's (erroneous) density miss-match between the Earth and Moon, the presence of four spherical shells was implied according to Halley since his observations indicated that the geomagnetic field had four magnetic poles. One pair of poles was embedded in the Earth's outer shell, and these poles were deemed to be fixed and stationary, the other sets of polar pairs were in motion, however, and it was the relative rotation and positions of the inner spherical shells that determined the actual properties of Earth's magnetic field over time. Halley's theory concerning Earth's interior became popular and widely known, and even as late as 1736, the somewhat somber-looking, octogenarian Halley, then 2nd Astronomer Royal, can be seen in his portrait by Michael Dahl (housed at the Royal Society's main office at Carlton House in London) holding a chart displaying his multi-shelled Earth.

Hooke's polar wander was an idea too far ahead of its time, and the best part of two centuries would pass before the basic idea of continental drift, as outlined by German meteorologist Alfred Wegener in 1912, was put in place. Using the very same ideas advocated by Robert Hooke, Wegener used fossil record patterns and similar rock types to match up the coastlines of the continents – assembling them all into a single super-continent called Pangaea (a name actually coined in 1927 and derived from the ancient Greek for 'entire Earth'). It is presently estimated that Pangaea began to break apart about 200 million years ago, having formed about 100 million years before the inevitable rifting set-in.

When they were first presented, Wegener's ideas were critically attacked and they were not generally accepted until the mid-1950s, when it was realized that the newly discovered phenomenon of sea-floor spreading offered a natural mechanism via which continents might move over Earth's surface. Circulation currents set in motion by deep mantle convection from within Earth's hot interior, it was further realized in the 1960s, could drive the process of plate tectonics, sea floor spreading and continental drift. The Earth is, as Robert Hooke firmly believed, a vibrant and dynamic body, ever changing both above and below ground.

While Hooke's earthquake model for landmass upheaval contains but a grain of the real truth within it, one can still ask if the Earth's spin axis physically moves when the Earth quivers. The answer to this question is, in fact, yes, although one must be careful to specify exactly what it is that is actually moving. The Earth is not a symmetric, uniformly layered, rigid body and accordingly its spin state is very difficult to define in any theoretical manner –

rather, it can only be measured. At play are multiple time-varying perturbations and interactions, and the consequence of these phenomena are measured through the motion of the spin axis about Earth's figure axis. The figure axis is the imaginary line that passes through the center of the Earth and about which its mass is in balance. It is the orientation of this axis that changes whenever the mass distribution within and upon the Earth changes. Indeed, the figure axis shifts continuously, at a current rate of a few centimeters per year, and this drives the displacement of Earth's spin axis, which is constrained to move around the figure axis. Continual shifts in the orientation of the figure axis come about due to an effect known as Chandler Wobble; a phenomenon relating to the redistribution of mass within Earth's core and mantle, and the slow rise of land (isostatic rebound) following the last glaciation cycle. The first, and perhaps least understood, of these terms was discovered by American astronomer Seth Chandler in 1891 and corresponds to what is know as free nutation. Where earthquakes fit into the picture is through there effects upon the landmass distribution close to Earth's surface. Indeed, the redistribution of landmass will affect both Earth's spin rate and the orientation of the figure and spin axis. In recent years several dramatic - and deadly - earthquakes have occurred, and for those recorded on February 27, 2010 [off the Chilean coast], and upon March 11, 2011 [off the coast of Japan], the figure axes was observed to shift by 8 and 15 centimeters respectively. These earthquakes also produced a measurable change of -1.26 and -1.6 milliseconds in the Earth's rotation rate.

Sympathetic Vibrations

It is perhaps only appropriate that the world's tallest building, the Taipei 101 Tower located in Taiwan, contains the world's largest pendulum. The colossal 660,000-kg bob of this record-breaking pendulum is housed between floors 87 and 91, and its function is vital to the building's stability. Composed of 41 massive steel plates the bob is attached to the building by four massive hawsers and its function is to stop the building from critical flexing during the commonly occurring earthquakes and typhoons that strike Taiwan. The bob is specifically a tuned mass damper, and its role is to stop the building from swaying too much during an earthquake or hurricane. Nominally at rest, the mass damper can be thought of as a giant plumb-line, the purpose now is not to define the local vertical, but to inhibit the motion of the swaying building. The bob does this by swinging in the opposite sense to the building and absorbing the vibrational energy of the building and literally, thereby, damping out its motion.

While all modern day skyscrapers contain some form of motion damping system, the massive pendulum installed in the Taipei 101 tower was necessitated by its location. Taiwan is situated close to the boundary between the Eurasian and Philippine tectonic plates, and it is one of the most active earthquake zones on Earth. Here the ground literally does move for you, often, and while such movements can endanger building integrity as well as threaten life and limb, they also afford geologists an incredible opportunity to understand the make-up of the Earth's otherwise invisible interior.

The seismic waves produced by landslides and earth tremors are generated according to two varieties: P-waves and S-waves. The P-waves are longitudinal waves that result in the compression and then dilation of the medium through which they are propagating. S-waves, on the other hand, are shear waves that 'shake' at right angles to the direction of propagation the medium through which they are moving. Unlike P-waves, which can propagate through any type of material whether solid, liquid or gas, the S-waves can only pass through solid material – liquids and gases being unable to support shear stresses. It has been through the deployment of recording devices (seismographs) to measure the P and S-waves generated by earthquakes that geologists have been able to map out the structure of Earth's interior.

In the modern era seismographs are produced in many different shapes and forms and they exploit various physical processes to detect and measure seismic waves. The earliest seismometers, however, were constructed around the inertial properties of a heavy plumb bob – the concept of inertia will be discussed later in Chapter 4. As illustrated in figure 3.9, however, the early seismographs consisted of a frame, a heavy plumb bob, with a scribe or pen attached to its lower surface, and a rotating paper drum. The plumb bob is isolated from the motion of the frame by a suspension spring, and its large inertial mass prevents movement even when the frame is vibrating – any small motions that the bob might make due to frame shaking is actively damped out by a magnetic coupling attached to the suspension wire. By isolating the plumb bob it is the relative motion of the frame that is recorded on the rotating paper drum. As the paper roll rotates and the frame shakes, so a time recording of the frames motion, driven by the seismic waves, relative to the fixed point of the plumb bob is recorded. In some sense we can think of the seismometer as being a reverse pendulum – while the position of the bob remains (almost) fixed the world and the seismograph frame oscillate around it. The schematic seismograph shown in figure 3.9 has been constructed to measure the horizontal shaking motion of the frame. To complement this design the vertical shaking motion of the frame could also be measured, in this latter case by mounting a pen on the equator of the bob and placing the rotating drum in a vertical position.

SHM and the Horizontal Pendulum

If a plumb-line is a stopped simple pendulum, then a plumb-line incorporating an elastic or stretchy string is the embodiment of new type of pendulum – in this case the pendulum will undergo vertical oscillations. For the vertical pendulum the restoring force acting to bring the bob back to its equilibrium position is the elastic pull of the support wire or spring. Hooke's law, which was first articulated by Robert Hooke in 1660, describes the key elastic properties of a wire (or spring), and states that the restoring force F_R varies in direct step with the extension y imposed such that $F_R = -k\,y$, where k is a constant[20]. With this linear restoring force, the equation of motion for a bob, of mass m, attached to the end of an extensible wire will be $m\,\ddot{y} = F_R = -k\,y$, where Newton's dot notation for time derivatives has been used. This particular differential equation has a simple solution in the form $y(t) = A\sin(2\pi t/T)$, where A is the amplitude of motion and T is the period of oscillation with $T = 2\pi\sqrt{m/k}$. Motion of this particular one-dimensional kind is called simple harmonic motion (SHM). The formula for the period of vertical oscillation is clearly similar to that of the simple pendulum, equation (1.4), but now we see that the mass of the bob enters directly into the period determination, while the length of the elastic wire or spring does not. In addition, the period of oscillation also depends upon the spring constant k: If k is very large then the period of oscillation will be very rapid, while if k is very small then the period of oscillation will be very long. In seismological work, there are distinct advantages to having a highly sensitive long period of oscillation pendulum that can respond to small vibrations. Such sensitivity can in principle be achieved by making a long, thin, loosely wound, narrow gage wire spring, but in practice this makes for an unwieldy pendulum configuration. To mitigate against this latter problem, seismometers often employ a horizontal pendulum which consists of a horizontal thin, low mass solid beam, hinged, and therefore allowed to swing like a gate, at one end, and a bob attached at the other. A spring, as shown in figure 3.10, is used to support the beam and hold it, when at rest, in the horizontal (equilibrium) position. In the most sensitive applications of horizontal pendulums the spring is specifically chosen to be, what at first seems to be a contradiction in terms, a zero-length spring.

A spring is a physical object, typically made from a coil of wire, which stores mechanical energy. If you stretch-out a spring, then it produces a restoring force, in accordance with Hooke's law, that will counteract the expansion. A zero-length spring is an especially manufactured spring such that if it could contract to zero physical length it would exert zero restoring force. If one performed a series of experiments, therefore, with such a spring a plot of the restoring force *versus* spring extension would yield a straight line going through the origin at (0, 0). Since springs cannot be physically reduced to zero length, however, they are manufactured with a built in tension – a tension equivalent to their physical length and one such that if they were (magically) contracted to zero length would produce zero force. The other way that this can be stated is that for a zero-length spring the restoring force (a constant of proportionality aside) is equal to its physical length. The idea of zero-length springs came about in the early 1930s when American engineers Lucian La Coste and Arnold Romberg (University of Texas at Austin) realized that they could be used to make ultra-sensitive seismometers and gravimeters.

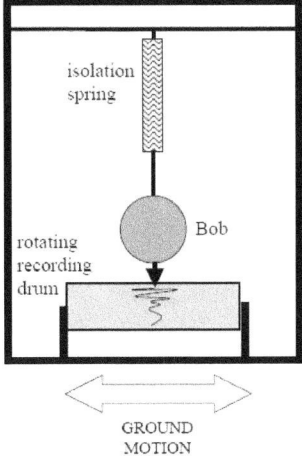

Figure 3.9. A schematic inertial mass seismometer. The plumb bob is isolated from frame movement by a suspension spring and a magnetic damping system. The rotating drum allows a record of the motion of the frame relative to the plumb bob to be recorded.

To see how zero-length springs help in the design of highly sensitive, long period of oscillation, horizontal pendulums, let us consider the equation of motion for the bob mass (as illustrated in figure 3.10). Accordingly, we have for small oscillations

$$m\ddot{y} + \left(k \frac{L_0}{L_{eq}} \sin^2 \phi_{eq} \right) y = 0 \qquad (3.2)$$

where the subscript eq indicates the equilibrium, system at rest, length of the spring, and L_0 is the un-stretched length of the spring. Equation (3.2) is precisely the equation expected for a SHM oscillator. The period of oscillation is therefore

$$T = 2\pi \sqrt{\left(\frac{L_{eq}}{L_0} \right) \frac{m}{k \sin^2 \phi_{eq}}} \qquad (3.3)$$

The equation for the period indicates that T can be made very large by making L_0 very small or by adjusting ϕ_{eq} to be close to zero degrees, or by a combination of both such adjustments. There are physical, engineering constraints that set a limit on how small the spring can be, and given that $\sin \phi_{eq} = D / L_{eq}$, for ϕ_{eq} to be close to zero requires that D, the distance between the spring attachment point and the hinge, becomes small. Once again there is a practical limit to how small D can be: as $D \to 0$ the whole hinge system becomes unstable. For the horizontal pendulum, therefore, a classic trade-off situation arises, and the engineer needs to find a combination of L_0 and ϕ_{eq} that produces the required sensitivity. In practice, a period of oscillation of $T \approx 30$ seconds is typically chosen, with ϕ_{eq} being set to an angle of 45 degrees, and L_0 being adjusted in line with the constant k appropriate to the wire being used to make the spring .

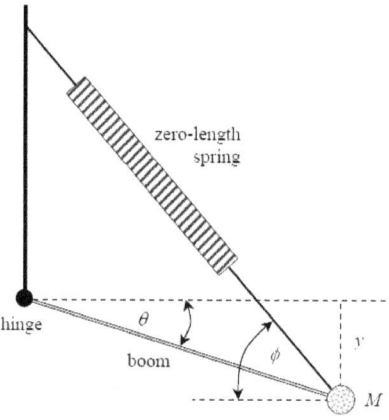

Figure 3.10: The horizontal pendulum

While networks of many oscillating horizontal pendulum can monitor even the smallest of terrestrial tremors, and reveal thereby the Earth's internal structure, a stopped horizontal pendulum – a horizontal plumb-line? – can be used to measure relative changes in the acceleration due to gravity. Again, with reference to figure (3.10), the torque at the support hinge due to gravity acting upon the bob is $\tau_{gr} = m\,g\,b\cos\theta$. The torque due to the spring, counteracting that of gravity on the bob, is $\tau_{sp} = k(L - L_0)\sin(\phi - \theta)$. The equilibrium position, $L = L_{eq}$, corresponds to a condition where there is no net torque, $\tau_{gr} = \tau_{sp}$. Using the identity that $\sin(\phi - \theta) = (D / L)\cos\theta$, the equilibrium condition becomes:

$$Dk\left(1 - \frac{L_0}{L_{eq}}\right) = mg \qquad (3.4)$$

The advantage of making L_0 as small as possible is now revealed, in the sense that for $L_0 \approx 0$, so equation (3.4) indicates that $g = D\,k\,/\,m$, and the instrumental variable at play is now the value of D. In the case of very small L_0, not only does the period of oscillation become very large, as indicated by equation (3.3), but the pendulum hovers on the brink of unstable equilibrium. The analogous situation is that of a mass placed on a perfectly smooth, friction-less horizontal plane. Under these conditions the mass will stay put wherever it is placed upon the plane, but if the plane tilts by just the smallest of amounts then the mass will begin to slide, in principle, all the way to infinity. The stopped horizontal pendulum, therefore, once set in its equilibrium position will be extremely sensitive to small changes in gravity. The relative change in the gravitational acceleration from one location to the next can now, in effect, be determined by simply monitoring the equilibrium position of the pendulum bar.

The Inside Story

The speed with which P and S-waves propagate through a solid varies with density and composition, and accordingly the depth of a specific region within Earth's interior can be gauged by studying the arrival times of seismic waves at widely dispersed seismograph stations. Some waves will propagate along the Earth's surface, others will be reflected from specific regions in the deep interior according to the wave's angle of propagation and the type of material being encountered, and some waves will propagate straight through the Earth's interior. If a liquid layer is encountered only P-waves will propagate through it, with the S-waves being terminated. It was by unraveling all these intricate details of travel times according to the seismograph to earthquake epicenter distance, as well as reflections and types of waves that can propagate that Irish seismologist Robert Dixon Oldham first deduced, in 1906, that the Earth must have a molten core. Specifically he noticed that for any given large earthquake there is a shadow zone region on the opposite side of the Earth (figure 3.11) where no S waves are detected. P-waves, on the other hand would be recorded in the shadow zone and this must mean that the S-waves have been terminated in a liquid medium. Further studies by the Dutch seismologist Inge Lehmann showed in 1936 that Earth's core is not one single molten sphere, but is divided into two zones with an inner (2000-km diameter) solid sphere of nickel-iron alloy surrounded by a molten nickel-iron alloy shell. The typical depth, density and composition of the Earth's inner regions, as deduced by the study of seismic waves, are shown in table 3.2. below.

Region	H (km)	Density (kg/m³)	Typical composition
Crust	30	2,500	Silicate rocks / basalts
Upper mantle	720	4,000	Olivine / pyroxene
Lower mantle	2,171	5,000	Magnesium / silicon oxides
Outer core	2.259	11,000	Sulfur / liquid nickel-iron alloy
Inner core	1,221	13,000	Sulfur/ solid nickel-iron alloy

Table 3.2. The main structural zones of Earth's interior as deduced from seismic studies. Moving from the surface crust to the central core the characteristic composition of material changes, the characteristic density increases dramatically and so too does the temperature. Data adapted from USGS.

Not only have pendulum seismographs mapped-out Earth's interior, they have also been used to study the internal make-up of the Moon and (less successfully) Mars[21]. Indeed, in the case of planet Mars the pendulum has gone where no human being has ever stood (so far at least). In relation to lunar exploration, one of the most important (and, as it turned out, most successful) outcomes of the Apollo landings of the 1970s was the establishment of a network of seismographs to study meteoroid impacts and moonquakes. The various instruments deployed[22] contained horizontal as well as vertical pendulums, with the response signal to seismic events being generated via the movement of a coil (situated close to the pendulum bob) moving through a magnetic field - this motion results, via Faraday's inductance law, in the generation of an electric current, and it is the variation of the electric current that is transmitted back to Earth for analysis. Meteorite impacts were regularly detected by the lunar seismic network, as well as moonquakes, and the collected data indicate that both P and S-waves can propagate through the Moon's interior, indicating that its core, which was once fully molten, must now be relatively cold and, at best, only partially molten in a relatively small central region.

The lunar seismic network revealed that the Moon, in spite of its small size, has a remarkably complex internal structure, and that the central core is off-set from the geometric center, being located closer to the lunar surface on the Earth facing side. In addition studies conducted from orbiting spacecraft revealed regions of high gravitational anomaly, with the mass concentrations (mascons) being located underneath a number of the lunar maria – these anomalies appear to be related to the intrusion of dense basaltic lavas. A whole series of moonquakes were detected by the lunar seismic network, some were deep-seated occurring at depths of about 1000 km below the surface, and these were most probably due to tidal flexing. Other quakes were detected much closer to the Moon's surface, and these were generally attributed to meteoroid impacts, or the thermal expansion and contraction of the lunar crust in those regions located close to the terminator – the solar illuminated / non-illuminated boundary line.

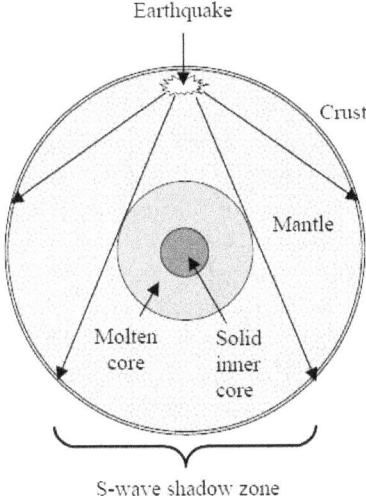

Figure 3.11. The S-wave shadow zone caused by the Earth's molten core. Although the arrows indicating the radiation of S-waves from the epicenter of the earthquake are shown as straight lines their actual paths will be curved because the speed of propagation will vary with depth and composition – see table 3.2.

The last Apollo lunar landing mission ended 40 years ago, and no human has walked upon the Moon since Eugene Cernan and Harrison Schmidtt successfully blasted from its surface aboard the *Challenger* lunar module. It is far from clear when the next humans will walk upon the lunar surface, although much recent interest and research has focused upon the possibility of returning to the Moon to establish a permanently staffed research station. To this end the remote satellite exploration of the Moon continues apace, and NASA, for example, has most recently initiated the *Gravity Recovery And Interior Laboratory* (GRAIL) mission. The GRAIL mission is based around two identical spacecraft, appropriately named Ebb and Flow, orbiting the Moon along near parallel polar orbits about 200-km apart. To map-out the Moon's gravitational field the small shifts in the distance between the two spacecraft, brought about by the gravitational force of the surface features located directly below their orbital tracks, are monitored. The distance between the two spacecraft is measured via a direct radio-wave link, and their separation is determined to an accuracy of a few thousands of a millimeter. The data from the GRAIL mission will eventually reveal a detailed structural map of the Moon's crust and underlying lithosphere.

A Small Spherical Correction

To finish this Chapter we shall begin to explore some of the 'real world' limitations of the simple pendulum analysis presented in Chapter 1 and as applied to SHM in this chapter. Moving beyond the correction due to the initial offset angle – as exemplified by equation (1.7) - we shall consider the effects of variable pendulum wire (spring) length, variable bob mass and spatial variations in the gravitational field. Given that this chapter has been concerned with the historical determination of Earth's shape, it seems only reasonable to first ask if its derived spherical(ish) shape might have some potentially observable affect upon the motion of a simple pendulum. The answer to this question is yes, and it relates to fact that the gravitational field is directed along radial-lines pointing towards the center of the Earth (here assumed to be a perfect sphere). This situation is in contrast to that assumed in deriving equation (1.4), and as illustrated in Figure 1.1, where the gravitational field lines are always taken to be parallel to the vertical defined by the pivot point and the middle of the bobs arc. This parallelism assumption essentially amounts to adopting a 'flat Earth' model for the gravitational field[23]. The affect of a gravitational field gradient can be studied via conservation of energy arguments (as discussion in Chapter 1), and we accordingly write the total kinetic plus potential energy of a pendulum of length L swinging just above the surface of a spherical Earth as:

$$E = \tfrac{1}{2}mV^2 - \frac{g\,R^2}{\sqrt{(R+L)^2 + L^2 - 2L(R+L)\cos\phi}} \qquad (3.5)$$

where m is the mass of the bob, V is the velocity, $g = GM / R^2$, with M and R being the mass and radius of the planet, and where G is the universal gravitational constant (to be discussed in detail in the next chapter). In deriving equation (3.5) the square root term is the distance of the bob from the center of the Earth for a pendulum opening angle ϕ - the derivation is, in fact, a straightforward application of the cosine formula. In the limit that $L << R$ (which, of course is the usual situation) the potential energy term in equation (3.5) will reduced to GM / R (see note 23), which is the correct 'flat Earth' approximation. Without reproducing all the details here, Lior Burko (University of Utah) has shown that under the conditions applicable to equation (3.5) so, to a first order approximation, the pendulum's period of oscillation becomes $T = 2\pi\sqrt{L/[g(1 + L/R)]}$. Comparing this result with the period derived for the ideal, 'flat-Earth', frictionless simple pendulum $T_0 = 2\pi\sqrt{L/g}$, it is apparent that for the spherical Earth model the pendulum moves as if it was embedded within an effective gravitational field $g_{eff} = g(1 + L/R)$. In other words, if one attempts to use the standard period of oscillation formula [equation (1.4)] to deduce the value of gravitational acceleration g at the Earth's surface, the end result will be an under estimate of its true value by a factor of order $(1 + L/R)$. Given that the Earth has an average radius of 6371 km, so the L/R term is typically going to be very small; for a 1-m pendulum, the correction amounts to a factor of 1.57×10^{-7}, but at 30-m the spherical Earth correction term is comparable to the finite amplitude corrections indicated in table 1.2. Pendulums having lengths of multiple tens of meters are not common, but they do exist. Indeed, as we shall see in Chapter 6, Foucault's famous 1851 pendulum experiment in the Panthéon in Paris, France had a length of 61-m. For the very long, simple pendulum the shape of the Earth, it turns out, does matter – this result is to be contrasted with the (hypothetical) infinite length pendulum considered in Chapter 1 (recall figure 1.6)[24].

Endnote: Three Pendulum Hybrids

Next to the gravitational parallelism approximation, one of the defining characteristics of the pendulum introduced in Chapter 1 was the constant length of the pendulum wire. This, of course, need not always be true and with the introduction in this chapter of thermal expansion effects, simple harmonic motion, Hooke's law and springs, we may now consider the intriguing dynamics associated with a few hybrid pendulums; first the Wilberforce pendulum and then the variable length and variable mass pendulums, all of which tell us a little bit more about the simple pendulum and how it works in the 'real world'.

The Wilberforce pendulum was first described by British physicist Lionel Wilberforce in the late 19th Century. It is an intriguing device that produces an initially unexpected display. Composed of a vertically hanging helical spring and a coupled end mass the Wilberforce pendulum is set in motion by gently twisting the end mass without stretching the spring or displacing it to one side. At first the end mass simply twists back and forth about the vertical axis (it is, in fact, behaving like a torsional pendulum – to be discussed in much greater detail in the next chapter). After a short while, however, something odd begins to happen. The twisting motion becomes less pronounced, and the mass begins to oscillate up and down in the vertical direction – just like a simple harmonic oscillator. Eventually, the twisting motion completely subsides and the motion is entirely in the vertical, up and down, direction. At this stage the behavior reverses, with the up and down amplitude subsiding and the twisting motion beginning again.

The Wilberforce pendulum is an example of a coupled mechanical oscillator, and it illustrates the point that, in reality and in spite of our Chapter 1 definition, there is no such thing as a constant length, mass-less pendulum support wire, and it also illustrates the point that the motion of a pendulum is

145

sensitive to the conditions under which it is set in motion. The behavior exhibited by the Wilberforce pendulum is entirely due to the construction of the spring; when the end mass twists about the vertical axis, the spring is slightly wound-up and then unwound, and this transfers energy into the translational (that is vertical) mode of oscillation. Gradually, energy is sapped from the rotational mode of oscillation and transferred into the vertical mode of oscillation and so on. The equation of motion for the Wilberforce pendulum can be written[25] in terms of the twist angle θ and the vertical motion Z:

$$m\ddot{Z} + kZ + \tfrac{1}{2}\varepsilon\theta = 0 \text{ and } I\ddot{\theta} + \delta\theta + \tfrac{1}{2}\varepsilon Z = 0 \qquad (3.6)$$

where m is the mass of the bob, k is the translational spring constant, δ is the torsional spring constant, I is the bob's moment of inertia, and ε is a mode coupling constant. A quick scan of equations (3.6) shows that if $\varepsilon = 0$, then we have two independently working simple harmonic oscillators. It is the additional ε-dependent terms that allows for the coupling and mode switching – unfortunately, there is no analytic solution to equations (3.6) and purely numerical methods must be employed to see how the system will behave under different coupling constants and spring characteristics – this is a topic we shall come back to in Chapter 6). The Wilberforce pendulum is a wonderful example of how strange and unexpected the behavior of a simple pendulum can be, and although in the modern era it is typically introduced as a device that shows mode switching in its behavior, the original design was used by Wilberforce to determine the mechanical properties of the wire (the k and δ constants in equation 3.6) out of which the spring was made.

The second hybrid pendulum to be considered is a modification of the Wilberforce pendulum in that the spring no longer undergoes a twisting motion, but it is allowed to move from side to side about the vertical axis. In this case the length of the pendulum is variable as the spring itself undergoes, say, SHM stretching and relaxing as it additionally oscillates from side to side

The equation of motion for this pendulum[26] over large angles of swing is written as

$$\ddot{\vartheta} + 2\left(\frac{\dot{L}}{L}\right)\dot{\theta} + \frac{g}{L}\sin\theta = 0 \qquad (3.7)$$

Again, a quick glance at equation (3.7) indicates that if the spring is no longer deemed to be variable in length $L(t) = $ constant, so $\dot{L} = 0$ and we recover (as expected) the equation for the simple pendulum (as derived in Chapter 1). Once again, there is no simple analytic solution to equation (3.7) so numerical solutions need to be sought if the detailed behavior of the pendulum is to be described. Indeed, the time variable history of $L(t)$ can take many different forms, and it need not, of course, be periodic. For example, the support wire might be engineered to lengthen or shorten in a linear fashion over time, $L(t) = L_0 (1 \pm \beta t)$, where L_0 is the initial wire length and β is some constant – such a length variation could be used to model the behavior of a crane lifting or lowering a swinging load. Numerical solutions to equation (3.7), when the pendulum length increases linearly with time ($\beta > 0$), reveal that the angle of oscillation decreases, but the period of swing increases as time advances.

Next to the crane hoist, a second everyday example of a variable length pendulum is that of the playground swing. In this case it is not the length of the chain that varies but the leg position of the person on the swing. By raising and lowering their legs, the swings occupant changes the location of the swings center of mass – and this is effectively a change in the pendulum's length (recall the discussion in Chapter 1). Without going through the mathematical details here (see note 25) it turns out, as all playground aficionados know, that the effect of rhythmically changing the height of ones legs, as the swing moves backwards and forwards, is to increase the amplitude of the swing. Indeed, all things being equal the energy in the system increases in a geometrical progression with the energy after the nth kick being $E_n = E_0(1 + k)^{n-1}$, where $k > 0$ is a constant relating to the effective variation in the length of the swing and E_0 is the initial energy. Provided one is happy to spend a lot of time on the swing, what this energy increase mode illustrates, is that even if the constant k term is very small (that is, if the swing occupant barely moves their feet) the energy will none the less grow as n becomes larger. This build-up of oscillation amplitude effect will be further discussed in Chapters 5 and 6, where it will be seen that even very small energy jolts, repeated over long enough intervals of time, can literally move a planet.

Our final hybrid pendulum example is a slight modification to the variable length pendulum just described. In this case, however, the effect of varying the mass of the pendulum bob is considered. The bob, for example, might be composed of a leaky water container, or the pendulum rod (as in early clock designs) might be made of wood which will have a variable mass according to the local humidity and its concomitant water content. At first thought, given the earlier discussion on dimensional analysis in Chapter 1, it might be argued that the mass of the bob is irrelevant, since the period of swing is simply related to the length of the pendulum – as shown in equation (1.4). What must be remembered, however, is that the length of a pendulum is not simply the length of its support wire, but it is the distance between the pivot point and the bob's center of mass, and if the bob's mass changes so too will the location of its center of mass and this effectively changes the length of the pendulum. The variable mass pendulum, therefore, has, in fact, characteristics in common with the mercury compensation and gridiron pendulums - although the latter are of a fixed mass. If, in contrast, the pendulum mass changes in such a way that the length of the pendulum remains constant (i.e., if, say, a spherical pendulum bob loses mass in a series of surface shells), then the equation of motion[27] will be the similar of equation (3.7), with

$$\ddot{\theta} + 2\left(\frac{\dot{m}}{m}\right)\dot{\theta} + \frac{g}{L}\sin\theta = 0 \tag{3.8}$$

where $\dot{m}$ is the rate at which the mass of the pendulum changes. Equation (3.8) is similar in form to the damped pendulum equation [equation (2.4)] introduced in Chapter 2 (see also Chapter 6 later). In this manner, the term associated with the first time derivative of θ in equation (3.8) will act to increase the period of oscillation over that of the constant mass, same-length simple pendulum, such that

$$T = T_0 \frac{1}{\sqrt{\left[1 - \frac{L}{4g}\left(\frac{\dot{m}}{m}\right)^2\right]}} \tag{3.9}$$

where T_0 is the period of a simple, frictionless, ideal pendulum of length L. That mass loss, or indeed, mass gain will always increase the period of the pendulum (at constant L that is) is indicated by the fact that the mass loss rate term $\dot{m}$ enters as a squared quantity and this dictates that the denominator of equation (3.9) will always be some number less than one. The limitations of the constant-length but variable mass approximation are inherent in equation (3.9) in the sense that the mass loss rate must be so limited that the square-bracket term in the denominator is neither zero nor negative. Numerical solutions for equation (3.8) reveal a somewhat counter intuitive result in the case of mass loss when $\dot{m} < 0$ and the $\dot{\theta}$ term is necessarily negative. What this reveals is that the pendulums amplitude of swing increases with time – until, that is, $m \approx 0$ (or more specifically when the mass of the bob compared to the mass of the otherwise ignored pendulum wire becomes comparable - at which point equation (3.8) is no longer valid). The mass loss effectively acts like a negative friction term allowing the pendulum to swing through larger and larger angles as time goes by[28].

The exact opposite effects to those seen with the simple pendulum come about for a spring undergoing vertical SHM and mass loss. In this case, as we saw earlier, the period of oscillation T is directly related to the square root of the mass m attached to the spring, hence $T \to 0$ as $m \to 0$. Likewise, the amplitude A of oscillation will decrease as the bob mass is reduced, since it is the mass of the bob that produces the spring extension (via Hooke's law): hence $A \to 0$ as $m \to 0$. An instructive experimental verification of this behavior has been presented[27] by Rafael Digilov and M. Riener who experimentally studied the behavior of a spring undergoing vertical SHM with a bob losing mass at a constant rate – a condition that was engineered by allowing sand to leak from the hollowed out interior of the bob (this mimics, of course, the steady flow rate of sand from the upper to the lower chamber in an hourglass – see note 12 in Chapter 2).

In the last two sections we have considered some of the small correction terms that act to change a pendulum's period of oscillation away from the 'ideal' given by equation (1.4). Indeed, as we shall see in Chapter 4 next, and in Chapter 6 later, the more the influencing factors that can act upon the motion of a pendulum are appreciated and analyzed so the further and further we move away from the ideal situation described in Chapter 1 - with respect to the teachings of Plato we are now entering the 'corruptible' domain of the real world[29].

CHAPTER 4

WEIGHING THE EARTH

He loved the mountains, or he had loved the thought of them
marching on the edge of stories brought from far away; but now
he was borne down by the insupportable weight of Middle-Earth.
He longed to shut out the immensity in a quiet room by a fire

The Lord of the Rings
J. R. R. Tolkien

In 1735 Pierre Bouguer noticed something strange; it was a small but unaccounted for anomaly. When making observations of star elevations form his camp situated close to the extinct volcano Mount Chimborazo[1] there was a systematic offset of several arc seconds in their supposed positions. This offset Bouguer correctly attributed to a displacement of his reference plumbline from the local vertical – a deviation caused by the gravitational attraction of the volcano. The mass of the volcano was literally pulling the plumb-line away from the vertical set by the rest of the Earth. Since the offset effect was systematic, however, it could be corrected for in subsequent calculations, but it was also realized that this tiny offset, under carefully controlled conditions, could be usefully used to determine the bulk density of the Earth, and then, at least in principle, the Earth's mass could be derived. While from the depths of Greek mythology we are told that in the aftermath of the Titanomachy, Zeus condemned Atlas to carry the weight of the celestial sphere upon his shoulders, the burthen of the Earth, it would appear, lies in the attraction of mountains. And, while the ancient Greek philosopher Archimedes (c. 287 BC – c. 212 BC) boasted that given a place to stand and a large enough lever he could move the Earth, the question now is how to determine the Earth's mass, and the answer, as with many things in life, is in the balance.

In the Balance

The enduring symbol of both the law courts and general commerce is that of the weigh scale – the evidence is literally and logically weighed, and the goods are proportioned out according to the infallible truth of the horizontal balance. Should the bar not rest steady and true, and tip to one side or the other, then a judicial verdict can be made – "guilty as charged"– or, an imbalance of goods to remuneration is revealed. Justice, power, and truth all held in the balance by the swing of a horizontal plumb-line: once again the mighty pendulum, albeit stopped and horizontal in this case, runs our society and oversees our daily lives.

At their most basic, weigh scales are a form of stopped horizontal pendulum. In this manner the bar, to which two pans are attached, functions as a rigid pendulum that is brought to rest about its center of mass. What are being compared in a set of weigh scales are the so-called moments on each side of the balance point or fulcrum (figure 4.1). The moment of some mass M situated a distance X from the fulcrum is simply MX, and for a beam to remain balanced – that is horizontal – the moments on each side of the fulcrum must be equal. In figure (4.1), the beam will balance only if $M_1X_1 = M_2X_2$. To construct an 'honest scale', say, for measuring out a quantity of apples, the idea is that two equal mass pans are situated an equal distance away from the fulcrum (corresponding to $X_1 = X_2$). A calibrated mass (say of 1 kg) is then placed in one pan and apples are placed into the second pan until the bar exactly balances. In the horizontal balance position, the two moments are equal, and since X_1 and X_2 are equal, we must have the result that $M_1 = 1$ kg calibration mass $= M_2 = 1$ kg of apples – as required by both the customer and shop owner, both sides being satisfied that they have neither paid out too much, or been short changed. If the balance bar is horizontal then the truth of the transaction has been made visible and both parties are certain of having being fairly served.

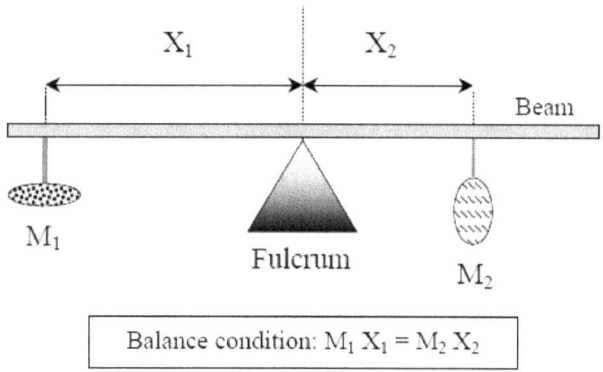

Figure 4.1. A balance pendulum. The working principle of the balance is to exactly match the moments on one side of the fulcrum to those on the other. If they match then the beam will come to rest horizontally.

As to whether Galileo was fairly served by the Papal authorities in his later life is an open question, but it is known that at the tender age of 22 years, he wrote a short article concerning *La Bilancetta* (*The Little Balance*). Within the pages of this work Galileo explores and then refines a story relating to the Greek mathematician Archimedes. As Galileo comments in his introductory paragraph, "as is well know to anyone who takes the care to read ancient authors that Archimedes discovered the jeweler's theft of Hiero's crown".

Remarkably, Galileo's comments hold true to this very day, and indeed, anyone who has taken even an introductory science class will have heard of Archimedes supposed streak through the ancient streets of Syracuse – it was his "eureka" moment. The story revolves around Archimedes's discovery of how fluid displacement can be used to determine the volume and thereafter the density (to be defined below) of irregularly shaped objects. The discovery, as such, does not appear in any of Archimedes known works, as Galileo noted, and it has even been questioned in modern times if the method could have been accurately applied in ancient time. What Galileo did in his *La Bilancetta*, however, was to show exactly how the Archimedean method might have been applied. The idea is to take a balance arm and attach to one end the mass of material for which a composition determination is to be made. This produces a moment of, say, M_1X_1 about the fulcrum point. Now, a counter-balance of mass M_2 is positioned appropriately on the other side of the balance arm in order that $M_1X_1 = M_2X_2$ - this is the balance position "in air". What Galileo suggests next is that the mass of unknown material be suspended in a beaker of water. Archimedes hydrostatic principle now dictates that the unidentified mass will weight less in water than in air, and the balance arm will accordingly move away from its equilibrium position. The counter-balance mass is accordingly adjusted to a new location X_3 at which the arm regains its equilibrium. Galileo then notes that in principle the measurable shift $X_3 - X_2$ in the counter-balance location can be used to determine the composition of the mass being investigated. This method works, provided that the device has already been calibrated according to the counter-balance shifts produced by various types of materials with known composition. It is a brilliantly straightforward scheme and one that highlights Galileo's early interest in physical experimentation – skills and interests that were to see further expression 52 years after the production of *La Bilancetta* in what is probably his greatest work, *Dialogues Concerning Two New Sciences* (published in 1638).

Weighing Gravity

In terms of performing balance experiments in extreme environments, the pains undertaken by Robert Hooke probably reign supreme. We have already recounted the details of Hooke's conical pendulum experiments in Chapter 1, and discussed how these resulted in his identification of a continuously acting central force. While Hooke identified the presence of such a force, he struggled to identify how it might vary with distance from the center. He knew it decreased, but was it a linear decrease or some other power law expression? Newton eventually solved this problem by showing that gravitational force must vary according to the inverse distance squared. In an attempt to measure the change of gravitational attraction with height, however, Robert Hooke took to climbing the rafters of the highest building that he could find, and in December of 1662 he made his way to the top of Westminster Abbey.

Once there he set up a balance system with two equal mass weights. One of the weights was attached to a 70-foot length of string, and while initially the equilibrium was determined with the weights being at equal, short distances from the balance arm, the system was then monitored as the weight on the long string was lowered towards the ground. Since the mass closer to the ground should weigh more than its more lofty companion, set at a fixed, short distance from the balance arm, it was expected that the system should move out of it equilibrium state. The experiment was inconclusive. The balance did move from its initial equilibrium, but it was not clear if this was due air movements in the Abbey or the long string becoming laden with moisture due to the damp air. Hooke repeated his balance experiment in the summer of 1664, working this time with a 200-foot thin brass wire from the top of St. Paul's Cathedral, but again, wary experimentalist that he was, Hooke was not convinced that any real difference in the weights had been detected. Further such experiments were carried out in a 315-foot deep well located on Banstead Downs, but again a successful conclusion could not be drawn. Once again, fate played Hooke a bad hand, and while the effect that he was trying to measure does assuredly exist, the experimental apparatus available to him in the mid-17th Century was simply not accurate enough to measure the change - a 200-foot difference in height, as realized in his St. Paul's Cathedral experiment, for example, amounts to a mere 0.002 percent variation in weight.

Working some 325 years after Hooke performed his inconclusive Westminster Abbey balance experiment, James Thomas (Lawrence Berkeley Laboratory) and co-workers set out to perform a similar such investigation[2] – and again, the aim was to see how the law of gravitational attraction might vary with separation. Specifically, in this latter case, however, the idea was to investigate the theoretical possibility that the gravitational acceleration law might undergo a short-range modification, the effect of which might be seen as a variation in the universal gravitational constant with $G = G_\infty(1 + \alpha e^{-r/\lambda})$, where r is the distance, α is a constant, λ is a length scale term and G_∞ is the gravitational constant at large distances. Various theories suggested that λ might be as large as a few hundred meters, and accordingly while the modification would have no effect upon astronomical interactions, where $r \gg \lambda$ and $G = G_\infty$, it should be manifest on scales varying from a few tens of meters to a few hundreds of meters. Interestingly, on distance scales where $r \ll \lambda$, the theories further indicated that $G = G_\infty(1 + \alpha)$, which suggested the existence of a 'fifth' fundamental (anti-gravitational) force.

In the experiment conducted by Thomas and colleagues, the acceleration due to gravity was measured with a set of highly sensitive La Coste-Romberg horizontal gravimeters, at a range of heights along a 465-m vertical tower located on the nuclear testing grounds at Jackass Flats in Nevada[2]. The tower experiment produced results that were consistent to within $(-60 \pm 95) \times 10^{-8}$ m/s^2 of those expected from standard Newtonian theory; leaving no room for any modification term – essentially, the data from the tower experiment implies that $\alpha = 0$; indeed, this is now the generally accepted result, and it is no longer thought likely that short-range modifications to the inverse square law (or an anti-gravitational, 'fifth' force) exist. This remarkable result not only vindicated the veracity of the inverse square law on scales between a few to 10^3 meters, it further reminds us that in the history of science it is more often than not advances in technology rather than changes in experimental methodology that enables a specific theory to be tested.

Packing It In

Before describing below how pendulum and plumb-line experiments can be combined to determine the mass of the Earth, we should first introduce a few definitions relating to the properties of matter, and remind ourselves of some key points developed in Chapter 3 concerning the internal structure of the Earth.

The density ρ is defined as being the mass of a certain amount of material m divided by its volume V; symbolically this is written, $\rho = m / V$. Density is a fundamental measure of material, and it varies from one substance to the next. The density of pure water, for example, is 998 kg/m^3, while that of oak is 650 kg/m^3, and that of lead is 11740 kg/m^3. For a composite body made up from many different layers of material, each with its own specific density, the bulk density is defined as the total mass divided by the total volume. The simplest composite model for the Earth is that of a sphere of radius $R_\oplus$ with a central core of radius R_C and density ρ_C, surrounded by an outer mantle of density ρ_M. If the core radius to total radius ratio is written as $x = R_C / R_\oplus$, then the bulk (or average) density can be written as $\rho = x\rho_C + (1 - x)\rho_M$. From this it can be seen that as $x \to 0$, so $\rho \to \rho_M$ and the sphere has a uniform density corresponding to that of the mantle material. As $x \to 1$, on the other hand, we find $\rho \to \rho_C$ and the sphere has a constant density corresponding to that of the core material. In general, however, $\rho_C \geq \rho \geq \rho_M$, and for a sphere $\rho = m / (4/3) \pi R_\oplus^3$. While the bulk density doesn't tell us much about the layers of different density within a planet, it is a measurable quantity (derived from the planet's mass and radius) and it does enable an estimate of the core radius to be made if appropriate densities are chosen for the interior regions. Below we shall see that the accepted value for the bulk density of the Earth is $\rho = 5515$ kg / m^3. If we assume that the Earth's core is composed of a nickel-iron alloy with a density of $\rho_C = 9000$ kg/m^3 and that its mantle is composed of rock with density $\rho_M = 3000$ kg/m^3, then $x = 0.419$, indicating that core must occupy about 4/10^{ths} of Earth's interior by radius.

In terms of mass, however, such a core would account for 11% of Earth's total.

Defining Up

The zenith, or the point on the sky directly overhead of an observer, is one of the cardinal references markers in observational astronomy. It is a point on the sky, however, that varies for every single observer (unless they are using the same telescope). The key attribute about the zenith is that it defines the local vertical and therefore, by definition, establishes the local horizon (as shown in figure 4.2). In addition, treating the Earth to be a perfect sphere, the zenith point on the sky is the extension of the line passing from the center of the Earth, through the observer's feet and on into the heavens. This vertical line from the center of the Earth, as Newton demonstrated in his *Principia* in 1687, and which had been well known to all builders since antiquity, was the same line traced out by a suspended weight. The stopped pendulum, or builder's plumb-line, points to the center of the Earth and defines the local vertical.

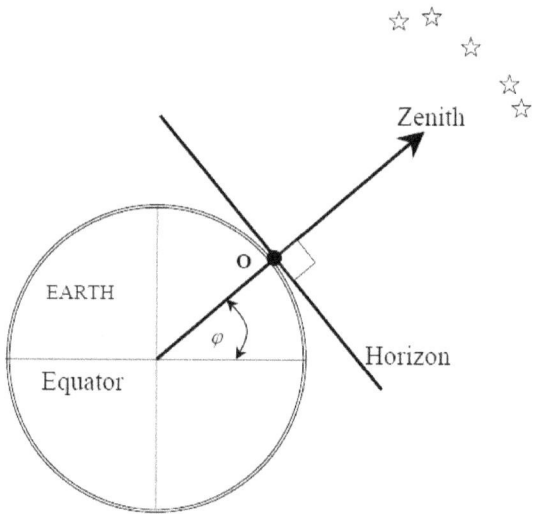

Figure 4.2: The observer's zenith and horizon. The astronomical horizon is set by definition to be the plane at 90 degrees to the observer's local vertical. The direction of the local vertical (zenith point on the sky) will change according to the observer's latitude (φ).

Most astronomical telescopes are mounted in such a fashion that they can be pointed to any location in the visible sky. There is one high specialized telescope, however, that is by its very design restricted to move through just a very few degrees along the meridian arc centered on the zenith point. Indeed, the very first custom designed telescope was a so-called zenith instrument.

The primary purpose of the zenith telescope is to measure, to the highest precision possible, the zenith angle of a star as it crosses an observer's north-south meridian. The ever inventive Robert Hooke knew this, and purpose built a telescope to determine the exact zenith angle of the star gamma Draconis (also known by the Arabic name *Eltanin*, which means *the serpent*) in 1669. His aim, no less, was to prove that the Earth must be in motion about the Sun. Since we will focus on the motion of the Earth in Chapter 5, it is just the properties of the zenith telescope that will be discussed below.

A zenith telescope can be thought of as an off-set stopped-pendulum. By suspending the telescope at a point close to the objective end of the telescope tube the optical axis defines the local vertical – the telescope, if you will, is acting like a plumb-line when left alone and simply hanging downward. Figure 4.3 shows the basic telescope design. The pivot point is set up so that the telescope can swing about P along a north-south meridian, and the zenith angle of a star is measured as it crosses the meridian by a scale set close to the observer's eyepiece. Since the telescope tube of a zenith instrument is typically very long, in order to accommodate the long focal length of the objective, the measurement scale can be very finely divided and consequently exceptionally small zenith angles can be measured. One of the great astronomical advantages of using a zenith telescope is that the starlight passes straight through the Earth's atmosphere and consequently there is no atmospheric distortion, or refraction terms to adjust for in the measurement.

Robert Hooke (unsuccessfully as it turned out) used a zenith telescope with a 36-foot radius in an attempt to measure the parallax motion of the star gamma Draconis, and with this instrument he estimated that he could measure zenith angles to an accuracy of at least one arc second – an outstanding achievement at that time. While all the details of the observational reductions need not occupy us here, the key point that should be made at this time is that by making a careful set of observations of the transit times and zenith angles of stars that pass overhead of an observer, a measure of their distance (as well as the gravitational attraction due to a mountain) can be made.

The Attraction of Mountains

The Reverend Nevil Maskelyne (1732 - 1811) cut to the very core of the problem when in the year 1772 he announced to the assembled Fellows of the Royal Society of London, "If the attraction of gravity be exerted, as Sir Isaac Newton supposes, not only between the large bodies of the universe, but between the minutest particles of which these bodies are composed,… then every hill must by its attraction, alter the direction of gravitation in heavy bodies in its neighborhood from what it would have been from the attraction of the Earth alone."[3]. If we imagine, therefore, setting up a plumb-line on a north-south line that cuts across a particular mountain, then when the pendulum is located to the south of the mountain, the gravitational pull of the mountain will cause the plumb-line to be tilted slightly towards the north.

Figure 4.4(A) illustrates the effect of the mountain on the plumb-line. Careful observations of the elevations of stars as they pass the observer's meridian can be used to determine the location of the zenith point on the sky, and this in turn establishes the local vertical independently of the gravitational effect of the mountain. The zenith point determined by the plumb-line, however, will be displaced southward (in our example) due to the northward attraction of the mountain. The tangent of the offset angle φ between the vertical set by stellar observations compared to the 'vertical' set by the plumb-line string can be related to the ratio of the distance d of the plumb-line from the center of the mountain (which we take to be a hemispherical dome) to the Earth's radius $R_\oplus$.. Assuming that the mountain and the Earth are made of material with the same density (an assumption that we will eventually see to be untrue) then[4] $\tan(\varphi) = \frac{1}{2}\,(d\,/\,R_\oplus)$.

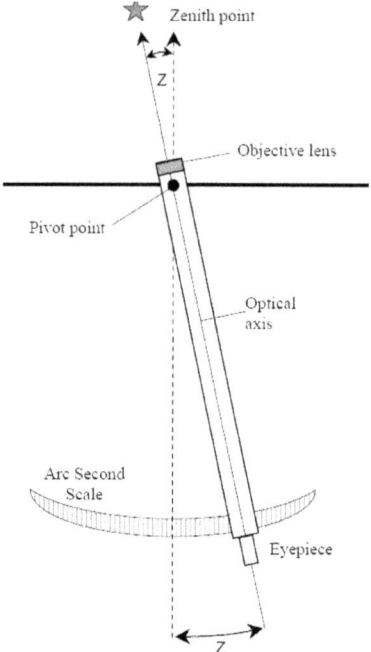

Figure 4.3. A zenith telescope. The telescope tube is attached to a pivot point, and the zenith angle Z is measured from the scale close to the observer's eyepiece.

Sir Isaac Newton, our ever present guide and pioneer, was fully aware of the gravitational influence of mountains, although he apparently failed to appreciate the useful consequences of trying to measure their effects. In article 22 of his *The System of the World*, Newton argued that when compared to the attraction of the Earth, "whole mountains will not be sufficient to produce a sensible effect." He then goes on to indicate that a three-mile high mountain "will not, by its attraction, draw the pendulum two minutes out of the true perpendicular." Converting miles into kilometers (our preferred *SI* units), we can calculate the deflection angle to be $\tan(\varphi) = \frac{1}{2} (3 \times 1.609) / 6371 = 7.5765 \times 10^{-4}$, which indicates that $\varphi = 1.3$ arc minutes. This value for the offset angle due to Newton's 3 mile high, hemispherical mountain is exactly the same as that determined by Maskelyne in 1772. Maskelyne knew, however, that the offset angle would probably be much smaller in any real-world situation. Indeed, Maskelyne quotes Bouguer as finding an offset angle of just 8 arc seconds from Mt. Chimborazo, but he also notes that this is much smaller than might have been expected. Maskelyne then goes on to estimate the expected offset angle for several of the highest mountain peaks in England's Lake District, finding that values of some 5 to 4 arc seconds should be realized. To these estimates Maskelyne comments that they "are not too small to be measured and demonstrated by an accurate zenith sector." The stage was therefore set for action. Maskelyne believed that if the right mountain could be found, then with the aid of the Royal Societies zenith telescope he could determine "a better idea of the total mass of the earth, and the proportional density of the matter near the surface compared with the mean density of the whole earth." [3].

Maskelyne Heads North

To measure the gravitational attraction of a single mountain is by no means straightforward. The difficulty being, of course, how to separate out which part of a mountain range is causing the attraction, and then exactly how much mass the mountain contains. Maskelyne considered several candidates for his experiment but eventually settled on Schaehallion[5], a 1081-m high mountain in the Scottish Highlands. In its shape and location Schaehallion was as near to perfect as one could hope for. It was massive, isolated from other surrounding hills, and had a reasonably regular shape. By being massive it should produce a large and hence more easily measured offset. By being isolated one could be reasonably sure that any offset was entirely due to the one mountain, and by being regular in shape a good estimate of its actual mass (or more specifically, the location of its center of mass) could be made. In his experiments in Ecuador, Bouguer had determined the attraction of Mt. Chimborazo by making two sets of zenith observations on the same parallel, but at different distances from the mountain peak. Maskelyne, however, was going to use a different method to determine the attraction of Schaehallion. His plan was to make zenith observations at two locations situated on a north-south arc on opposite sides of the mountain. The zenith sector observations would provide the apparent difference in the latitude of the two observing sites. These observations would then be combined with a comprehensive geodetic (triangulation) survey that would provide the true latitude difference between the observation points. The offset angle corresponding to the gravitational attraction of Schaehallion would then be half the difference of the two latitude measures.

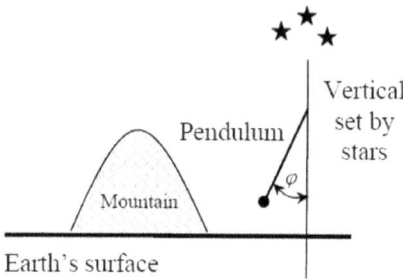

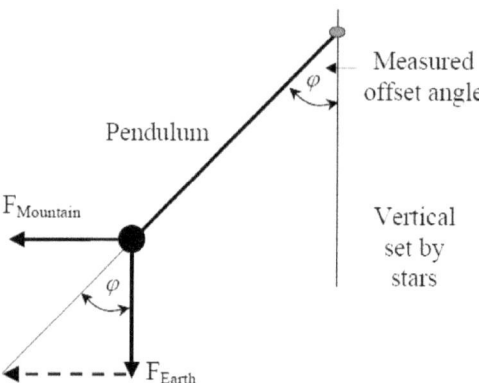

Figure 4.4. The gravitational offset due to a mountain. (A) The gravitational influence of the mountain causes the plumb-line to tilt away from the local vertical determined by the stars. (B) The offset angle φ is related to the gravitational force $F_{Mountain}$ exerted by the mountain and the gravitational force F_{Earth} exerted by the rest of the Earth.

Financed through a grant from the Royal Society, and with a 10-ft radius zenith sector in tow, Maskelyne headed north to central Scotland in the summer of 1774. Hundreds of astronomical observations were made of many different stars and the apparent latitude difference between the two observing stations was determined to be 54.6 arc seconds. The detailed land survey of Schaehallion further revealed a true latitude separation of 42.94 arc seconds between the observing locations. Maskelyne had his result, and on 26 July, 1775 triumphantly presented his findings to the Royal Society. The gravitational attraction due to Mt. Schaehallion had deflected the plumb-line, defined by the optical axis of the zenith sector, by (54.6 − 42.94) / 2 = 5.8 arc seconds. The day following his report, the Council of the Royal Society met and voted to award Maskelyne the society's highest honor − the Copley Medal − for his work on the lonely Scottish mountain. The final step in the calculation was to determine the bulk density of the Earth. This task was given to Professor Charles Hutton of the Royal Military Academy in Woolwich. Being paid £150 for his labors, Hutton commented that "these calculations were naturally and unavoidably long and tedious". Indeed, his 1778 final report[3], published in the Philosophical Transactions of the Royal Society, consisted of 99-pages of detailed analysis and tables of reductions. In conclusion, however, Hutton found that "the ratio of the mean density of all the matter in the Earth, in comparison with the density of the matter of which the hill is composed ... is equal to the ratio 9 to 5." By then assuming that the material out which Schaehallion was composed had a density similar to that of "common rock" (i.e., $\rho_{Mountain}$ = 2500 kg/m³), so the bulk density of the Earth came out to be ρ_{Earth} = 4500 kg/m³. Hutton took the circumference of the Earth to be 131,284,080 feet (about 40,000 km) and consequently the mass of the Earth to be 5 x 10²⁴ kg. Hutton's value for Earth's mass is certainly good to order of magnitude, but as we shall see below it is about 17% on the small side with respect to the modern day value.

The methods by which the structure of the Earth's interior can be measured were described at the end of Chapter 3, but one important result that Hutton discussed in his 1778 paper was that concerning the physical extent of the Earth's inner core. Hutton noted that the density of 'metals' (e.g., iron) was about ten times that of water. Given, therefore, the known size of the Earth and its bulk density (ρ_{Earth} = 4500 kg/m³), so Hutton argued that about "one-half of the matter in the whole earth must be metal .... or nearly 2/3 of the diameter of the earth is the central or metalline part". This is, in fact, not a bad result although a little on the high side in comparison with the modern day estimate of the core mass (which is reckoned to about 1/3ʳᵈ of the Earth's total mass).

It seems not a little remarkable that while he was engaged in the detailed geodetic survey of Schaehallion Maskelyne made no attempt to study the rocks out of which the mountain was made. Such a lithological survey was eventually made, however, in the summer of 1801 by John Playfair and Lord Webb Seymour. Urged, in fact, to undertake the survey by Hutton, Playfair eventually presented the outcome of his investigations to the Royal Society in a paper published in 1811. The new survey data didn't change the earlier results by a great amount, but they did raise the estimate for the Earth's bulk density to $\rho_{Earth} = 4700$ kg/m^3, and this subsequently resulted in a small increase in the mass estimate for the Earth.

Airy Down a Coalmine

If Maskelyne's method of using zenith sector observations to weigh the Earth can be thought of in terms of a stopped pendulum result, then the series of experiments performed by George Biddell Airy are a direct pendulum measurement[6]. Rather than attempt to measure the deviation of a plumb-line due to a large mass (i.e., a mountain), Airy chose to exploit one of Isaac Newton's famous results concerning the gravitational attraction at any point within a hollow shell. Specifically Newton showed that there is no net gravitational force resulting from the material that constitutes a spherical shell. Figure 4.5 indicates the idea and the practical consequence of this theorem: if a measurement of the gravitational force is made at some point in Earth's interior (assumed to be a homogeneous sphere) at distance $0 < r < R_\oplus$ from the center, where $R_\oplus$ is Earth's radius, then the gravitational attraction is simply that due to the mass interior to the sphere of radius r. In other words, the gravitational force exerted by all the material in the shell (of thickness $R_\oplus - r$, and mass M_{shell}) exactly cancels out[7].

A young Airy first attempted to perform a series of pendulum experiments along the lines of Newton's shell cancellation method in 1826 and again in 1828. While he initiated pendulum experiments at the top and the bottom of the Dolcoath mine[8] in Cornwall, England, the experiment was never completed due to flooding in the mine and the loss of data in a fire. Twenty-six years later, however, Airy tried the pendulum experiment again, this time at the Harton Colliery in South Shields, England. The mineshaft gave a shell thickness of $R_\oplus - r = 1200$ feet (or about 366-m), and it was found that the pendulum at the bottom of the mineshaft (point P in figure 4.5) gained 2.5 swings over the pendulum at the top of the mine (point Q in figure 4.5) during a period of 24-hours. Counter to expectation, Airy's measurements indicated that the acceleration due to gravity was actually greater at the bottom of the mineshaft than at the top - a result that actually shows the Earth is not a homogeneous sphere. In spite of this unexpected result, Airy was able to show that Earth's bulk density has a value of some $\rho_{Earth} = 6560$ kg/m^3, which suggests that Earth's mass is of order 6.8 x 10^{24} kg. While Maskelyn's mass estimate was a little on the low side, Airy's result is, in fact, on the high side (about 14% too big) with respect to the accepted modern day value: $M_\oplus = 5.9742 \times 10^{24}$ kg. With one result being too small and the other too large, the Earth's bulk density must therefore lie somewhere between 4500 and 6560 kg/m^3. The decisive experiment that pinned down the correct value for the Earth's bulk density was performed in 1798, not down a coalmine, nor upon the flanks of a large mountain, but in a small private laboratory located on the outskirts of London, England.

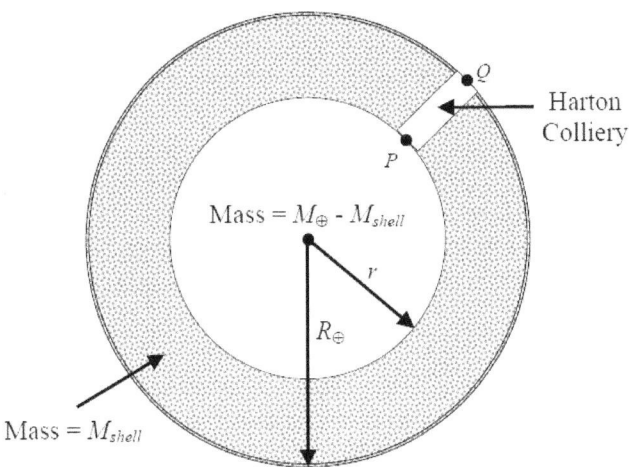

Figure 4.5. Airy's method for finding the mass of the Earth. The gravitational influence of the material in the outer shell exactly cancels, and consequently the pendulum period at P is dependent upon the mass $M_\oplus - M_{shell}$, while the period of the same pendulum at point Q is dependent only upon the Earth's mass $M_\oplus$.

The Cavendish Experiment

Henry Cavendish (1731 - 1810) was a remarkable scientist but a very strange man[9], and in addition to determining the definitive value for the Earth's bulk density, thus allowing its mass to be enumerated, he also discovered hydrogen, the lightest, most common, and most elementary of all atoms. Working over a century after the initial publication of Newton's *Principia*, Cavendish measured an effect that Newton thought was immeasurable. Indeed, in the *Principia* Newton dismissed the very idea that laboratory scale measurements relating to gravitational attraction could be made. In his account of the *System of the World*, Newton sets out the following problem: imagine that two spheres, each 1-foot in diameter (0.152 m), are hung from very thin strings with their surfaces just ¼ of an inch (6.35 x 10⁻³ m) apart. If there are no opposing forces (such as air currents) then what is the time required for the two spheres to draw together and for their surface to touch? Newton claimed that "they would not ... come together by the force of their mutual attraction in less than a month's time". This time estimate for coalescence, however, represents one of those rare circumstances where Newton made a large numerical error. Indeed, he was out by a factor of nearly 10,000 and the actual time for the two spheres to come together under their own gravitation interaction will be less than about ten minutes[10]. What this correct calculation implies, of course, is that careful laboratory-based studies of gravitational effects are entirely possible, albeit extremely exacting.

The classic Cavendish experiment, as it has become known, was actually designed and developed by the Reverend John Michell (1724 – 1793). Michell is perhaps better known in modern times for his suggestion that the escape velocity of some very massive stars might exceed that of light, and therefore their presence would only be realizable through their gravitational effect on other less-massive stars. Such dark stars would now be called black holes (although the latter objects have very different origins and properties to the stars envisioned by Michell). By scientific inclination, Michell was a geologist and he developed the idea of a torsion balance to determine the bulk density of the Earth. As illustrated in figure 4.6, the torsion balance is a horizontally twisting pendulum. Indeed, the torsion balance is, in some approximate sense, an amalgamation of the simple and the conical pendulum – it has a rotational arm and it swings horizontally from side to side. The oscillation period of the torsion pendulum, unlike its simple and conical cousins, however, has a period that depends upon the shape of the bob (specifically its moment of inertia), which according to the application can be a bar, a flat disk, or a dumbbell shape. In the case of the torsion pendulum illustrated in figure 4.6 the beam AOB twists back and forth about its rest position, with the restoring force being provided by the twisting of the suspension wire. One of the most important characteristics of the support wire in a torsion balance is that the restoring force responds linearly with respect to the angle that it is twisted through: $F_{wire} = k\varphi$. Indeed, it was just this property of the ultra-fine, ultra-pure quartz crystal fibers produced by Charles Boys in the late 19th century (recall Chapter 2) that enabled researchers to turn the torsion balance into a refined and highly accurate scientific instrument.

Central to the torsion balance experiment for determining the bulk density of the Earth is the placement of two large masses (M) close to the two smaller masses (m) suspended from the end of the balance arm AOB. Figure 4.7 illustrates the idea for the experiment, and by making the arm supporting the two larger masses (M) moveable about the center point O, their physical distance from the smaller masses can be varied.

The Cavendish experiment begins by determining the period of oscillation T_0 of bar AOB with the large masses M moved to a distance where they can have no gravitational effect upon the smaller masses m. Next the large masses are brought in to position, and the bar AOB is once again set in motion. The new period of oscillation T will now be slightly different from T_0 due to the gravitational interaction between the bar end masses m and the larger masses M. Finally, with the large masses M still in the position shown in figure 4.7, the bar AOB is allowed to come to rest and the physical separation r between the centers of m and M, and the angle φ are measured.

The point of the oscillation measurements is to determine the ratio of the restoring force F, due to the torsion wire, and the weight $W = mg$, where g is the acceleration due gravity at the Earth's surface. It turns out that $F / W = (\varphi / l)(T_0 / T)^2$, where l is the half-length of bar AOB. At this stage, Cavendish did something rather clever and he introduced into his analysis a fictitious sphere of water of radius R_W = 6-inches (0.1524 m – for the sake of consistency with Cavendish's original paper I will use the 6-inch value rather than its *SI* equivalent in the analysis that follows). Having introduced the idea of the water sphere, Cavendish then observed that the mass of each of the lead spheres M being used in his experiment would be 10.64 times greater than that of the imaginary sphere of water. From this observation, Cavendish reasoned that the force of attraction due to the lead sphere would be 10.64 $(6 / r)^2$ times greater than the force due to the water sphere if the mass m were situated on the water sphere's surface. Finally, by setting the gravitational force between m and M to be equal to the restoring force F then $F / W = (\varphi / l)(T_0 / T)^2 = 10.64 (6 / r)^2 [6 \rho_W / R_\oplus \rho_\oplus]$, where $R_\oplus$ is the Earth's radius, $\rho_\oplus$ is the Earth's bulk density, and ρ_W is the density of water. The only quantity not known in the final equation turns out, therefore, to be the Earth's bulk density – everything else is known beforehand or is measured during the experiment.

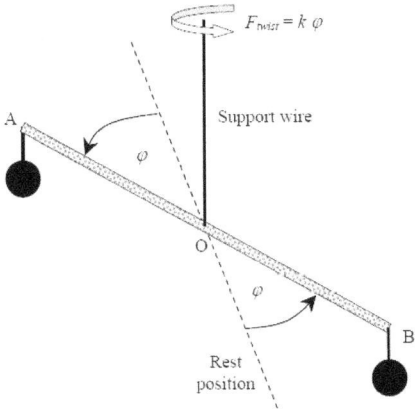

Figure 4.6. The Torsion balance

The 66-year old, reclusive and pathologically shy Cavendish began collecting data from his torsion balance experiment on August 5th, 1797. He continued to test, refine and gather data through September and October of that same year, after which he called a halt to his experimentation. In May of 1798 he performed some additional experiments and then being finally satisfied read the results of his study to the assembled Fellows of the Royal Society on June 21st 1798. His final conclusion was that the bulk density of the

Earth was some 5.48 times greater than that of water (in other words, $\rho_\oplus = 5469$ kg/m³). The accepted modern day value for the Earth's bulk density is $\rho_\oplus = 5150$ kg/m³, which indicates that the accuracy to which Cavendish worked was remarkably good, his value coming out at a little over 6 % on the large side.

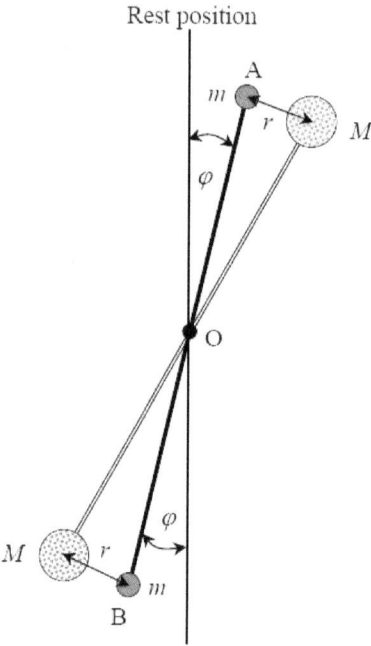

Figure 4.7. Configuration for the moveable, large attracting masses in the Cavendish Experiment. This view is from above, looking down along the support wire attached to point O. The 'rest position' line corresponds to the orientation of bar AOB when the large masses M are far away and have no effect on the bar orientation.

The Measure of Big-*G*

In modern-day texts the Cavendish experiment is usually introduced as the means by which the universal gravitational constant *G* is determined. Historically, however, this interpretation is something that neither Isaac Newton nor Henry Cavendish would have readily appreciated. Indeed, in modern textbooks Newton's formula for the force F_{grav} resulting from the gravitational attraction between two spherical masses m_1 and m_2 separated by a distance r between their centers is written as

$$F_{grav} = G\frac{m_1 m_2}{r^2}$$
(4.1)

but, of course, this is an equation that Newton never actual wrote down (at least symbolically as such). Equation (4.1) is really the mathematical shorthand for Proposition VII, Theorem VII of Book III of Newton's *Principia Mathematica*, combined with Corollary II to theorem VII. In this proposition Newton argues "there is a power of gravity pertaining to all bodies, proportional to the several quantities of matter they obtain", to which in the corollary he adds, "The force of gravity towards the several particles of any body is inversely as the square of the distances of places from the particles". In modern-day usage, the constant G in equation (4.1) accounts for the "proportional to" statement in theorem VII. Rather than work in absolute terms, that is use something like equation (4.1), Newton and his contemporaries always worked in ratios, one force compared to another, and under these circumstances any constant terms (such as G) cancel out. The other point, of course, is that a very precise measurement of a small angle is required in order to determine G, and this was something that Newton thought impossible – as indeed it was with the experimental equipment available in his time. Remarkably, writing 163 years after the first appearance of the *Principia*, however, British mathematician George Stokes (University of Cambridge) was to comment in 1850 that, "the great importance of the results obtained by means of the pendulum has induced philosophers to devote so much attention to the subject, and to perform the experiments with such scrupulous regard to accuracy in every particular, that pendulum observations may justly be ranked among the most distinguished by modern exactness". These statements bear great testament to the improvements directed towards experimental techniques in the 19th Century, and they also underscore the central importance of the pendulum in the advancement of the physical sciences.

Building upon Stokes' eulogy, modern-day versions of the Cavendish experiment do away with the idea of an imaginary water sphere, and simply measure with exquisite accuracy the twist angle φ directly (Figure 4.8). The displacement angle is typically measured with the help of a laser beam that is directed towards and then reflected from a small mirror attached to the sup port wire. Once the angle of deflection has been measured and k determined then G is derived through the relationship:

$$G\frac{m_1 m_2}{r^2} = k\varphi \qquad (4.2)$$

Many torsion pendulum experiments have been designed, built and operated during the past century but it is still the case that the universal gravitational constant is the least well determined of all the fundamental constants of nature. The presently accepted value for is $G = (6.67428 \pm 0.0010) \times 10^{-11}$ N m² kg⁻². The problem inherent in determining a precise value for G stems directly from the extreme weakness of the gravitational force, and with the manufacturing problems associated with the production of highly homogeneous masses and uniform torsion fibers.

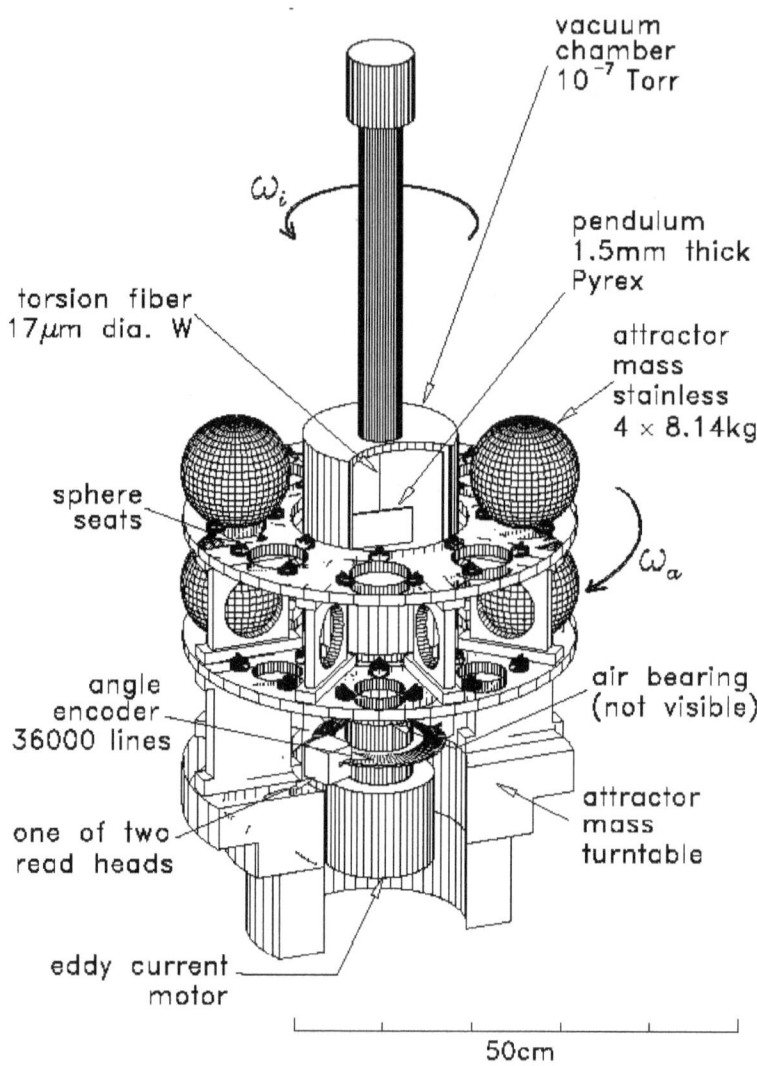

Figure 4.8. A schematic diagram of the high precision torsion balance apparatus for measuring the gravitational constant as developed by the Eöt-Wash Group at the University of Washington. For experiment details see J. H. Gundlach, A rotating torsion balance experiment to measure Newton's constant (*Measurement Science and Technology*, 10, 454-459, 1999).

A Pendulum That Can Never Be: III

In our minds eye let us imagine a continuation of the Airy mineshaft experiment. Indeed, let us imagine that a gang of over exuberant miners have bored a hole all the way to the center of the Earth and out again, exiting on the opposite side of the Earth to which they started. If we now further imagine a small observation craft of mass m, bristling with scientific instruments, being constructed and dropped down the borehole what then, we may ask, will happen. The equation of motion for the observation craft will be

$$m\frac{d^2r}{dt^2} = -\frac{GM(r)m}{r^2} \tag{4.3}$$

where $M(r)$ is the mass interior to radius r. Now, assuming the Earth is a sphere of constant density ρ, so $M(r)$ can be expressed in terms of the density and the volume such that $M(r) = \rho\frac{4}{3}\pi r^3$. Substitution in equation (4.3) now reveals that inside of the (constant density) Earth the acceleration varies linearly with the radius (see figure 4.9)

$$m\frac{d^2r}{dt^2} = -\frac{Gm}{r^2}\rho\frac{4}{3}\pi r^3 = -\frac{4}{3}G\pi\rho m r \tag{4.4}$$

and equation (4.4) is the familiar pendulum equation from Chapter 1. Initially, therefore, the observation craft will have zero velocity, but as it descends down the borehole it will pick-up speed, accelerating according to the gravitational attraction of the sphere of the Earth below its specific location r – as per equation (4.4). At the Earth's center $r = 0$ the acceleration due to gravity will go to zero, but it is at this very point that the observation craft will have its maximum velocity. The craft will therefore sail on through the center of the Earth and head on outwards to the borehole's second opening, now, however, the motion of the craft is being retarded due to the gravitational attraction of the Earth's mass below its specific location – it therefore begins to slow down as it approaches the second opening. Upon reaching the extreme end of the borehole, the observation craft will come to a stop, only to thereafter fall back down the borehole, once more heading for the center of the Earth. This motion will be repeated indefinitely (assuming, as it is a mind's eye experiment, that there is no frictional or air-resistance effects), and the observation craft will behave like a pendulum, moving back and forth across the Earth's interior with a period T_{cross}. From equation (1.4) we have, therefore, that $T_{cross} = 2\pi\sqrt{R/g}$ where $g = GM/R^2$ is the acceleration due to gravity at Earth's surface. Putting in representative numbers it is revealed that $T_{cross} = 2534$ seconds, or about 42 minutes[11].

Clearly our minds eye experiment is rather fanciful, but a scaled down version of the experiment has been proposed as a means of determining the gravitational constant G. The experiment would be performed in space and a solid, constant density sphere with a borehole drilled through its middle would be isolated within the spacecraft. A small pellet would be released down the borehole and its period of oscillation T_{cross} would be measured. If the test sphere were made of a material with density ρ, it turns out[12] that $G = 3\pi/\left(2\rho T_{cross}^2\right)$. While this envisioned space experiment is certainly possible (monetary issues aside), the biggest challenge would be, once again, the production of a sphere with a truly uniform density distribution.

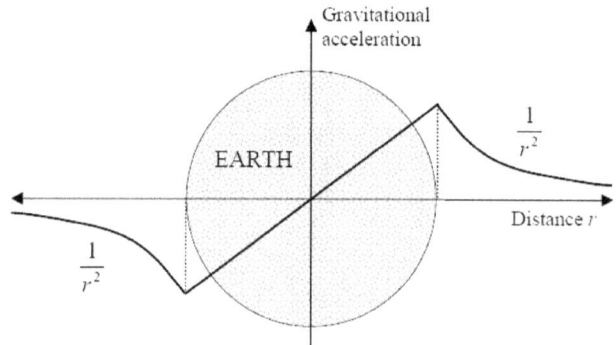

Figure 4.9. Variation of the acceleration due to gravity approaching and then crossing the interior of the Earth. Outside of the Earth the gravitational force will vary according to the inverse square of the distance. Interior to the Earth (assuming a constant density) the gravitational acceleration will vary linearly with the radius. The circle for the Earth is only intended to illustrate the boundaries at which the gravitational force law changes. The change of sign at the origin reflects the idea that a particle (say a miniature black hole) approaching with some initial velocity from the left along the central axis will first experience an acceleration towards Earth's center, but a deceleration after passing right through it.

The Roots of Mountains

As we saw in Chapter 3, Sir George Everest spent twenty-five years of his life, from 1818 to 1843, with the geodetic survey of India. In 1847 he published the results of his labors and described the measure of an 11.5 degree meridian arc that stretched from Cape Comorin in the far south of India, to the Himalayan Mountains in the far north. It was a tremendous accomplishment, undertaken by many surveyors working under Everest's command, under harsh conditions and debilitating circumstances. While Everest recognized that there was a small difference between his triangulation deduced length for the 11.5 degree meridian arc and that implied by astronomical observations, he attributed the difference to errors in the geodetic measurements. The Venerable John Henry Pratt (1809 – 1871), Archdeacon of Calcutta reinterpreted Everest's measurements, however, and found that the gravitational influence of the Himalayas was, in fact, very much smaller than might be expected – indeed, it was only about 1/3rd of expectation. This result, Pratt argued, indicated that the Earth's crust must vary in density, with lower density material being found under the mountains, and higher density material underlying the lowland (non-mountainous) regions. It was this variation in the density of the underlying crust, Pratt argued, that resulted in the reduced gravitational attraction produced by the Himalayas.

In contrast to the constant thickness, but variable density crust picture espoused by Pratt, Airy suggested that Earth's crust varied in thickness and that mountain's actually put-down roots. In this manner, Airy imagined mountain ranges, made of lower density material, to be floating on a higher density crust, the large weight of the mountains, however, caused the crust to spread, and it is this displacement of the higher density crust that neutralizes the attraction due to the mountains (recall the mountain height calculation presented in Chapter 3). Figure 4.10 illustrates the essential difference between the two ideas being put forward by Pratt and Airy. The theory that Earth's crust physically accommodates, or dynamically compensates, for the formation of mountain ranges is known as isostasy[13].

Airy's isostatic model is essentially a statement of Archimedes' principle of hydrostatic equilibrium, in which a floating body (the mountain) displaces its own weight within the medium supporting it (the mantle). Ships, for example, float because they can displace their own weight in water. Archdeacon Pratt's model, on the other hand, implies that the vertical expansion of the crust takes place with no change in mass, requiring that the density of the mountain range must vary inversely with height – that is, the higher it is, the lower its density must be. Both of the models described by Airy and Pratt can explain, at least to a first approximation, the observations with respect to the pendulum attraction of mountains – the true story is somewhat more complicated, however, than either model suggests.

Before we move on to consider the moving Earth, there are still a few unresolved issues concerning the physics of gravity that should be looked at first. One of the problems that not only baffled Newton, but still baffles physicists to this very day, is how a gravitational presence is transmitted. And, coupled with this issue, another question that intermittently arises is, "are there short-range variations in Newton's formula [equation (4.1)] related to the composition, and/or density of the masses that are interacting gravitationally?"

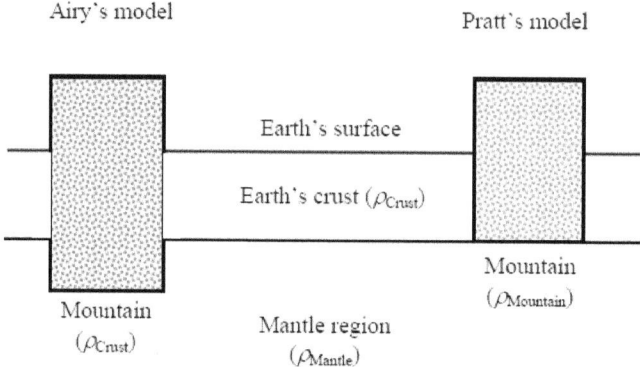

Figure 4.10. Idealized isostacy for rectangular mountains. Under Airy's scheme (left) all of the Earth's crust has the same density, but in regions were the terrain is elevated the crust is thicker. Archdeacon Pratt's scheme (right), on the other hand argues that the density of the crust varies and that in those regions where the terrain is elevated the density is lower. Under Airy's scheme the 'roots of a mountain' push into the Earth's mantle. In this figure it is intended that $\rho_{\text{mountain}} < \rho_{\text{crust}} < \rho_{\text{mantle}}$.

Gravitational Aether

In the every day world of human experience a common sense notion suggests that forces must be transmitted by or through something. If the force can be represented as a small particle then we can imagine that a process analogues to that of 'throwing' and 'catching' the particle would 'transmit' the required force – just think of catching a ball in some sports game, where the energy that has been put into the ball's motion is transmitted to the catcher's hand, who thereby feels the 'presence' of the 'thrower'. In this sense, the gravitational force is transmitted by particles called gravitons. The other way that force can be transmitted is through a transmitting medium. One example of this corresponds to two people each holding one end of a long rope. If one person waves their arm up and down, starting a wave to move along the rope, the other person will eventually feel the presence of the companion when the wave reaches the end of the rope that they are holding, and their arm begins to be forced up and down in sympathy with the motion of the sender. If we now imagine the rope being expanded along its diameter and flattened down, a sheet will develop, and by flapping the sheet so a force can be transmitted from one place to another. In this latter, so called, field sense it is not the whole rope or sheet that physically moves (as in the particle case), but it is a wave that passes along the rope, or over the sheet that transmits the required interaction information. By way of analogy, we are all probably familiar with magnets, and magnetism, and magnetic interactions are generally thought of as field (or in our case sheet) interactions. Indeed, when we use a compass for navigation, we are seeing an effect caused by Earth's magnetic field. So, the question arises, is gravity transmitted via discreet particles (such as gravitons) or via a transmitting medium or field of some sort.

Isaac Newton initially believed that there was a gravitation transmitting medium – a gravitational aether. This aether, Newton argued, permeated the small pores that exist in all objects whether a solid, or a liquid, and it was the contraction and dilation of this medium that transmitted the gravitational force between two objects. Not only this, but gravity apparently acted instantaneously over exceptionally large distances – as witnessed by the fact that the planets had stable orbits about the Sun. In the penultimate paragraph of Book III of the *Principia*, however, Newton, eventually concedes that he has, "not been able to discover the cause of those properties of gravity from phenomena," and to this he adds his famous line, "and I frame no hypothesis" [which in the original Latin was written *hypothesis non fingo*]. Newton was criticized by some for this apparently defeatist attitude and, further ridiculed by others for his absolute reliance upon experimentation and observation. It is also a somewhat strange statement for Newton to make since he readily 'framed' hypotheses in many other areas. The key experiment that convinced Newton that there was no gravitational aether was a straightforward pendulum experiment performed in the early 1680s. The experiment is described towards the end of Section VI in Book II of the *Principia*. Specifically, what Newton did was to construct a pendulum some 11-feet (3.35 m) in length with a bob made of a small wooden box. According to his account, which he acknowledges was written from memory (the original paper being lost), Newton carefully determined the displacement position of the pendulum after various numbers of oscillations. After this calibration step he filled the wooden box with lead and repeated the experiment. What he found was that there was no sensible change in the pendulums motion. If there was a gravitational aether, Newton argued, the pendulum, when the box was full of lead, should have been slowed more rapidly than when it was empty. The reasoning behind this expectation being that the aether would have to interact with more matter (since lead is very dense it must have many more constituent 'particles' per unit volume than air) and accordingly it should have an additional drag-like effect on top of that due to air-resistance alone. In the end, Newton could only offer the following comments in the *General Scholium* of Book III in the *Principia*, "...to us it is enough that gravity does really exist, and acts according to the laws which we have explained." And, indeed, Newton is absolutely right, and there have been no unequivocal experimental circumstances (well, see below) where his formula for the gravitational force [equation (4.1)] has given the wrong answer under circumstances where it is expected to apply.

While Newton was unable to find an explanation for the mechanism that transmits gravitational force, the modern day physicist is not that much better equipped to answer the question either. Certainly Einstein's general relativity theory (introduced in 1915) relates gravitation to a manifestation of the curvature of 4-dimensional spacetime, and this theory is thoroughly successful under the domain in which it should apply. By explaining gravity in terms of curvature Einstein essentially nullified it as having to be a force that acted between two objects, but and in spite of this brilliant move, there is still no answer to the fundamental question which asks, "why and how does matter (and energy) curve spacetime in the first place." In the quantum mechanical world of the incredibly small, a mass-less particle called the graviton has been proposed as the carrier of the gravitational force, but to date physicists have not been able to build a universally accepted theory of quantum gravity, and nor have they been able to device an experiment that would clearly indicate the existence of gravitons. This latter lack of physical understanding, however, is exactly what makes the field of quantum gravity one of the most vibrant and exciting areas of theoretical and experimental research in the current era.

The Weak Equivalence Principle

The discussion in Chapter 1 introduced mass as a fundamental measure of the quantity of matter within an object. Physicists, however, distinguish between two fundamental kinds of mass; the inertial mass and the gravitational mass, each of which is introduced in a distinct and slightly different manner. The inertial mass $m_{inertial}$, is employed in dynamic-style calculations and it is a measurement of an object's resistance to a change in its velocity. In this fashion, the greater the inertial mass of an object, so the greater is its resistance to any velocity change due to an applied force. This behavior is embodied in Newton's second law of motion, which states that the acceleration a an object of inertial mass $m_{inertial}$ experiences in response to some applied force F is such that $F = m_{inertial}\, a$. In contrast to the inertial mass, the gravitational mass $m_{gravity}$ is introduced as a measure of how strongly an object interacts with a gravitational field. For a given gravitational field g a larger gravitational mass experiences a greater gravitational acceleration than a smaller gravitational mass, and the force F experienced is given by the expression $F = m_{gravity}\, g$. Clearly, the life of the physicist would be much more complicated if there really were two different types of mass, and the weak equivalence principle importantly stipulates that the two masses are identical and accordingly $m_{inertial} \equiv m_{gravity}$. The word equivalence is introduced in order to indicate that a gravitational force is equivalent to any other type of net force that results in an observable acceleration.

One of the first experiments to test the idea of equivalence (although not known by that term) was supposedly performed by the ever irascible Galileo circa 1590. The term supposedly is used here since there is no physical evidence to indicate that Galileo actually performed the experiment[14], but the story goes that he dropped two unequal masses from the top of the Leaning Tower of Pisa and noted that they struck the ground at the same time (figure 4.11). This observation is sometimes called the universality of free-fall, and it embodies the idea that the path of a falling object depends solely upon its initial velocity and its initial position and is independent of the object's mass and composition. The idea of equivalence can be illustrated within the framework of Galileo's supposed experiment by imagining what a small insect attached to one of the masses will see. Since the two masses are free-falling at the same speed at any instant, the little insect will perceive the second mass to be simply hovering in space, as if the Earth's gravitational attraction had simply disappeared, in other words the motion and acceleration that the little insect undergoes while attached to its host rock cancels out any observable effects resulting from the Earth's gravitational field.

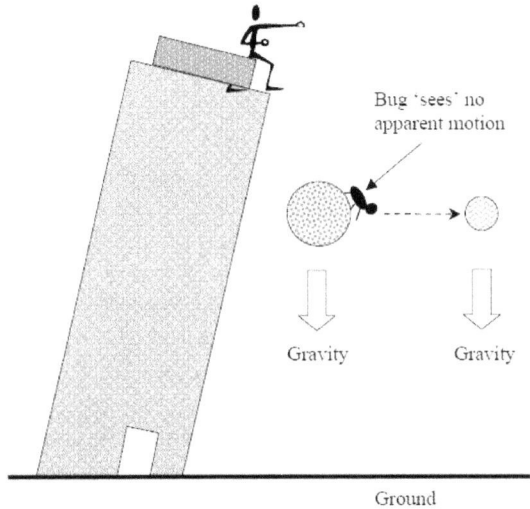

Figure 4.11. Galileo's supposed Leaning Tower of Pisa experiment. A little insect attached to one of the dropped spheres 'sees' the companion sphere simply hover in space as if the Earth's gravitational field has disappeared. This effect witnessed by our hapless insect comes to an abrupt end, however, when the two spheres hit the ground!

Since the principle of equivalence has such profound consequences, and is the cornerstone Einstein's general theory of relativity, the question automatically arises as to its correctness, and one can ask if there are ways to test its claim. It turns out that the key to testing the equivalence principle is to think a little more about what it is that the test bodies are made of. Indeed, embodied within the weak equivalence principle is the notion that the motion of an object is independent of its composition. In this manner, let us imagine that we take a steel ball-bearing of gravitational mass m and construct by careful machining a second sphere with exactly the same gravitational mass but made out of, say, wood. The wood sphere will be much larger than the steel ball-bearing (because of its lower density), but if the two spheres are placed in a pan balance then the balance arm should remain horizontal because the gravitational field of Earth acts upon both masses equally. The question now is whether the inertial mass of the steel ball-bearing and the wooden sphere are the same. That is, will the wooden sphere be accelerated at the same rate as the steel ball-bearing under the action of the same applied force? The test for this was first developed by the Hungarian physicist, Baron Loránd Eötvös in 1908.

The Eötvös experiment, as it has become known, is a modified Cavendish-style torsion balance experiment which uses the rotation of the Earth to determine the equality of inertial masses. The key idea is to orientate the arm of the torsion balance in an East-West direction (see figure 4.12) and to suspend from the ends of the beam two identical gravitational masses. Since the Earth is spinning (as we shall see in the next chapter) the masses on the torsion balance are subject to two forces. The first is the gravitational force due to the Earth, but this, recall, will be equal by the design of the experiment. The second force is due to the Earth's spin – a centripetal force, the details of which will be discussed in Chapter 5. Now, if the centripetal force on the steel ball bearing (say) was greater than that on the wooden sphere, the balance arm would tip, and this would indicate that their inertial masses were different. If this imbalance was truly present then the arm would also rotate slightly until the torsional force in the suspension wire matched the difference in the centripetal forces acting upon the two masses. The reason behind this rotation will be explained in the next chapter, but for the moment the key point is that the small angular twist in the suspension wire is something that can be measured. In a series of experiments Eötvös was able to show that the inertial and gravitational masses of materials such as tallow, copper, asbestos, water, wood, magnalium and copper sulfate were the same to a precision of 5 parts in one billion. A refined, modern day version of the Eötvös experiment has improved upon this accuracy and finds that the inertial and gravitational masses of copper (Cu) and beryllium (Be) are the same to a precision of one part in 100 billion.

Torsion balance experiments, such as the Eötvös experiment, are so-called null experiments in that they balance the acceleration of one body against that of another and look for small departures from equilibrium: the idea of the Eötvös experiment being that if the weak equivalence principle is true then nothing should happen. A dynamic version of the Eötvös experiment, however, was developed at Princeton University in the mid-1960s by a group of researchers under the directorship of by Robert H. Dicke. In the Princeton experiment the balance arm was orientated in a North-South direction (figure 4.13), and the idea was to see if a difference in the Sun's gravitational influence on the suspended masses could be detected. Specifically, as the Earth spins on its axis any difference between the Sun's gravitational interaction with the two masses will result in a 24 hour modulation, or oscillation, in the orientation of the balance arm as seen in the laboratory. The

Princeton group found no modulation in the torsion balance orientation, and concluded that the Sun's gravitational acceleration on identical aluminum and gold masses was the same to one part in one hundred billion.

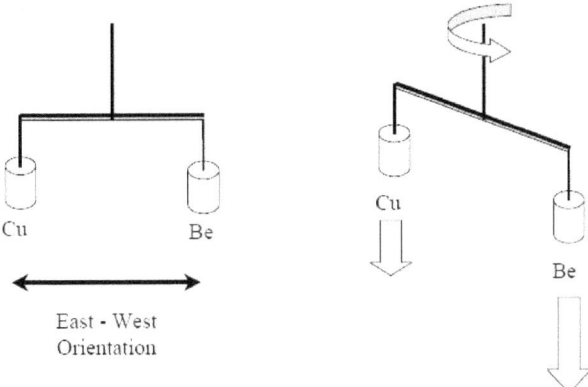

Figure 4.12. The Eötvös experiment is designed to test the equivalence of the inertial and gravitational mass. In the arrangement shown here the two masses are assumed to be made of copper (Cu) and Beryllium (Be), and the diagram illustrates the situation (see the block arrows) where the centripetal force on the beryllium mass is greater than that on the copper mass. With one mass being higher than the other, a small difference in the centripetal acceleration will result and this will cause the balance arm to twist.

The two experiments described in this section are dependent upon knowing that the Earth spins and that this spin results in the generation of observable accelerations. In our modern world it is hard to imagine that the Earth doesn't spin – indeed, we have been told this fact throughout our lives. But how do we really know that during the course of the day it is the Earth that is spinning from West to East that produces the motion of the Sun through the sky, rather than the Sun moving from East to West around a stationary Earth? It is not an obvious observation which motion is true, at least to our human senses, and while Copernicus could argue in 1543 that the Earth was in motion about the Sun and also spinning on its axis, he had absolutely no demonstrable proof of his conjecture. The definitive, experimental verification that the Earth must be spinning was presented in 1851 (more than 300 years after the publication of Copernicus's text) and yes, you guessed it, as we shall see in the next chapter, at the heart of the experiment was the mighty pendulum.

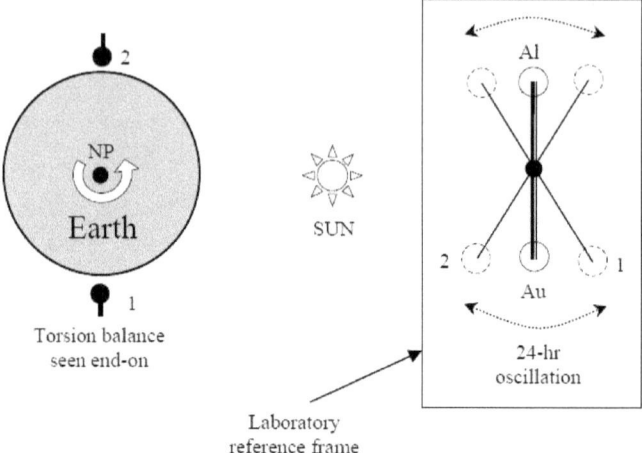

Figure 4.13. The Roll-Krotkov-Dicke experiment was designed to see if a gold (Au) mass experiences a greater solar gravitational acceleration than an aluminum (Al) mass.

How Constant a Constant is Big-G?

There are relatively few areas in which the domains of geology and cosmology overlap, but one such cross-over field is that relating to variable G-cosmologies. In these specific cosmological models, which are admittedly no longer in popular vogue, the absolute constancy of the universal gravitational constant [G in equation (4.1)] is questioned. While Einstein's general theory of relativity admits no possible variation in G once its value has been fixed, it is possible to construct alternate cosmological models in which the gravitational 'constant' varies slowly with time, or more specifically with the age of the universe. Clearly, if G does vary then it must either increase or decrease as the universe ages, and accordingly there should be distinct and quite opposite consequences to separate between alternative models. Since the size of Earth's orbit about the Sun is related to the gravitational constant, a value of G that increases with time will result in Earth's orbit contracting; if on the other hand G decreases with time then Earth's orbit will expand. Not only this, however, the physical size of Earth will decrease if G increases with time, and its physical size will increase if G is getting smaller. The Sun will also be effected since its luminosity varies as G to the power 7.5, and accordingly it would have been much brighter in the past if G was larger and less bright if G was smaller. American physicist Edward Teller was one of the first scientists to study the solar-terrestrial consequences of a variable gravitational 'constant' and he noted[15] that if G was as much as 10% larger some 200 to 300 million years ago then Earth's oceans would have long ago boiled away, and all life would have been extinguished. Clearly such a situation didn't come about, since we are here to ask the question, but of course, Teller's result simply asserts that the change in the gravitational 'constant' has to be less than about 10% on timescales of order a few millions of years. Teller's constraints on the allowed variations in G were confirmed observationally by Irwin Shapiro and co-workers in 1971. Using a series of atomic clocks and high-powered radar systems Shapiro and co-workers studied the orbital motion of the planets Mercury and Venus, and concluded that the fractional variation of the universal gravitational constant could not be greater than $(dG/ dt) / G = 4 \times 10^{-10}$ per year – where (dG/ dt) is the time variation of the gravitational constant. More recent studies based upon the lunar ranging project (forming part of the Apollo lunar landing program discussed in Chapter 3) have shown that an even stronger limit on the variation of the gravitational constant can be set with the fractional variation $(dG/ dt) / G < 10^{-12}$ per year.

The idea that the universal gravitational 'constant' might vary with the age of the universe was first considered in the mid-1930s. Pioneering astrophysicist Edward Milne, for example, described in 1935 a cosmological model in which the gravitational constant increased over time. His motivation for this idea was an apparent abhorrence for Einstein's theory of general relativity and its prediction of curved spacetime. Milne's theory was complex and never popular and was soon cast aside as untenable. The variable G cosmology developed by Paul A. M. Dirac, long-running candidate for the strangest physicist award[16], in 1937, however, proved more interesting[17] and has been revised in various forms over ensuing decades. Dirac's motivation was based upon dimensional analysis, as described in Chapter 1, and specifically the construction of very large dimensionless numbers out of the fundamental constants such as the electron charge, the mass of the neutron, the mass of the electron, the speed of light and the age of the universe. Dirac realized that the two numbers N_1 = age of the universe / the light crossing time of the atom $\approx 6 \times 10^{39}$, and N_2 = electrostatic force between the proton and an electron / the gravitational force between the proton and an electron $\approx 2 \times 10^{39}$, were to within a factor of three equal. This is a remarkable result, and completely unexpected since it invokes a commonality between the incredibly small world of the atom and the staggeringly large cosmos. In light of such numbers Dirac formulated his famous large number hypothesis (LNH). The LNH states that any two of the very large dimensionless numbers (such as N_1 and N_2) occurring in nature are related by a simple mathematical relation in which the coefficients are of order unity. Now, given $N_1 \approx N_2$ and by its very definition N_1 varies according to the age of the universe (it becomes larger with time), then the LNH indicates that one of the terms in N_2 must also change with time. Dirac chose G to be the time variable term in N_2, and in order to satisfy the LNH its value will have to decrease as the universe ages.

While the gravitational constant was the first fundamental constant to be defined, it is far from the first fundamental constant to have its constancy questioned. Indeed, Dirac's LNH can also be interpreted in terms of the time variation of the speed of light, or the electron charge or the electron mass. Exploring these latter possibilities has lead to the construction of various non-standard cosmological models but none, so far, has proven especially popular, or generally convincing. At the present time it seems reasonably safe to conclude that there is no strong observational evidence to indicate that the gravitational constant varies with time and the age of the universe.

That the gravitational constant does not vary (at least significantly) as the universe ages has potential consequences for our very existence. This result can be cast in terms of the dimensionless gravitational coupling constant $\alpha_G = Gm_p^2/\hbar c \sim 10^{-39}$, where m_p is the proton mass, $\hbar$ is Planck's constant divided by 2π, and c is the speed of light[18]. The importance of this term resides in the fact that it can be shown[19] that the characteristic mass of a star is dependent upon the ratio of fine structure constant (see below) and α_G. Not only this, however, Brandon Carter (Laboratoire Univers et Théories, Meudon) has shown that should the value of α_G be just a little larger than is observed, then the energy transport in all stars would proceed via convection and supernova explosions, critical for the formation of all atoms other than hydrogen and helium within the universe, would not occur. In short, if there were no supernova, we would not exist, and nor would planet Earth[20].

While the characteristics of stars are ultimately determined by the value associated with α_G, it also turns out that the so-called natural units, derived from various combinations of the fundamental constants of physics, by Max Planck are also dependent upon the value of the universal gravitational constant G. Specifically, the Planck mass $m_{Planck} = \sqrt{c\hbar/G} \sim 10^{-8}\,kg$ specifies the hypothetical mass of a miniscule black hole having an event horizon with a radius equal to the Planck length: $l_{Planck} = \sqrt{G\hbar/c^3} \sim 10^{-35}\,m$. It is upon scales of order l_{Planck} and smaller that both general relativistic and quantum mechanical principles must be applied in describing the behavior of a physical system. To date no theory has been fully developed to assimilate these two diverse domains, and a working model for quantum gravity still eludes the grasp of modern scientific investigations.

While the workings of quantum gravity are played-out upon the unimaginably small Planck length scale, the Planck mass, in odd contrast, is almost tangible and approximately corresponds to the mass of a very small grain of sand. Just as remarkably, it is also quite likely, though not yet proven observationally, that Planck mass black holes do exist in the universe - they might also be produced[21], albeit only very rarely, in the proton-proton beam collisions being engineered at CERN's Large Hadron Collider (LHC). There is no existential threat posed to the Earth should the LHC manufacture even multitudes of miniature black holes, or even if Earth chances to encounter a cosmic, so called, primordial black hole (PBH)[22]. This is simply because such miniature black holes are so exceptionally small that they cannot effectively accrete ordinary matter. Indeed, in the case of PBH's they will likely pass straight through the Earth (this situation is reminiscent of the pendulum that can never be III discussed earlier – see also figure 4.9). Interestingly, however, the passage of a miniature black hole should generate a signal detectable by surface seismic arrays (recall Chapter 3). Yang Luo and co-workers at Princeton University have recently[23] considered what the seismic signal might look like should the Earth encounter a PBH, and find, remarkably, that it should have a distinct and easily recognized form. The key characteristics would be that the seismic signal would be detected simultaneously over the entire Earth's surface. To date no such event has ever been recorded, but it is rather pleasing to think that the greater body of the Earth itself has the potential to act as a bell-like, seismically ringing detector for some of the smallest, most extreme physical objects ever (hypothetically) created within the history of the universe.

An Aside on the TOE and the "Australian Dipole"

The great motivating drive behind much of 20th and 21st Century physics has been and continues to be the search for an overarching theory of everything (TOE) – a theory that unifies all four of the fundamental forces that enable matter to exist and objects to move. The four forces in question are the electromagnetic force, the weak force, the strong force and gravity. The first of these forces enables electrical and magnetic interactions to occur between atoms, the second and third allow for the existence of atomic nuclei (and their occasional transmutations), while the fourth controls the dynamics of large-scale accumulations of atoms (that is, ordinary matter). The characteristic properties of these four forces vary dramatically, and while the reach of gravity and electromagnetic waves is essentially infinite, the domain of the weak and strong forces is restricted to the miniscule scale of the atomic nucleus. The strength of the forces also vary dramatically, with, for example, the electromagnetic force acting between an electron and a proton (at any separation) being some 10^{39} orders of magnitude greater than their mutual gravitational force[24]. Indeed, for all of its infinite reach, gravity is by far the weakest of the fundamental forces.

Finding the appropriate mathematical and physical architectures to unify the fundamental forces, into one grand TOE, has proved to be a great intellectual challenge, and it has resulted in many remarkable suggestions and postulates concerning the very make-up of matter, space and time. String theory[25], for example, requires the existence of multiple additional dimensions, beyond the four (three space and one time dimensions) that we are familiar with – and since these additional dimensions must be hidden from our direct observational gaze they must be wrapped-up upon themselves to sizes smaller than the Planck length itself. Just as remarkable as the requirement of extra-dimensions, many of the possible versions of string theory predict both spatial and temporal variations in dimensionless constants such as the fine structure constant $\alpha = e^2/\hbar c \approx 1/137$ (where e is the elementary charge carried by a single proton) and the electron to proton mass ratio $\mu = m_{electron}/m_{Proton} \approx 1/1836$. The fine structure constant α characterizes the strength of the electromagnetic interaction, while μ essentially characterizes the ratio of the weak force to the strong force. In the standard model of particle physics the values of α and μ are deemed to be absolute constants, but the important observational question is, are they really constant over all space and over all time[26].

One way to test the constancy of the fine structure constant is to look at the spectra of distance quasars. These active galactic nuclei are located at vast distances away from the Earth, and they are powered by the accretion of matter onto supermassive black holes. John Webb (University of New South Wales, in Australia) and co-workers[26, 27] have made detailed studies of such spectra over the past decade, and they have found evidence for both a spatial and temporal variation in the fine structure constant. Their method entails making measurements of what is called the Lyman alpha[28] forest. The key point of such observations is that given the great distance to any quasar there will inevitably be numerous intervening galaxies and intergalactic hydrogen clouds along the observational line of sight. Each one of these structures will imprint a Lyman-α absorption line on the quasar spectrum, and by analyzing the characteristics of the numerous Lyman-α lines a measure of the fine structure constant can be deduced. Since each cloud is at a different distance from us, and therefore has a different age (that is look back time), so the spatial and temporal variations, if they exist, of the fine structure constant can be studied. While the results are still preliminary, Webb's research group has found evidence that the fine structure constant does vary with respect to direction and age of the universe, being larger in the past in one direction and smaller in the past in the opposite direction. Dubbed the "Australian Dipole", the variation reported by Webb and co-workers[27] suggests that $\delta\alpha/\alpha_0 = (1.10 \pm 0.25) \times 10^{-6} \, r \cos\psi$, where $\delta\alpha/\alpha_0$ is the relative variation in the fine structure constant at distance r, measured in billions of light years, with ψ being the line of sight angle relative to the axis of the "Dipole", and where α_0 is the value of the fine structure constant measured on Earth (located, of course, at $r = 0$) at the present time.

It is far too early to tell, yet, if the Australian Dipole is a real feature of our universe. But, if it is a real phenomenon, then the consequences for physics and cosmology will be profound. It is a high stakes game, and the mantra that extraordinary claims require the support of extraordinary evidence holds true, with the collective scientific community, for the present, holding its final judgment in reserve. If, again, the results for a spatially and temporally vary-ing fine structure constant are verified, then the implication is that other di-mensionless 'constants' such as μ will also vary in the same direction and sense. Time, the relentless and impartial judge, will eventually tell how these extraordinary observations and their implied potential for new and exotic physics is going to play out.

The Nanomechanical Whisker

This chapter began with a discussion of the ordinary balance – the horizontal stopped pendulum. It is a device that is as old as civilization itself, and it is a device that has seen relentless adaptation and incredible improvements in its sensitivity. Indeed, as we have seen, Galileo described in his *La Bilancetta* (The Little Balance), published in 1586, how through Archimedes principle a bal-ance could be calibrated to determine chemical composition. Robert Hooke was also to be found, in 1662, precariously perched amongst the rafters of Westminster Abbey, trusty balance in hand, in an attempt to determine the manner in which gravity varies with distance – the balance and the pendulum, one and the same, all being put to use for commerce and fundamental scien-tific investigation. In the modern era the principles and practices have hardly changed, but the scale of application has plumbed ever-smaller quantities and physical scale. To this latter end, one of the smallest and most sensitive weigh scales ever to be constructed was recently fabricated in the laboratory of Adrian Bachtold and co-workers at the Institute of Nanotechnology in Barce-lona, Spain[29]. This particular device is made of a carbon atom nanotube just 1 nanometer wide and 150 nanometers long – indeed, a veritable infinitesimal whisker from Charles Dodgson's vanishing Cheshire cat.

Carbon nanotubes are, as the name suggests, minute cylindrical structures composed of a trellis-like network of carbon atoms arranged in rings, and key to their application as high sensitivity mass sensors is to set them oscillating. Once set in motion, in a high vacuum environment, the nanotube will adopt some very specific resonant frequency f_0, typically a few gigahertz, and this can be controlled and monitored by the experimental arrangement. If some additional small mass is then added to the nanotube it will show a measurable shift Δf in its resonant frequency. This change in frequency is directly linked to the change Δm in the mass of the system, with $\Delta f = -\left(f_0/2m_0\right)\Delta m$, where m_0 is the initial mass of the nanotube. By carefully monitoring, therefore, how the frequency of a nanotube changes it has been possible to determine the masses of various kinds of molecules, biological cells and even individual atoms. The nanomechanical resonator, which at its heart is a pendulum-like flexing beam, is indeed exquisitely sensitive, and the research group in Barce-lona[29], have recently announced state-of-the-art measurements to the level of several yoctograms. This lowest 'officially recognized' unit in the *SI* system corresponds to a minuscule 1 yg = 10^{-24} grams, and Bachtold and co-workers were able to obtain precise mass measurements of the xenon atom and the naphthalene molecule ($C_{10}H_8$). The resolution of the Barcelona nanotude weigh scale is an incredible 1.7 yg, which corresponds to the mass of a single proton.

Once again, we find that the pendulum, our mighty pendulum, under the transformed guise of a weigh scale, has enabled humanity to explore the fundamental workings of the universe, to determine the mass and make-up of the Earth, and even directly measure the masses of molecules and individual atoms.

CHAPTER 5

ALL IN A SPIN

Even the Pendulum is a false prophet. You look at it,
you think it's the only fixed point in the cosmos, but
if you detach it from the ceiling of the Conservatoire
and hang it in a brothel, it works just the same.

Foucault's Pendulum
Umberto Eco

Published as he lay half-paralyzed[1] and upon his death bed in the spring of 1543, Nicolaus Copernicus argued in his great work *De revolutionibus orbium coelestium* that the Earth had to undergo three different kinds of rotational motion; a yearly orbital motion about the Sun; a diurnal spin motion about the axis running between the north and south celestial poles (recall figure 2.1), and a third, annual, rotation that kept the direction of the Earth's spin axis pointing in the same direction as it orbited the Sun (figure 5.1). The reasoning behind the third rotation was a consequence of the elementary observation that from one day to the next the north celestial pole remained a fixed point on the sky. If this third rotation did not take place, Copernicus correctly reasoned, then the location of the celestial poles would vary during the course of the year – indeed, they would move along the circumference of 23½-degree radius circles on the sky.

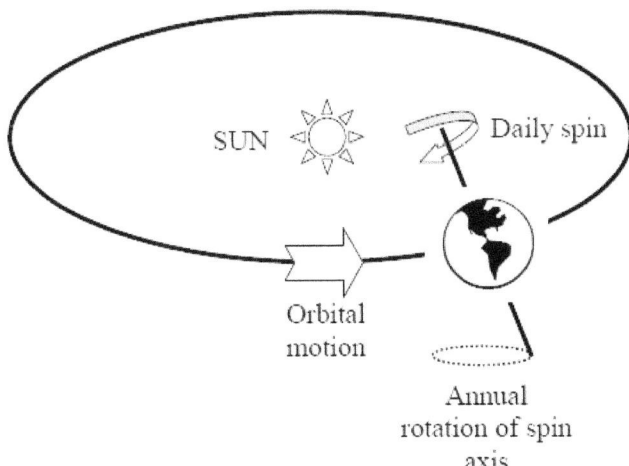

Figure 5.1. The Earth's three rotational motions as introduced by Copernicus in 1543.

While Copernicus could offer no proof that the Earth orbited the Sun, the definitive detection of stellar aberration by James Bradley in 1729 indicated that it must, indeed, be moving, traveling on its rounds at a sprightly speed of about 30-km/s. Rather than relating to its spatial, that is orbital, motion, however, the two other Earthly rotations introduced by Copernicus relate to its bodily spin. As we shall see in this chapter, it required the construction of an enormous pendulum to experimentally demonstrate the existence of Earth's spin, but an understanding of spinning objects in general was required before an appreciation the third rotation could be made – indeed, as we shall see, this latter rotation essentially takes care of its self.

What Goes Up.....

French Minim monk and polymath Marin Mersenne (1588 – 1648) had a certain penchant for performing experiments that made lots of noise. To this end, for example, Mersenne estimated the speed of sound, in 1637, by having a friend fire a pistol into the air at a measured distance away from him. By timing the interval between the discharge flash and the arrival of the sound he was able to estimate that sound traveled at a speed of about 450 meters per second - not bad for a first attempt, but a little on the high side, since the modern day value is more like 330-m/s. Mersenne's procedure works, of course, because the speed of light, conveying the discharge flash information, travels much faster than the sound wave; indeed, it travels 900,000 times faster, although the finite speed of light was not remotely known when Mersenne performed his experiment. Within a year of performing his pistol shot experiment Mersenne investigated a long-standing and rather bothersome scientific problem, but this time he employed a much higher caliber weapon.

It was on a beautifully clear, late-spring morning in 1638 that Mersenne along with his engineer assistant, and superintendent of fortifications, Pierre Petit, tramped into the fields beyond the outskirts of Paris and there, at a safe distance from any buildings and spectators, set up their borrowed cannon (figure 5.2). The hefty bore of the cannon was pointed skywards, straight into the air. Amidst a great concussion of sound and a billowing plume of smoke, Mersenne fired the cannon, and a massive iron cannon-ball careened upwards into the azure skies. The two observers soon lost sight of the speeding sphere, and in fact it was never seen again. Dumfounded at the loss of their cannon-ball, Mersenne and Petit retuned to Paris; the experiment hadn't exactly gone as planned – but then what was it that they were trying to measure? The aim of the experiment had been to demonstrate the spin of the Earth.

The cannon experiment conducted by Mersenne and Petit wasn't based upon a new idea, but it did have a few refined features compared to earlier attempts at measuring the Earth's rotation. The essential argument behind the test is as follows; if the Earth is spinning, say from West to East and an

Figure 5.2. *Retombera – t – il?* reads the legend (Will it fall back down again?). A woodcut illustration of the cannon experiment conducted by Father Mersenne and Pierre Petit.

object is thrown vertically upward, then in the time that it takes the object to go up and then fall down again, the Earth, and the observer standing on it, will have been carried eastwards, and accordingly the object should fall to the West of the observer. The argument is simplicity itself – but entirely wrong, as we shall see. By using a high velocity cannon-ball it was argued that the flight time, up into the sky and back down again, should be appreciably long and therefore the westward displacement from the observer large and easily measurable. The large mass of the cannon-ball should also reduce the effect of wind gusts upon its motion. Well, that was the theory. Figure 5.3 (a) illustrates the (physically incorrect) situation in which the cannon ball is fired upwards at position 1 and simply falls back down again after some specific flight time T. In that same time T the observer and cannon have moved eastward to point 2 due to the spin of the Earth and accordingly the cannon-ball falls to the West of the observer. The physically correct picture, however, is shown in figure 5.3 (b). Since the cannon-ball is initially at rest in the cannon it has at the moment of launch an eastward velocity equal to the Earth's spin velocity (this situation will be described more fully in a moment), so rather than simply traveling upwards and then falling back down again, our birds-eye view of the experiment will show the path of the cannon ball [the dotted line in figure 5.3 (b)] as a parabolic curve. Indeed, in a perfect world in which air resistance plays no role, the cannon-ball will fall back into the bore of the cannon out of which it was fired – presumably with none too pleasant effects. It was perhaps just as well that the experiment conducted by Mersenne and Petit was a failure.

While Mersenne's vertical-cannon experiment didn't yield the result that was expected, Mersenne himself was an important figure in Renaissance science. He was literally the central communications hub of a vast network of

European scientists. He wrote nearly endless numbers of letters to all practi-tioners of philosophy and mathematics, and he passed-on new scientific ideas and information. It was Mersenne that was responsible for making Galileo's work more widely known outside of Italy, and he openly encouraged the young Christiaan Huygens in his many mathematical and experimental works. Likewise, Mersenne was a good friend of and correspondent with the great French philosopher Rene Descartes. Indeed, the cannon experiment was performed in the hope of helping Descartes prove his vortex theory of the universe, one of the consequences of which required the Earth to be spin-ning. Ironically, Mersenne and Petit had misunderstood the experiment as outlined by Descartes, who actually suggested that the cannon should be fired horizontally along a north-south line. In this situation the cannon-ball should then fall to the East of the firing line.

The cannon and cannon-ball experiment by Mersenne and Petit was a bold (albeit loud and incorrectly conceived) attempt to demonstrate the spin of the Earth. In order too make further headway, it was soon realized, if the spin of the Earth really was to be demonstrated then a more subtle experi-mental approach was required. Acting upon instructions provided by Isaac Newton (Descartes arch nemesis) Robert Hooke was the person who eventu-ally performed the more refined experiment, replacing the massive cannon-ball with a small-bore musket bullet.

A Technical Aside on Spin Dynamics

Since the Earth spins as a solid body once every sidereal day, we can define its angular velocity in terms of the number of degrees (actually we shall use radians – recall figure 1.2) that it spins through per second. Accordingly, $\omega = 2\pi / (24 \times 60 \times 60) = 7.27 \times 10^{-5}$ radians per second – although the conven-tion is to simply express the units as 'per second'. The actual instantaneous velocity tangential to the Earth's surface at a distance R away from its spin axis can now be written as $V_T = \omega R$. Since it is the distance from the spin axis that counts, rather than the Earth's physical radius, the tangential velocity at a latitude φ will be $V_T = \omega R \cos \varphi$. At the equator $\varphi = 0$ and the tangen-tial velocity attains its maximum value. As the latitude increase towards either one of the poles, where $\varphi = \pi/2$ radians $= 90$ degrees, however, the tangen-tial velocity becomes smaller and smaller, disappearing altogether at the poles since at these locations the displacement from the spin axis is zero. It is es-sentially because of the cos φ term, in fact, that the launch complexes for commercial, military reconnaissance and scientific rockets are placed as close to Earth's equator as possible (or at least as southerly as national boundaries allow).

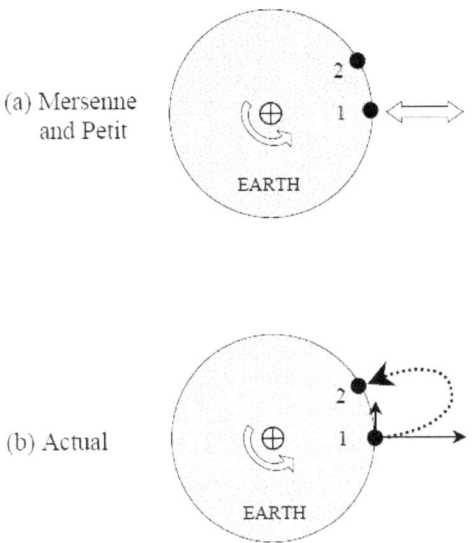

(a) Mersenne and Petit

EARTH

(b) Actual

EARTH

Figure 5.3. (a) The expected result of Mersenne's cannon experiment in which the cannon-ball simply goes up and falls down again, but lands to the West of the observer. (b) Since the cannon has an initial eastward velocity (due to the Earth's spin), the actual flight path of the cannon ball (as seen by an observer stationed above the Earth's North Pole) is a parabola (dotted line). The vertically launched cannon-ball will, in a perfectly mathematical world in which there is no air resistance, fall back in to the bore of the cannon from which it was fired. Note that in each diagram the scale has been highly exaggerated.

In addition to describing the velocity of an object, we shall also be interested in annotating the change in an objects velocity with time – the acceleration. The equation that describes the acceleration of an object moving close to Earth's (rotating) surface can be written as the sum of three competing terms:

$$\text{Acceleration} = \text{Newton} - \text{Coriolis} - \text{Centripetal} \qquad (5.1)$$

The 'Newton' term in equation (5.1) corresponds to Newton's second law of motion which states that the acceleration a produce by a force F acting upon an object of mass m is $a = F / m$. This equation applies whether we are dealing with a rotating system or not. The 'Coriolis' term is named after the French engineer Gaspard-Gustave Coriolis who first realized, in 1835, that an additional acceleration term must be included when the equation of motion is expressed in a rotating frame of reference. The Coriolis acceleration is a, so-called, fictitious acceleration that comes about when a freely moving object is viewed from within a rotating frame of reference (such as that corresponding to an observer on Earth's surface). The classical example of the Coriolis Effect is that demonstrated by large bodies of water and air. Rather than an air flow flowing along a straight-line path from a high pressure to a low pressure region, as it would on a non-rotating planet, the flow is either towards the right or towards the left according to whether the observations are being made in the northern or southern hemisphere. This effect is particularly evident in the airflows associated with cyclones (regions of low pressure) where the sense of rotation is counterclockwise in the northern hemisphere, and clockwise in the southern hemisphere. Indeed, this effect is incorporated in the rule-of-thumb law described by the 19th century Dutch meteorologist C. H. Buys Ballot: if you stand with your back to the wind in the northern hemisphere, then the region of low pressure is to your left. In the southern hemisphere the region of low pressure will be to your right. The magnitude of the Coriolis acceleration term is the product $2\,\omega\,v$, where ω is Earth's angular velocity and v is the velocity of the object relative to Earth's surface.

The final 'Centripetal' term in equation (5.1) is another fictitious acceleration that arises in a rotating reference frame. A playtime example of this effect is that experienced by a child and their parent when enjoying the thrill of a merry-go-round ride. As the merry-go-round begins to turn the riders experience a sense of being pushed outward – this is often described as the centrifugal force. The outward acting centrifugal force is the response (described according to Newton's third law of motion: forces always act in pairs that are equal and opposite), however, to the inwardly acting centripetal force. We can appreciate the origin of the centripetal acceleration by considering the motion of a ball at rest on a rotating table as viewed from a distant fixed (so-called, inertial) reference frame. To the distant stationary observer the ball appears to be moving in a circle, and accordingly an inwardly acting (or centripetal) force is required to maintain such a motion. The magnitude of this acceleration term is given by the product $\omega^2\,r$, where r is the distance from Earth's spin axis and ω, again, is the Earth's spin rate in radians per second.

A Letter from Newton

Robert Hooke really was, as we have already seen, a remarkably man, and while clearly a gifted experimenter he was typically far too busy, with a multitude of conflicting projects, to complete any one of them satisfactorily. We saw this in Chapter 5, for example, when discussing his unlucky, lens breaking, attempt at measuring stellar parallax – if he had only been granted more time, we may safely assume then he would likely have refined and repeated his observations with a new lens. Another experiment in which Hooke ran ahead of his results, and which could have used more time and greater refinement was that in which he attempted to measure the diurnal spin of the Earth. This particular experiment resulted in an apparently happy exchange of ideas between Hooke and Isaac Newton, but it also planted the seed for what latter became a bitter, and for Hooke, a devastating feud.

In 1677 Hooke became Secretary to the Royal Society of London, and one of his assigned duties was to write to Isaac Newton, at Cambridge University, in the hope of encouraging him to resume his philosophical correspondences with the Society[2]. Newton replied that his, "affection to philosophy being warn out", and that he was accordingly interested in other areas of study. This being said, Newton also wrote, "I shall communicate to you a fancy of my own about discovering the Earth's diurnal motion". Hooke was no-doubt delighted at Newton's "fancy" since here was the outline of a fundamental experiment that he knew he could perform. He was also delighted, as his future behavior revealed, since he spotted a mistake in Newton's reasoning[3].

Newton's idea for demonstrating the spin of the Earth was based upon the notion of dropping a small weight (a bullet was actually advocated) from a height of some "20 to 30 yards" and comparing its path with the vertical set by a plumb-line. Newton correctly reasoned that the bullet should fall to the East of the point directly below the release point (as set by the plumb-line), since being released at a height H above the Earth's surface it would have a greater horizontal velocity than that of the ground (see figure 5.4).

The experiment described by Newton was exactly the kind that Hooke excelled at, and indeed, within a few months of receiving the letter Hooke had obtained a series of experimental results (to be discussed shortly). If Newton had simple described the experimental procedure for demonstrating Earth's diurnal motion in his missive, then his relationship with Hooke might have taken a very different path. In grand philosophical fashion, however, Newton imagined what path the falling bullet would follow if it could pass unimpeded through Earth's interior under the influence of gravity alone. To this end he drew the path of the bullet as a spiral coming to rest at the Earth's center (see figure 5.5). It was this diagram, and the interpretation of the bullet's imaginary path through the Earth, that resulted in the vitriolic events that ensued. Indeed, as his subsequent power and influence grew within the Royal Society, Newton single-handedly contrived to write Hooke out of scientific history[4]. Such, for so history tells us, are the not uncommon actions of many a spurned genius.

At Garraways
Thomas Garraway opened his famous coffee-house in 1669, and all manner of hot beverages were sold there for the next 200 years. Located in Change Alley, Cornhill in the City of London, Garraway's was the social hub at which the merchants from surrounding businesses could make their transactions and catch-up on local gossip. For the price of a penny or two, any well-dressed and respectable-looking gentleman could gain access to its bustling rooms. And, once inside, amidst the din of mingled voices and the fug of

tobacco smoke, a patron could order a dish of coffee and thereafter either join-in the evanescent conversation or quietly peruse a copy of the local newspaper. Garraway's was a place to learn, listen and communicate. Hooke's rooms at Gresham College were just a 5-minute walk away from Garraway's and many of his intimate diary accounts[5] simply read "at Garraways" – it was his home from home. Indeed, the coffee house was a regular meetinghouse for many members of the Royal Society and it was amidst the noise and aromatic fog of Garraway's that Hooke must have assuredly enjoyed many a conversation about the latest scientific news.

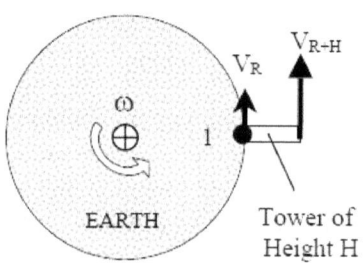

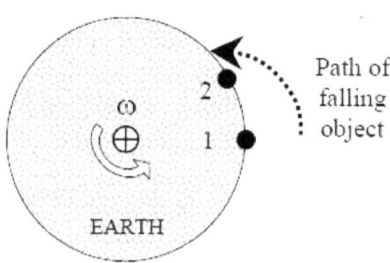

Figure 5.4. The flight path of a weight (scale greatly exaggerated) dropped from a tower of height H as seen by an observer located above the Earth's North Pole. The upper figure indicates the initial horizontal velocities of the observer (at the base of the tower) and the weight at the top of the tower. Upon being released (at point 1) the path of the weight is shown by the dotted line (lower diagram), striking the Earth's surface at a point eastward of the observer, who in the same time, has moved with the spin of the Earth to point 2.

Not only was Garraway's a place to discuss new scientific ideas, it was also a place where experiments could be performed – presumably under the bewildered gaze of the local merchants. On one occasion Hooke, and a number of his fellow Royal Society companions, dissected, examined and sketched a large porpoise while in the coffee house - an exercise that may not have been so pleasing to near-by patrons. On January 16th, 1680, however, Hooke and his long-time assistant and friend Harry Hunt, strode enthusiastically to Garraway's with the intent of performing an experiment to demonstrate the spin of the Earth. Hooke recorded in his diary for that day, "at Garways [sic], tryd fall of bullet in the hall with Hunt". The experiment was evidently undertaken in the front hallway amidst the hustle and bustle of coffee-sipping patrons, and it represents a nice early example of taking science to the public arena. To begin with, Hooke and Hunt set-up a 27-foot (8.23-m) long plumb-line above a tray of soft tobacco-pipe clay. The plumb-line marked the vertical path against which the trajectory of the dropped bullet would be measured. The aim was to test Newton's assertion that the weight should fall to the east of its release point – that is to the east of the central point, indicated by the plumb-line, of the target tray.

The experiment, according to a letter written by Hooke to Newton the day after the Garraway's experiment, was a great success. Hooke wrote, "I am now perswaded the Experiment is very certaine, and that it will prove a Demonstration of the Diurnall motion of the earth as you have very happily intimated…". What Hooke had physically measured was the eastward deviation of the falling weight compared to the plumb-line vertical, finding a shift of about ¼ of an inch (6.35 millimeters). While Hooke, and indeed Newton were apparently convinced by the experimental results obtained at Garraway's, in the light of more modern understanding we now know that Hooke was deceived. His falling weight experiment, while more subtle than Mersenne's and Petit's cannon extravaganza, was still not sensitive enough to truly demonstrate the rotation of the Earth. The actual eastward movement that will result from a 27-foot (8.23-m) drop should amount to 0.5 millimeters (at the latitude of London), and accordingly Hooke over-measured the effect[6] by a factor of about 13 times. The spin of the Earth introduced by Copernicus in 1453, long held to be true by astronomers, had still not been demonstrated, at least to everyone's satisfaction, by experimentation. Indeed, the first successful performance of the critical Earth-rotation-proving experiment was not to be performed until 148 years after Hooke's death.

In principle there was nothing wrong with the idea behind the experiment conducted by Hooke; it was just a question of the accuracy to which measurements could be made. Theoretically, the longer the drop, so the further the displacement should be, and accordingly perhaps the ultimate version of the falling bullet experiment was that conducted by the German chemist Ferdinard Reich in 1832. Echoing Airy's pendulum experiments within mine shafts (recall Chapter 4), Reich set up a 520-foot (158.5-m) free-fall drop in an abandoned mine near Freiberg, in Saxony. The average eastward displacement from 106 trials was determined to be 1.12-inches (2.84-cm), which is in remarkable agreement with the expected displacement of 1.08-inches (2.74-cm).

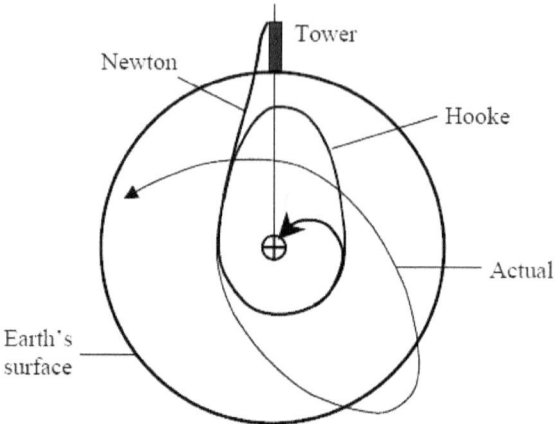

Figure 5.5. The path of a weight dropped from the top of a tower and its subsequent unimpeded passage through the Earth. In Newton's originally communication to Hooke November 24th, 1679 he drew a spiral track for the weight's path. Hooke argued that the path should really be an ellipsoid-like orbit (similar to the orbit of a planet about the Sun). The actual hypothetical path that the imagined bullet would take is shown by the lopped path – in the latter case the spin of the Earth his been included.

Flamsteed Checks for Consistency

The 27-foot plumb-line set up at Garraway's was not the first time that Hooke had experimented with long pendulums – all be it a stopped pendulum in the coffee-house experiment. Indeed, some ten years prior to his Earth rotation experiment Hooke had produced and then demonstrated before the assembled Fellows of the Royal Society a pendulum with a length of 14-feet (4.27-m) having a bob of mass 3-pounds (1.36-kg). This giant pendulum had a 2-second period and a mere ½-inch (1.3-cm) amplitude of swing. The long length of the pendulum wire and its short amplitude of swing were adopted in order to reduce the circular error (as described in equation 1.7) and to thereby eliminate the need for introducing cycloidal cheeks (recall figure 2.7) to ensure an isochronal swing. In addition to its inherent accuracy the characteristics of Hooke's great pendulum was such that it could power a clock on one winding for over a year. Indeed, the very first astronomical research program, by the first Astronomer Royal, John Flamsteed (1646 – 1719), at the newly opened Greenwich Observatory in England was to determine the isochronal spin of the Earth. To demonstrate this property Flamsteed used a pair of especially commission long-pendulum, year-going clocks built along the designs described by Hooke by master-craftsman Thomas Tompion[7].

The foundation stone of the Royal Greenwich Observatory was laid on August 10th, 1675. John Flamsteed, the 'astronomical observer' to King Charles II, drew-up a horoscope for the occasion but qualified it with the comment *Risum teneatis amici* (My friends, can you keep from laughter?). The Greenwich Observatory was constructed with the aim of solving the longitude problem described in Chapter 2, and within a few months of moving in to his new home Flamsteed set about showing that Earth's spin rate (or the Equation of Natural Days as Flamsteed called it) was constant. The twin clocks used to verify the isochronal nature of Earth's spin were commissioned and paid for by Sir Jonas Moore (a close friend and supporter of Flamsteed who otherwise had arguments and vitriolic disagreements with just about everyone else; Newton, Halley, Hooke and the Royal Society included), and they were designed to operate with 13-foot pendulums. The clocks were built into the walls of the Great Room (now known as the Octagon Room), with the pendulums connecting to the clock mechanisms from overhead rather than swinging from below. To monitor Earth's rotation rate Flamsteed made observations of the Sun, to determine the moment of local apparent noon (when the Sun was due South – recall Chapter 2), and he permanently mounted a 6-foot focal length telescope to one of the observatory walls (making it thereby a transit telescope) to specifically monitor the transits of the star Sirius (the brightest star, after the Sun, in our sky). The transit times of Sirius, which is bright enough to be seen during the day, gave Flamsteed a direct measure of the sidereal day. The going of the clocks was generally highly satisfactory, and in spite of numerous teething problems Flamsteed can be found writing to Sir Jonas Moore on March 7th, 1678 with the conclusion that "the clocks have proved that rational conjecture a very truth".

For all practical purposes Flamsteed, with his two long-pendulum, year-going clocks and his observations of the Sun and Sirius transits had shown that the Earth spins at a constant rate at all times of the year. Flamsteed's demonstration of the isochronal spin of the Earth was not, in fact, brought into question until the 1940s when the newly developed quartz driven electrical clocks revealed that small, indeed miniscule fractions of a second, variations in the Earth's spin rate do actually exist. With the isochronal spin of the Earth established to his own satisfaction, however, Flamsteed thereafter embarked upon his key observational programs to map out the heavens and to determine the Moon's motion around the zodiac. Sadly, as a consequence of several very public disputes and a near-complete lack of funding, Flamsteed refused to publish his astronomical observations, and upon his death in 1719 his wife even removed all the instruments (including the Tompian clocks) from Greenwich Observatory, claiming that they had all been paid for out of the family's pocket. The astronomical instruments were never seen again and are now lost. The two clocks, however, have survived to this day; one presently being on display at the British Museum and the other being located at Holkham Hall in Norfolk. Flamsteed's successor as Astronomer Royal, Edmund Halley inherited the mere shell of a building and had to equip the Observatory with new instruments over again.

Continued improvements to the mechanical design and construction of clocks during the 18th century resulted in the eventual abandonment of long-pendulum driven clocks, but while their days of keeping track of the heavens were numbered, the mighty pendulum was destined to return, in the third decade of the 19th century, to prove, once and for all time, that the Earth is most definitely spinning, and the person to make this classic demonstration was the great French experimenter Leon Foucault (1819 – 1868).

The Pendulum at the Panthéon

The demonstration of the Foucault pendulum at the Panthéon in Paris (figure 5.6), in 1851, was nothing less than a brilliant success. It was the biggest show in town, and simply everyone, who was anyone, had to be seen enjoying the spectacle[8]. In a startlingly straightforward fashion a large, but entirely standard simple pendulum, was used to reveal the spin of the Earth; well, if you had an hour or two to spare that is. Copernicus was once again vindicated by actual experiment – albeit some 282 years after the publication of his great thesis. Astronomers had long accepted before Foucault's time that the Earth rotates, but here was the final physical proof of the fact. As with many complicated questions the answer is often simple, provided one looks at the problem in the right way, and this is where the genius of Foucault came into its own.

The Foucault pendulum only makes sense as a demonstration of the spin of the Earth provided its motion is interpreted correctly. Once again, it is a relative motion that is observed and, accordingly the tricky part is to correctly determine which object is actually moving and which is staying still.

The pendulum that Foucault constructed was massive. A one-foot diameter bob made of iron was suspended on a 200-foot long wire from the central dome of the Pantheon. Great care had been taken to ensure that the bob could swing freely without experiencing any external influences, and that the pendulum could be set in motion without receiving any lateral deflection. Once the pendulum was set in motion the bob passed over a slightly raised circular rail that was 12-feet in diameter. A small ridge of sand was built up around the rim of the rail so that a pin attached to the underside of the bob could just scrape the sand away. As the bob passed over the circular rail, the attached pin would leave a telltale groove in the sand ridge indicating the pendulum's plane of motion. After a few minutes, however, it becomes clear that the sand grove is getting wider and after several hours the grove has turned into a very definite gap. To the observer it appears that the plane of the pendulum is rotating in a clockwise direction. What is really happening, however, is that the Earth (and the observer attached to it) is spinning beneath the plane of the pendulum.

Figure 5.6. Foucault's 200-foot (60.96-m) long pendulum in the Panthéon.

By measuring how rapidly the length of the sand furrow on the raised circular rail is growing the apparent rotational period of the pendulum can be found. In Paris, the plane of the pendulum takes about 32-hours to complete one revolution. The seemingly odd 32-hour rotation period is a consequence of the latitude at which the experiment was being performed. Indeed, the full theory behind the pendulum indicates that the observed rotation period P should vary according to the relationship $P_0 / \sin \varphi$, where P_0 is the Earth's sidereal period and φ is the latitude at which the experiment is being conducted.

Foucault's very public demonstration of the spin of the Earth created quite a stir amongst his Parisian audiences, and similar such experiments were being enthusiastically repeated throughout all major cities of the world. Indeed, a pendulum mania swept the globe with savants and lay-scientists a-like performing experiments at many different latitudes[8]. Although small by the standards of Foucault's original experiment the world's biggest Foucualt pendulum is presently located at the Oregon Convention Center in Portland. This particular pendulum is 70-feet (21.3-m) long, has a swing of 15-feet (4.6-m), and its 900 lb (408-kg) bronze-bob takes nearly 34 hours to complete one revolution about its starting axis. Perhaps the most difficult conditions under which the Focault pendulum experiment has been performed, however, are those described by Mike Town, John Bird and Allan Baker working in the harsh conditions at the South Pole[9] during the winter of 2001.

A Technical Aside: The sin(φ) Term

The dependency of the Foucault pendulum's rotation period upon the sine of the observer's latitude is at first glance a little odd. The term, however, has several clear-cut implications. For example, on the equator where the latitude by definition is zero degrees, the rotation period of the pendulum is infinite – in other words it won't deviate from its starting position at all. At either of Earth's poles, however, where the latitude is ninety-degrees, the pendulum's period of rotation is exactly one sidereal day[9]. At latitudes between the equator and the poles, therefore, the rotation period of the pendulum must always be greater than one day.

In order to visualize why it is that the sin(φ) term appears in Foucault's famous equation, first take a look at figure 5.7 (a). In this diagram we have constructed a cone that is coaxial with Earth's spin axis having a surface constructed from the tangents corresponding to all possible observers situated around a constant latitude φ. The radius of the base of the cone is $R \cos \varphi$, where R is Earth's equatorial radius. The length of the side of the cone is $L = R \cot \varphi$, where $\cot \varphi = 1 / \tan \varphi = \cos \varphi / \sin \varphi$ is the cotangent of the angle of latitude. Now, imagine that the cone has been cut along one tangent line and unfurled to produce a flat disk, of radius L, with a triangular wedge cut out of it - this is shown in figure 5.7 (b). The angle V in figure 5.7 (b) corresponds to the sum of all the angles between the adjacent tangent planes touching the Earth along the parallel of latitude φ. The plane of oscillation of the pendulum corresponds to any straight line that is drawn on our flattened out cone, and it is this plane that remains fixed with respect to the stars. If the pendulum is set in motion at point A on the Earth, then it will again be at point A [that is at point A' in figure 5.7 (b)] one sidereal day later, however, the plane of oscillation of the pendulum will not have rotated through 360 degrees (for an observer located on the Earth) until the Earth has rotated through the additional angle of (360 - V) degrees.

Now, the circumference of the circular disk is $2 \pi L = 2 \pi R \cot \varphi$, and the arc length along the constant parallel of latitude is $2 \pi R \cos \varphi$. Given that the Earth spins at a constant rate, so the ratio of times to complete the two arc lengths will be $2 \pi R \cos \varphi / P_0 = 2 \pi L / P = 2 \pi R \cot \varphi / P$, where P_0 corresponds to one sidereal day and P is the rotational period for the plane of the pendulum relative to the Earth. If we now cancel constant terms, we recover Foucault's equation that $P = P_0 / \sin \varphi$.

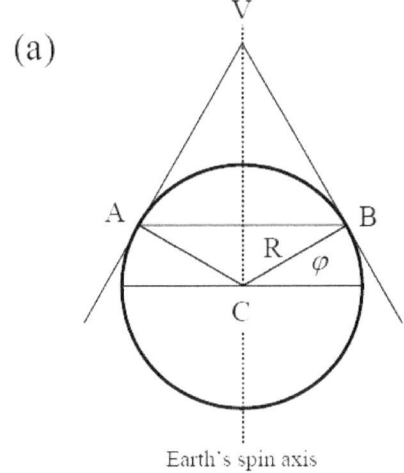

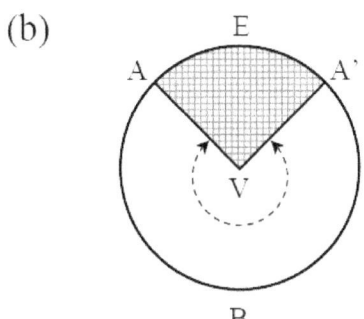

Figure 5.7. (a) The coaxial cone constructed with sides corresponding to the local horizontal planes for all observers at latitude φ. **(b)** The development of the coaxial cone.

The apparent oddity of the pendulum's rotation time is now revealed. Close to Earth's poles the cone automatically flattens down into a near circle and the angle V approaches 360 degrees, with the extra time for the pendulum plane to complete one rotation is reduced to near zero. At the very poles themselves (where φ = +/- 90 degrees) the angle V = 360 degrees and the Earth literally rotates beneath the plane of the pendulum once every sidereal day. Close to the equator, in contrast to the situation at the poles, the angle V becomes smaller and smaller, and the additional rotation time required for the pendulum plane to complete one revolution, relative to an observer on Earth, becomes longer and longer. At the actual equator, the cone degenerates into a cylinder and the period of rotation of the pendulum plane relative to an observer on the Earth becomes infinitely long.

It is perhaps worth noting that whereas the Foucault pendulum demonstration works best at either of Earth's poles, these are the two locations where Newton's falling bullet experiment would be least effective at showing Earth's spin. Alternatively, the amount of the eastward drift in Newton's falling bullet test works best at the equator, where the Foucault pendulum rotation time is infinite and therefore not physically measurable.

The Bravais Pendulum

As a footnote to history, at almost the very same time that Foucault was performing his celebrated pendulum experiments at the Panthéon, so fellow French physicist Auguste Bravais (primarily remembered today for his work on crystal lattice theory) was developing another experiment to demonstrate the spin of the Earth. Bravais, however, based his measurements on the motion of a conical pendulum (recall figure 1.9). Specifically, what Bravais realized was that given that the Earth is spinning then the travel times for the pendulum as it moved from east to west and then from west to east should be different – the difference being related to the angular velocity of the Earth at the location of the experiment $\Omega = \Omega_e \sin \varphi$, where Ω_e is Earth's equatorial angular velocity and φ is the latitude at which the experiment is being conducted. Bravais argued that if the Earth was at rest then the pendulum would oscillate with a constant angular frequency ω_0 (technically, this is the angular frequency in the inertial frame of reference). With the Earth spinning, however, the apparent angular frequency ω with which the pendulum will move should vary according to its direction of motion: when moving west to east $\omega = \omega_0 - \Omega$, and when moving east to west $\omega = \omega_0 + \Omega$. With this realization in place Bravais[10] setup a 10-m long, 6.5-seconds period conical pendulum at the Paris Observatory and in 1851 reported that, "when the pendulum rotates from west to east, the angular velocity at Paris is retarded 11'.43 per second of time; on the other hand, when the motion is from east to west, the velocity is increased by the same amount". While in several important ways the Bravais pendulum offers a better, indeed, more accurate, means of measuring and demonstrating Earth's rotational motion, it lacks any obviously observable effect for the by-stander to see (e.g., there is no rotation of the plane of oscillation to view) – the Foucault pendulum, in contrast, was a large, bold and brassy affair, and as a public viewing spectacle the Bravais pendulum simply could not compete. Not only did the Bravias pendulum experiment fail to catch the public's imagination, it also appears that the scientific community shunned it (for no good reasons) and as far as can be determined very few similar such experiments were ever conducted.

A Centrifugal Correction and Spin Apart

The acceleration experienced by an object in a rotating frame of reference is described in equation (5.1) by a term corresponding to Newton's second law of motion, a Coriolis term and a centripetal acceleration term. For a plumb-line at rest the Coriolis term will be zero and the bob will be influenced by only three forces: the force per unit mass that the tension in the support wire must supply, the force of gravity due to the Earth and the centripetal acceleration term caused by Earth's spin. Accordingly, the equation of motion for the plumb-line is

$$\text{Acceleration} = \text{Tension} + \text{Gravity} - \text{Centripetal} = 0 \qquad (5.2)$$

The acceleration, of course, is zero because the bob on a plumb-line isn't actually moving with respect to the Earth's surface. The tension provided by the plumb-line, therefore, is equal to the reduced gravitational acceleration g_{eff} = Gravity - Centripetal.

At Earth's equator the centripetal force acts vertically in opposition to gravity, and its magnitude is given by the term $C = \omega^2 R$, where ω is Earth's angular spin rate, and R is Earth's equatorial radius. Now, with reference to figure 5.8, the centrifugal force acting at some latitude φ (along Oc in figure 5.8) will be $c = C \cos\varphi$, and the component of the centrifugal acceleration acting vertically against gravity (along OB in figure 5.8) will be $c \cos \varphi$. The net gravitational force acting upon the plumb-line bob will be, therefore, $g_{eff} = g(1 - c/g)$, where $c = \omega^2 R \cos^2\varphi$, and where g is the acceleration due to gravity in the case of there being no rotation (i.e. when $\omega = 0$).

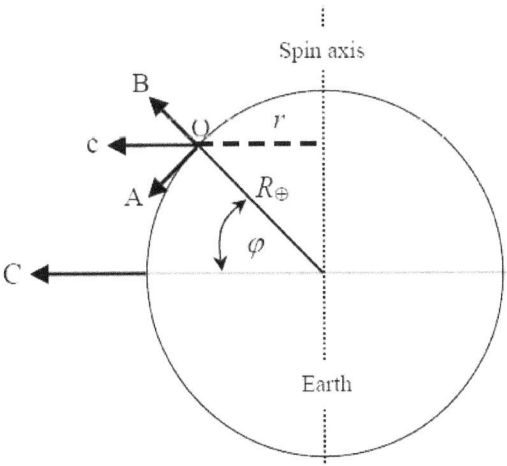

Figure 5.8. Resolving the vertical and horizontal components of the centrifugal terms.

If a pendulum is being used to determine the local effective gravitational acceleration g_{eff} therefore, a centripetal correction term, which is latitude dependent, will need to be applied against the observations. Substituting for ω = 7.272 x 10^{-5} radians per second and R = 6371.0-km, the centripetal correction term comes out to be δ(centripetal) = $\cos^2\varphi$ / (291.175) - this amount being added to g_{eff} to find the actual gravitational attraction due to the Earth alone. The correction is clearly largest at the equator where φ = 0 degrees, and zero at the poles where φ = +/- 90 degrees. At the equator the centripetal correction to any pendulum measurements of gravity amounts to a factor of about 0.3%.

The component of the centripetal acceleration acting at right-angles to gravity (OA in figure 5.8) is given by the term C $\cos\varphi$ $\sin\varphi$ = ½ C $\sin 2\varphi$. Interestingly this term tells us that the surface (or local horizon) of a large body of still water, at latitude φ, is determined by the local gravitational attraction and a centripetal modification term which acts towards the equator. This observation further tells us, perhaps counter-intuitively, that the surface of a still body of water is not actually perpendicular to the local zenith (excepting when the body of water is located exactly on the equator or at either of the poles).

Another consequence of the reduction in the effective gravity by centripetal acceleration is that if the Earth is imagined to be continually spun-up ever faster and faster then a limit will eventually be reached when g_{eff} =0. Once this condition is achieved then any loosely attached matter on Earth's surface will weigh absolutely nothing and be spun-off into space. The condition g_{eff} = 0 is realized when ω_{crit}^2 $R \cos^2\varphi$ = g = $G M$ / R^2. This critical rotation rate is usually expressed in terms of Earth's mean density ρ and accordingly $\omega_{crit} \approx$ (2 / $\cos\varphi$)$\sqrt{G\rho}$, where G is the universal gravitational constant. At the poles, where φ = 90 degrees ω_{crit} is infinitely large, but at the equator, where φ = 0 degrees, ω_{crit} =$2\sqrt{G\rho}$. Using the mean density determined for the Earth by Astronomer Royal, George Biddle Airy (recall Chapter 4) we find ω_{crit} = 1.3 x 10^{-3} radians per second. The critical spin rate is therefore about 18 times faster than that currently experienced by Earth.

As we shall see later on, Earth's rotation rate is actually slowing down as the solar system ages, although it is unclear what Earth's initial rotation rate might have been shortly after it formed some 4.5 billion years ago. Indeed, some astronomers have suggested that when it had finally assembled the Earth's spin rate might have been very close to its critical, fly-apart, limit. That the Earth spins much more slowly now is most probably due to a cataclysmic collision with a Mars-sized planetesimal[11]. This collision not only slowed Earth's rotation rate it also resulted in the formation of the Moon.

The Eötvös Effect: A Final Correction

When pendulum observations of the local gravitational acceleration are made from, say, an ocean-going ship moving at a speed v relative to the Earth's surface in an East-West direction, then a Coriolis correction term must be applied (as well as the centripetal one described earlier) against the observations (recall equation 5.1). The acceleration due to gravity measured from our supposed ship-borne experiment will now be given by the expression $g_{Cor} = g_{eff} - 2\,\omega\,v\,\cos\varphi$, where φ is the (constant) latitude along which the observations are being made. It is the second term in the expression for g_{Cor} that Hungarian born physicist Baron Lorand Eötvös, who we encountered earlier in Chapter 4, first observed and explained in the early 1900s. Indeed, in 1908 an experiment to specifically test his prediction was carried out aboard two ships, one traveling eastward and the other westward, on the Black Sea. The expectation was that since the relative velocity will vary according to whether the ship is moving eastward or westward, so the deduced values for the gravitational acceleration should be different. Just as he had predicted, Eötvös found that the gravitational acceleration measured from the westward moving ship was smaller than that determined from the eastward moving ship. If, for the sake of argument, we imagine that the two research ships, located now for our mathematical convenience at the equator, are both traveling at a speed of 5 knots (≈ 2.6 m/s), then the eastward bound ship will have a velocity $v_E = 465 + 2.6 = 467.6$ m/s, where the 465-m/s term is the equatorial spin velocity of Earth. The westward bound ship will have in contrast a speed $v_W = 465 - 2.6 = 462.4$ m/s. As a result of the two ships having these different velocities with respect to the Earth, the two deduced values for the gravitational acceleration will differ by a term of order $\delta_{Eotvos} = 2\,\omega\,(v_E - v_W) = 7.56 \times 10^{-4}$ m/s^2 which is a small but certainly measurable effect.

With the Eötvös effect described, we may now pull together all the correction terms involved in the determination of the gravitational acceleration at a specific location on Earth with a simple pendulum. The various correction terms discussed in this chapter as well as in Chapters 1, 2 and 3 can be arranged into expressions that relate to the terrain [Bouguer's correction, and possibly a mountain offset correction term], the height/depth [free-air correction], the temperature [table 3.1], the circular correction [equation (1.7)], and the speed of the vessel from which the measurements are being made [the Eötvös effect], as well as latitude dependent terms relating to the oblate shape of the Earth [as expressed in equation (3.1)] and Earth's rotation [the centripetal effect]. Accordingly:

$$g(\varphi) = g(\text{measured}) + \delta(\text{temperature}) + \delta(\text{circle}) + \delta(\text{height}) +$$
$$\delta(\text{mountain}) + \delta(\text{oblate}, \varphi) + \delta(\text{centripetal}, \varphi) +$$
$$\delta(\text{Coriolis}, \varphi) + \delta(\text{Eötvös}, \varphi, v) \qquad (5.3)$$

It is only through the careful consideration of all of these correction terms that geoid maps, such as that shown in figure 3.8 can be constructed, and the multiple number of correction terms that need to be considered in the construction of such maps and models reminds us of the fact that even the simplest of experiments (i.e., the setting up of a pendulum) can have very complex interpretations – in short, we are reminded that life and the interpretation of the world around us is complicated.

The Gyroscope and the Spin That Isn't

The Foucault pendulum experiment while simple in its appearance, is for the layperson far from an obvious demonstration of Earth's rotation – it only makes sense once the audience has been told what is going on, and it also relies on the audience being prepared to accept that the rotation of the pendulum plane is not a result of some slight perturbation induced at the time of setting the pendulum going, or as a result of air currents. Not only this, as British mathematician J. J. Sylvester pointed out in the *London Times* newspaper on 11 April, 1851, for the general observer the notion that the rotation period of the pendulum should be longer than one day (set according to the sine of the latitude term) is a "hard tax upon their faith to believe". Indeed, the British Astronomer Royal, George Biddell Airy, repeatedly dismissed, in his private correspondence with other scientists, Foucault's pendulum experiments as a 'fraud', and argued that while the mathematical analysis was perhaps of some interest, its practical application was nil. Publicly, however, after conducting his own experiments, Airy eventually sided with Foucault's interpretation although he continued to grumble that the "difficulty of starting a free pendulum, so as to make it vibrate at first in a plane, is extremely great"[8]. Criticisms such as those given by Airy can rightly be leveled against any experiment subject to subtle starting and maintenance conditions, and it was with such thoughts in mind that Foucault set about developing a second, more compact, more direct experimental demonstration of Earth's rotation – one, indeed, that did not depend upon the observers latitude. What caught Foucault's attention was the then new experimental device called the gyroscope. In fact, Foucault actually coined the name gyroscope in 1852 through the concatenation of the two Greek words *gyros* (= rotation), and *skopeein* (= to see).

Many readers are probably familiar with toy gyroscopes and their amazing ability of spin to stabalize what are otherwise unstable objects. A typical model gyroscope consists of a heavy fly-wheel disc that is made to spin withinin a surrounding frame. The spin axis of the system passes at right angles through the center of the fly-wheel (along its support axle) and once the fly-wheel is set in motion the orientation of its spin axis remains fixed in space. The orientation of the spin axis remains fixed because of the conservation of angular momentum[12]. This behaviour indicates that provided there is no net twisting effect (or torque) acting upon the gyroscope then the direction in which its spin axis points will not change, and accordingly any observed shift in the direction of the spin axis must be an apparent motion (as with the Focault pendulum experiment) due to the movement of the observer about the fixed rotational plane of the gyroscope.

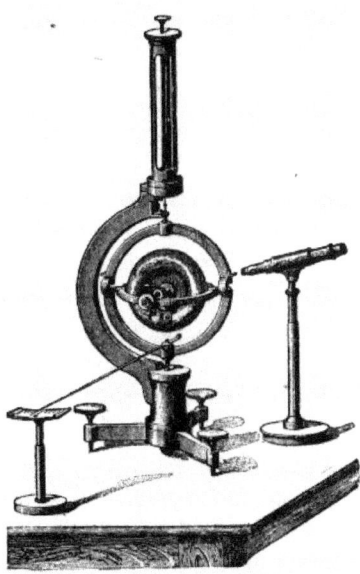

Figure 5.9: Foucault's gyroscope experiment to demonstrate the spin rotation of the Earth. The fly-wheel of the gyroscope is shown as the central 'donut shaped' wheel and it is constrained to move within a low friction gimbaled system. The orientation of the flywheel's spin-axis will remain fixed in space, and accordingly the Earth's rotation will be seen as a motion of the pointer (seen to the lower left) or through the microscope (shown to the right).

Working with instrucment maker Gustave Froment, Foucault constructed an exquistely balanced, gimble mounted gyroscope (figure 5.9), and set its fly-wheel spinning. Although the gyroscope fly-wheel would only spin for about ten-minutes at a time it was enough to reveal the steady rotation of the Earth at a rate corresponding to 360 degrees of rotation per sidereal day. Importantly the apparent rotation direction of the gyroscope's spin axis is the same (that is, counterclockwise) irrespective of the spin direction of the fly wheel.

In the modern era the spin-axis stability properties of the gyroscope are used to control the orientation and stablity of all manner of land, air, space and sea vessels. For our story, however, the gyroscope also solves the mystery of the third spin introduced by Copernicus in 1543 (recall figure 5.1). In order to make sure that Earth's spin axis always pointed in the same direction, that is to the same point on the celestial sphere, as it orbited the Sun, Copernicus introduced an annual rotation term. Treating the Earth as a giant gyroscope, however, we now see that no such annual rotation is required; the orientation of the spin axis is automatically maintained simply because the Earth spins.

Inconstant Spin, and Spin Down

Prior to the mid-nineteenth century it was taken for granted that Earth's spin rate was constant. John Flamsteed, as we saw earlier, had carefully compared the transit times of bright stars against his two long-pendulum, year-going Tompian clocks and had concluded by 1678 that there was no measurable variation in Earth's rotation rate. With time, of course, astronomical measuring techniques as well as clock precision improved and discrepancies in Earth's spin rate were eventually revealed.

The first hints that Earth might not be spinning at a constant rate appeared in the early 1800s and followed from highly accurate observations made of the positions of the Moon and planets. In particular the so-called lunar acceleration was shown to be in part due to variations in universal time (UT – recall Chapter 2). Indeed, the astronomical demonstration (based upon Moon position data) that the Earth must have a non-constant rotation rate only appeared a few years before the first quartz crystal regulated clocks were developed. By the close of the 1940s several observatories had studied transit time variations with state-of-the-art crystal oscillator controlled timing instruments and had successfully shown that Earth's spin rate did vary. Using data gathered between 1943 and 1945, with two intermittently running quartz clocks located at the Potsdam Observatory, Friedrich Pavel and Werner Uhink obtained the first indication that Earth's spin rate varied at the millisecond level per year level. Combining four years worth of data obtained with four continuously running quartz clocks, however, the first definitive demonstration of millisecond rotation rate changes was provided in 1950 by observers working at the Royal Greenwich Observatory in England. By the mid-1950s it had also been shown that there was a seasonal variation in Earth's spin rate, with a gradual increase being apparent from a minimum attained in July to a maximum achieved in January each year.

The eventual verification of changes in Earth's spin rate is perhaps not an especially great surprise – it was really a case of the precision catching up with the observations. With the equipment that was available to him in the 17th century, the ever fastidious Flamsteed stood no chance of detecting time variations as small as a few thousandths of a second, and yet the idea that Earth's rotation rate should be slowing down had been set in place with the publication, in 1687 of Newton's *Principia*. Specifically Newton had explained the fall and rise of the oceans tides through the gravitational pull of the Sun and Moon. The rhythmical rising and falling of the great ocean masses should accordingly result in the gradual reduction in Earth's spin rate. Modern-day observations with atomic clocks have fully confirmed this prediction (and more, as we shall see shortly). Indeed, atomic clock based observations reveal a fortnightly variation in Earth's spin rate that is modulated by the Moon moving between perigee and apogee within its orbit. The deduced long-term trend in the slowing down of Earth's spin rate amounts to about 2 milliseconds per century, although much larger, shorter-timescale fluctuations are also observed.

The steady tidally driven reduction in Earth's spin rate, which can alternatively be expressed as a steady increase in the length of day (LOD), is confirmed by the study of ancient eclipse times. Indeed, it is now clear that the lunar and solar eclipses observed by the ancient Babylonians, some two thousand years ago, systematically occurred five to six hours earlier than would be expected if Earth's spin rate had been constant and maintained at its present value.

In addition to the long-term near linear increase in the LOD by about 2-ms/cy, a whole host of other fluctuations in Earth's spin rate are now known. There are irregular fluctuations of 4 to 5-ms on decadal timescales, as well as semi-regular variations on timescales of about 5 years. The decadal variations are believed to be caused by angular momentum transfer between Earth's inner core and the solid mantle – when the mantle gains angular momentum the spin rate necessarily increases and the LOD must accordingly decrease. The shorter-term variations relate to what is known as polar motion, which is in turn comprised of an annual component and a 14-month duration, so-called, Chandler wobble[13]. The latter wobble is believed to result from variations in the mass distribution of Earth's oceans. Additional short-term variations relate to the gravitational production of deformations in the Earth's mantle by the Moon, and seasonal variations in the mass distribution of the atmosphere. In the modern era variations in the LOD and polar wonder are continuously monitored by the International Earth Rotation and Reference System Service (IERS), located at the Observatoire de Paris in France.

Moon Drift and the End of All Things

It is a well-known principle that there is no such thing as a free lunch and that change is always a process of give and take. That Earth's rotation rate is slowing down (or alternatively stated the LOD is steadily increasing) as a result, at least in major part, because of a gravitational interaction with the Moon, then some compensatory lunar effect must be taking place. Indeed, the inescapable consequence of the conservation of angular momentum[12] dictates that as Earth slowly spins down so the distance to the Moon must increase in compensation. That the Moon's mean distance from Earth is increasing was confirmed by the laser ranging experiments conducted as part of the Apollo Moon landing missions in the 1970s (recall Chapter 3), and the present rate of increase amounts to about 3.8 cm per year. The Moon is getting further and further away from us; in fact, at about the same rate as our fingernails grow.

That the Moon's orbital distance is increasing has a number of interesting consequences, and at least one potentially deadly outcome. As the distance to the Moon increases so its gravitational interaction with the Earth will become weaker and weaker resulting, eventually, in the breakdown of the synchronization between its spin and orbital period. Our very distant descendents might accordingly live to see the entire Moon's surface from Earth, rather than just the one hemisphere that we see now. In addition, as the Moon's distance increases so the spin-down effect it induces will weaken and the LOD will begin to stabilize. Perhaps more alarmingly for the future of life on Earth, however, is the fact that as the Moon's gravitational influence wanes, so too will its stabilizing affect upon Earth's equatorial bulge and the orientation of its spin-axis. Detailed numerical studies[14] have shown that without the Moon's gravitational influence the direction in which Earth's spin axis points (i.e., the angle of the obliquity of the ecliptic) can change considerably. Rather than being constrained to vary within just a few degrees, as at present, it will be free to move through many (perhaps 10 or so) degrees, on timescales of perhaps a few millions of years. Such dramatic changes in the obliquity will potentially wreak climatic havoc, with regions that were once situated in temperate zones finding themselves in near artic domains. The concomitant consequences for large swaths of life on Earth would be devastating – animals and plants that were well adapted to one region will not likely survive in another, and the timescale for evolutionary adaptation will be relatively short. There is little doubt, however, that life will continue to survive and thrive in one form or another - even when the pendulum Earth is no longer Moon stuck. The end of the biosphere, when all surface life will assuredly die off, will not be caused by a retreating Moon, but rather by an ever brightening Sun[15], and this inevitable heat-death is set to occur some two to three billion years from the present. Indeed, the continued warming, eventual evaporation and loss of the oceans through a moist greenhouse phase will bring to a close the 'erratic pendulum' that is complex life, and the multitudinous microbes, buried deep under ground, will once again inherit the Earth. Even the reign of the microbes will not last for ever, however, and as the Sun swells dramatically in size during its final giant phase, some eight billion years from the present, so planet Earth will be consumed within its fiery outer layers. This, of course, is exactly how it should be with respect to our literary pendulum metaphor, with life, once given the impetus of birth, being subjected to the vigorous swings and roundabouts of evolution, eventually coming into decay - the pendulum-like to and fro of existence being brought to a timeless standstill.

The Inverted Pendulum

Within the pages of his *Natural History*, an encyclopedic tome written in the first century A.D. by Gaius Plinius Secundus (Pliny the Elder: 23 – 79 A.D), it is explained that the Moon can induce insanity and a state of mental unrest when it reaches its most resplendent and full phase. This notion appears to trace back to at least Aristotle (4th Century B.C) who argued that since the brain, the center of all emotions, is mostly made of water, so the Moon should affect its function and state of stability, just as it affects the tides in the oceans. The idea of lunacy is as old as philosophical reasoning itself, and indeed it is a lunatic fringe idea – there is absolutely no evidence, counter to numerous urban myths, to indicate that an increase in states of mental disorder, crimes and/or homicides occur when the Moon is full. Using the term lunacy in its more derogatory 'to be an idiot' form, however, it would appear, at first glance, that one would indeed, be a lunatic to contemplate the construction of an inverted (or vertical) pendulum – at least, that is, in the simple form described in figure 1.1. Strings and thin wires, to begin with, have virtually no compressional properties that might allow them to support a bob located above the attachment point[16].

For all of its apparent contradictions, the vertical or inverted pendulum, as we shall see, has a number of interesting properties. To make such a pendulum one must first replace the support wire or string of the simple pendulum by a thin lightweight bar – this will provide the resistance against compression needed to support the vertically lofted bob. Secondly, and this is the more interesting point, one must find a way of stabilizing, even controlling, the motion of the bob once it has been released.

For the vertical pendulum (perhaps better described as the inverted plumb-line) supported loosely at one end, there are just two equilibrium locations: one with the pendulum bar pointing straight up and the other with the pendulum bar pointing straight down. Only the latter of these two equilibrium points is stable, however, with even the very smallest of motions destabilizing the bar when set in its upright position. The situation is similar to that of trying to balance a pencil upon its sharpened end. For a few brief seconds it might seem that a balance has been successfully achieved, but it will soon fall away from the upright orientation, finding a more stable balance position, resting lengthwise upon the supporting table. The upright equilibrium position of the pencil, as well as that of the vertical pendulum, teeters on the very edge of instability – only the very freezing of time itself allowing stability to be maintained. At issue, of course, is the fact that there is no restoring force to keep the vertical pendulum vertical – even the very smallest of perturbations sets in motion an unstoppable avalanche of change, driving the bar to its lower stable equilibrium point. The upper equilibrium positions of the balanced pencil and the loosely supported vertical pendulum are both subject to the destabilizing control of the butterfly effect – a subject that we shall pick up upon again in Chapter 6.

The equation of motion for the loosely supported inverted pendulum is very similar to that for the simple pendulum shown in figure 1.1, except in the case that the angle position θ is now measured from the vertical above the pivot point. Accordingly we have an equation of motion given by $\ddot{\theta} - (g/L)\sin\theta = 0$. To see that this configuration must result in an unstable equilibrium, we can use the small angle (in radians) approximation $\sin\theta \approx \theta$ and then multiply the equation through by $\dot{\theta}$. Solving now for θ reveals that: $\theta = \theta_0 \exp(t/\tau)$, where $\tau = \sqrt{L/g}$, and we see that for any small initial displacement $\theta_0 > 0$ (our butterfly wing beat) that the angular acceleration is positive, and that the angular displacement grows exponentially with time – we also see that since τ is directly proportional to the square root of the length of the pendulum bar L, so tall inverted pendulums fall more slowly than short ones. In terms of recreational activity, this latter point explains why it is easier to balance, say, a broom-handle upon the end of ones finger, than, say, a small pencil since the response time required to adjust for the pencil's movement is very much shorter than that required to compensate for the broom-handle's motion.

Generally speaking, the loosely supported inverted pendulum provides us with very little of actual interest. It has one stable and one unstable position. It is either up, temporarily, or it is down permanently – and that's it. The vertical pendulum can be transformed into something much more interesting, however, if it is allowed to have an active support – as in the case of the recreational balancing just mentioned. Indeed, circus performers, tightrope walkers and riders of unicycles and Segway PTs have long appreciated this stabilizing trick. By purposefully allowing the pivot point of the inverted pendulum to move appropriately the perturbations that it provides will miraculously transform the unstable, vertical equilibrium state of the pendulum into a stable one – motion at or about the pivot point effectively counteracting the instability caused by the perturbing butterfly effect. There are three ways in which an inverted pendulum can be stabilized. The first method allows for the horizontal back and forth sliding motion of the pivot point, the second allows for vertical up and down perturbations of the pivot point while the third introduces an active rotational torque at the pivot point to counter the pendulum's motion[17]. Methods one and three require active feedback mechanisms to control the pendulum's motion, and they are now classical demonstration experiments in engineering control theory. The second method is slightly more intriguing, however, in that the system becomes autonomously stable. For the vertically oscillating pivot point the equation of motion is written as

$$\ddot{\theta} - \left(\frac{g}{L} - \frac{A}{L}\omega^2 \sin\omega t \right)\sin\theta = 0 \qquad (5.4)$$

where it has been assumed that the pivot point undergoes simple harmonic motion with amplitude A and frequency ω. Equation (5.4) has no simple analytic solution, but it can be studied under various limiting conditions. One approximation, in which the pivot oscillates rapidly, was that studied[18] by the Soviet physicist Piotr Kapitza in 1951. Under the condition that $\omega \to \infty$, with $A\omega$ remaining finite (that is the pivot point undergoes very rapid but relatively small amplitude oscillations), so equation (5.4) becomes

$$\ddot{\theta} - \left(\frac{g}{L} - \tfrac{1}{2} \left(\frac{A}{L} \omega \right)^2 \cos\theta \right) \sin\theta = 0 \qquad (5.5)$$

Equation (5.5) indicates how the vertical position ($\theta \approx 0$) of the inverted pendulum can be made stable in the sense that $A\omega$ is essentially engineered so that $A\omega = \sqrt{2gL}$. Under this constraint the angular acceleration is zero and the bar simply oscillates, up and down, in the vertical position. Remarkably, even when the inverted pendulum is displaced away from the vertical position, stable back and forth oscillations can be engineered by the appropriate adjustment of A and ω. The inverted pendulum in this latter situation is the genuine, vertically flipped counterpart of the simple pendulum[19].

While it might seem remarkable enough that a pendulum can be made to stand vertically, it further transpires that the whole process can be extended to make a multiply-linked pendulum stand vertically. This generalized n-link vertical pendulum result was first studied[20] in the early 1990s by David Acheson (Oxford University) and Tom Mullin (University of Manchester). Interestingly, the new results by Acheson and Mullin build upon ideas first discussed by Christiaan Huygens in his *Horologium Oscillatorium* (published in 1673). Specifically Huygens suggested that a vibrating musical string might be modeled as a series of linked masses with each mass undergoing simple harmonic motion. Huygens idea was taken-up by Swiss mathematician Johann Bernoulli (1667 – 1748) in 1727, when he formally studied the mathematical properties of a string loaded with equally spaced, equal mass beads. It was Johann's son, Daniel, however, who in 1733 really got to grips with the problem, and showed that an n-linked vertically suspended chain could under go n different patterns of motion each of which has its own specific frequency of oscillation. In the lowest frequency f_L mode the entire set of links swings as if it is one simple pendulum; the highest frequency f_H mode, however, corresponds to the situation where every link is moving in the opposite direction to its two end neighbors – like a stylized zigzag streak of lightning. While Bernoulli considered the chain to be hanging vertically downwards, the remarkable result found by David Acheson in 1993 was that the linkages behave in exactly the same way when standing up-right as they do when hanging downward. Furthermore, to stabilize the vertical n-linked pendulum, Acheson was able to show that the displacement D of the base support must satisfy the condition $D < 0.023g / f_H^2$, where g is the acceleration due to gravity. Since f_H is typically very large, the displacement D is generally going to be very small. In addition, however, vertically stability also requires that the frequency f with which the base plate oscillates must satisfy the condition $f > 0.072g /(f_L D)$. These two conditional results reveal that to make a long chain stand vertically upright the base plate must undergo very rapid, but very small amplitude oscillations. At first glance it might be thought that the proof of stability for an n-linked vertical pendulum might offer some hope for perfecting the *magic* behind the Indian rope trick[19]. The problem one encounters, however, is that a rope, perhaps counter to expectation, cannot be directly approximated as a multiply-linked chain of pendulums (rope is made of flexible intertwined strands rather than simply connected individual links). Indeed, the mathematical constraints determined by Acheson indicate that as n becomes very large (conditions that make a chain something like a flexible rope) so the base plate must oscillate very, very rapidly and the displacement must be very, very small. The *magic* behind the working of the Indian rope trick, it would appear, is to make the rope's base plate oscillate infinitely fast while not actually moving!

The stabilization of the n-linked vertical pendulum by the rapid oscillation of its base plate is an autonomous process – once the plate has the right oscillatory characteristics stability is automatically achieved. The active torque vertical pendulum, however, applies a more directed approach to the problem in that each joint in an n-linked chain is controlled via some form of motor. This approach has obvious computer control characteristics and such systems are seen in robotic arms, such as the Canada Arm on the International Space Station, and within the animal kingdom – the human hand and arm, for example, can be thought of as a 3-linked pendulum composed of the humerus, radius (and associated ulna) and the hand. The control and stability of the arm and hand is further maintained by active muscular control, and sensory feedback, about the various bone joints. The simplest active torque control system has just one link ($n = 1$) and its equation of motion is easily obtained as

$$mL^2\ddot{\theta} - mgL\sin\theta = T \qquad (5.6)$$

where m and L are the mass and length of the pendulum link (the mL^2 term is the link's moment of inertia), and T is the torque applied at the pivot point. In principle this pendulum configuration can be stabilized at any angle θ – one simply has to apply the appropriate motor torque at the pivot point such that the arm moves from some initial angle θ_0 to some final angle θ in an allocated time interval Δt. As we would expect, however, if the torque is set to $T = 0$, so equation (5.6) reduces to that for the unstable, freely articulated simple pendulum.

In addition to the application of an active torque and/or linear displacement about the pivot point, the inverted pendulum can also be stabilized through rotation of the bob. In this particular device the pendulum bar is free to rotate about the pivot point but the bob is replaced by a spinning disk (see figure 5.10a). This form of inertial wheel pendulum was first developed and described by Mark Spong (University of Texas at Dallas) and co-workers[21] in 2001. The pendulum's equation of motion is written as

$$M_T \ddot{\theta} + M_D \ddot{\eta} - \overline{m} g \sin \theta = 0 \qquad (5.7)$$

where M_T is the moment of inertia of the entire pendulum, M_D is the moment of inertia of the rotating disk, and $\overline{m} = m_B L_{CM} + m_D L$ with m_B and m_D being the masses of the bar and the rotating disk, L_{CM} is the distance from the pivot point to the systems center of mass, and L is the length of the bar. In addition to equation (5.7) the system must also satisfy the coupled equation: $M_D (\ddot{\theta} + \ddot{\eta}) = T$, where T is the torque provided by the motor to drive the rotating disk. Again, there is no simple analytic solution to equation (5.7), but detailed numerical integration of the equations (and also through experimentation with physical models) reveal that by varying the speed at which the disk rotates (that is by controlling the torque T) so the position angle of the pendulum relative to the vertical can be controlled, and indeed, the bar can be made to stand vertically, with the rotating disk being placed directly above the pivot point.

Rather than replacing the pendulum bob with a rotating disk in order to open-up the vertical stable state, Katsuhisa Furuta (Tokyo Institute of Technology) and colleagues[22] developed an inverted pendulum device attached to a rotating disk or radius arm (figure 5.10b). Since being introduced in 1992, the Furuta pendulum has inspired the generation of a vast number of research related articles and engineering projects – the key issue, once again, being the dynamical control of the rotating platform so that the pendulum bar, freely articulated about its pivot point, can remain in a stable up-right position.

The inertial wheel, inverted and Furuta pendulums bring us to a new landscape where the simple pendulum is no longer the obvious paradigm, and the system dynamics are non-linear in their responses to input variations. Such pendulums require active control and continuously working feedback systems if any semblance of stability is to be maintained. They are pendulums located on the border of chaos – the boundary beyond which predictability fails us, and where the butterfly effect reigns supreme. All is not lost, howev-

er, and as we now enter Chapter 6, in order to explore the world of unpredictability and chaotic motion, our progress will continue to be guided by the
steady and comforting swing of the mighty pendulum.

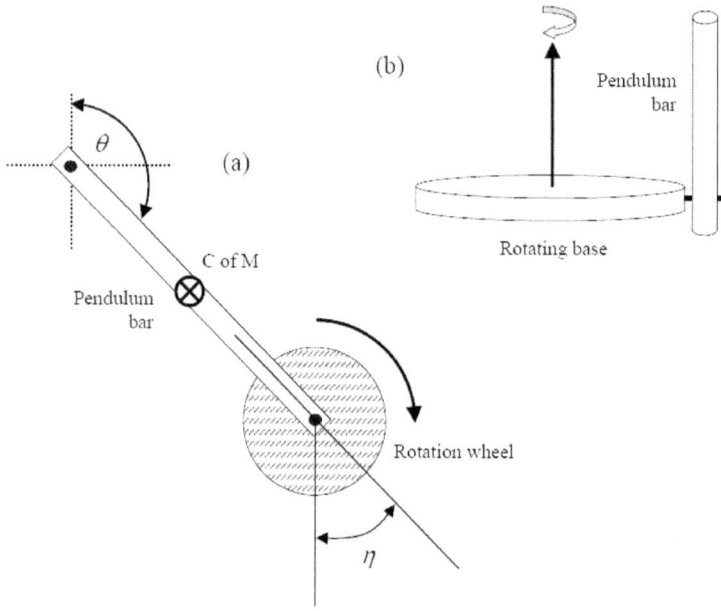

Figure 5.10. (a) The inertial wheel, and (b) Furuta rotational pendulums.

CHAPTER 6

FROM BEAUTY, TO RESONANCE, TO CHAOS

Whence is it that Nature does nothing in vain;
and when arises all that Order and Beauty
which we see in the world.

Optiks, or a treatise of the reflexions, refractions, inflections and colours of light (1704)
Isaac Newton

The Harmonograph

Art and science often merge to produce phenomena of great inherent beauty. One happy example of such synergy concerns the harmonograph, a simple pendulum device that can produce an endless variety of beautiful patterns and curves[1]. The simplest harmonograph is a pendulum suspended in such a way that it is free to swing in any direction – just like Foucault's pendulum. Indeed, the harmonograph is to some extent a deliberately spoiled Foucault pendulum experiment in which the bob is set in motion with both a lateral as well as a radial push. Under these circumstances the pendulum will swing backwards and forwards making graceful arcs and loops as it proceeds. Figure 6.1 shows the mechanical arrangement for a typical harmonograph device.

The twin elliptical pendulum illustrated in figure 6.1 brings together the combined motions of two conical pendulums. The table, upon which the harmonic figure will be produced, is attached to the upper end of the main pendulum rod, and this pendulum is further supported in a gimbaled mount that allows it to move freely in any direction. A second deflector pendulum is attached below the main pendulum bob via a length of thin wire. The pen arm is allowed to move up and down, but must at all times remain in contact with the table that is attached to the main pendulum. By varying the direction of rotation of each pendulum as well as the length of the wire supporting the deflector pendulum, an infinite variety of beautifully complex figures can be produced. Indeed, it is remarkable how by simply combining the output of two pendulums that so much variety can come about (as discussed earlier in Chapter 1). If we think of the table upon which the harmonograph design is being produced (figure 6.1) as a Cartesian (X, Y) graph, then the essential motion that is being mapped out is the sum of four sine terms, each with a different amplitude, angular frequency and phase. Coupled to this, since we are also dealing with a mechanical device there will also be a damping term associated with each sine term resulting in the point (0, 0) being the harmonograph's final resting location. Figure 6.2 shows a simulated harmonograph design, revealing at once the simplicity of the equations, the complexity of the resulting design and a picture of some inherent graceful beauty.

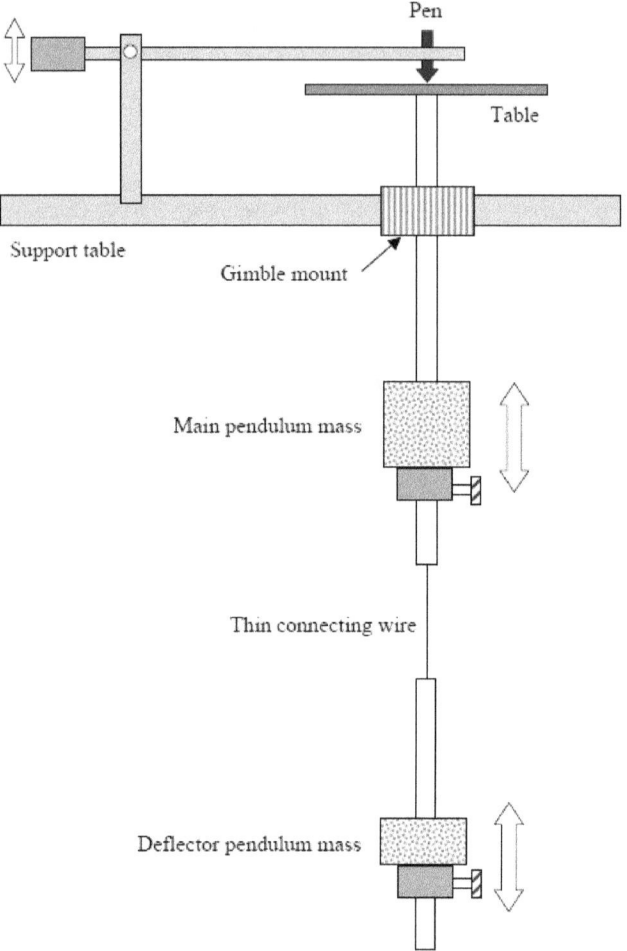

Figure 6.1. A twin elliptical pendulum harmonograph.

The harmonograph is the twin pendulum set free. The steady and predict-able paths of the simple and conical pendulums are entirely overwhelmed within the harmonograph's domain, with the highly restricted order of the arc and circle being replaced by the tangled grandeur of numerous foliated and intersecting curves. Indeed, the harmonograph reminds us that order is a restricted, rare and even fragile state of existence. For all of the inherent beauty within the harmonograph designs they are the veritable parable of the frightening infinities that so scared French mathematician and philosopher Blaise Pascal – the designs literally straddle the boundary that separates har-monious order from disharmonious randomness and chaos. There are no (presently) known physical applications for the harmonograph and it illustrates no specific fundamental phenomena found in nature. The harmonograph, however, is a wonderful invention and the source of an infinite variety of visually inspiring figures – with the harmonograph the pendulum simply writes and having writ moves on.

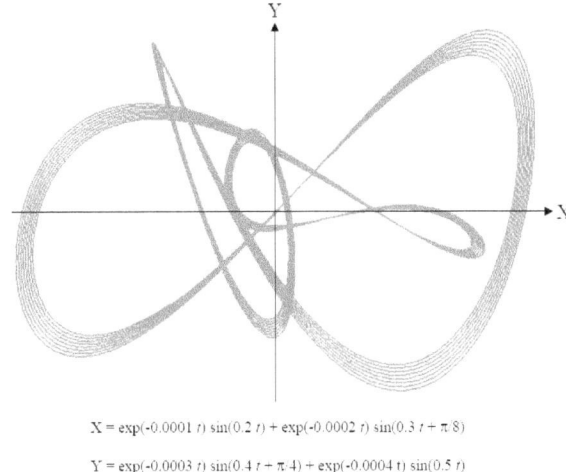

$$X = \exp(-0.0001\ t)\ \sin(0.2\ t) + \exp(-0.0002\ t)\ \sin(0.3\ t + \pi/8)$$

$$Y = \exp(-0.0003\ t)\ \sin(0.4\ t - \pi/4) + \exp(-0.0004\ t)\ \sin(0.5\ t)$$

Figure 6.2. A simulated harmonograph design. The design assumes that each coordinate is the sum of two sine terms, one for each pendulum, the amplitudes of which decay over time. The specific equations for the X and Y coordinates used to generate the figure are shown at the bottom of the diagram – the exponential terms are included to simulate frictional decay.

Lissajous Figures

The mathematics behind the ordered, that is, the periodic and repeating curves produced by a harmonograph are described in terms of so-called Lissajous figures. Jules Antoine Lissajous was a talented French mathematician, and the figures that now bear his name were first used to study sound waves in 1857. Lissajous experiments combined the use of a narrow beam of light, and two mirrors each one of which being attached to a tuning fork. The light beam was directed to strike the mirror attached to the first tuning fork which was mounted vertically, and from there the beam was reflected on to the mirror attached to the second tuning fork mounted horizontally. After reflecting off the surface of the second mirror the light beam was allowed to hit a screen (or viewing eyepiece). If neither tuning fork is in motion then the light beam will produce a small bright dot on the screen. Once the tuning forks are set in motion, however, the light beam traces out an intricate two-dimensional curve, or Lissajous figure. By choosing various combinations of differing frequency tuning forks Lissajous was able to study musical harmonies and harmonics visually – indeed, Lissajous had literally turned sound into a dancing geometry of light, and for this advancement he was duly awarded the prestigious Lacaze Prize of the French Academy of Sciences in 1873.

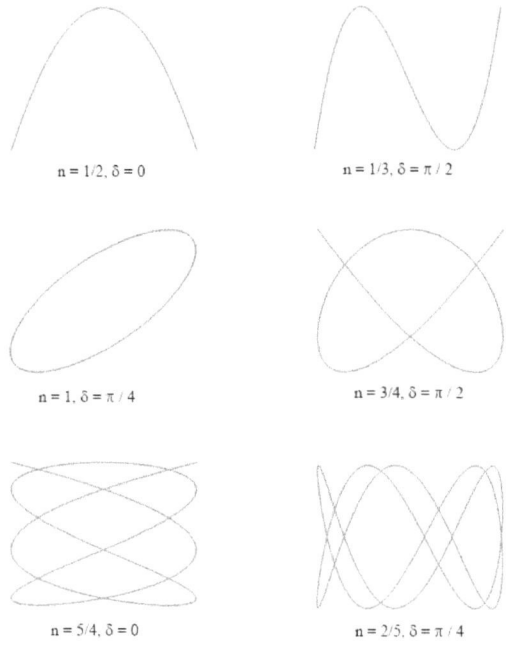

Figure 6.3. A selection of Lissajous figures: $A=B=1$; $a=n$, $b=1$.

Working in 1815, American mathematician Nathaniel Bowditch first studied the mathematical equations that describe the form of Lissajous figures[2]. Since we can think of Lissajous figures as resulting from the combined motion of two coupled pendulums, one swinging horizontally in the X-direction, and the other swinging vertically in the Y-direction, the figures are mapped-out over a rectangular grid of width $2A$ and height $2B$, where A and B are the amplitudes of swing associated with the generating pendulums. The location of each (X, Y) point on the Lissajous curve, situated within the rectangular grid, can be expressed as a function of time t through the parametric equations $X = A \sin(a\, t + \delta)$ and $Y = B \sin(b\, t)$, where a and b are constants that define the angular frequency of each pendulum. The term δ is an additional constant, or phase offset term, that allows the two pendulums to swing as if governed by different time streams. By studying the characteristics of the parametric equations, Bowditch found that a number of special case solutions existed. For example, if the curve is to be closed, that is periodically retrace its path, then the ratio of the angular frequencies a / b must be rational – that is expressed as a ratio of two integers. If, $a / b = A / B = 1$, then when $\delta =$

π / 2, the resultant Lissajous figure will be a circle. On the other hand, if a and b are not equal integers and $\delta = \pi$ / 2 then an ellipse will be produced. If the phase offset $\delta = 0$ and a / b is rational then the Lissajous figure becomes a straight line. A selection of Lissajous figures are shown in figure 6.3.

Dean's Two-Point Pendulum
The inspiration behind Bowditch's analysis on the motion of coupled pendulums was a paper published by James Dean[3] (University of Vermont) in 1815. As an intriguing thought exercise, Dean was interested in picturing how the Earth would appear to move across the sky for an observer located on the Moon. Specifically, Dean wanted to understand the effect that the Moon's librations would have upon the appearance of Earth's path. That the Moon's 'face' undergoes a small back and forth oscillation (libration) has been known since antiquity, and it accounts for the fact that more than half of the Moon's surface can be seen and mapped from the Earth. Dean suggested that a pendulum experiment could be used to illustrate the effect of the Moons libration on the observed path for the Earth. Rather than being a simple pendulum arrangement, however, Dean's pendulum has two attachment points (see figure 6.4).

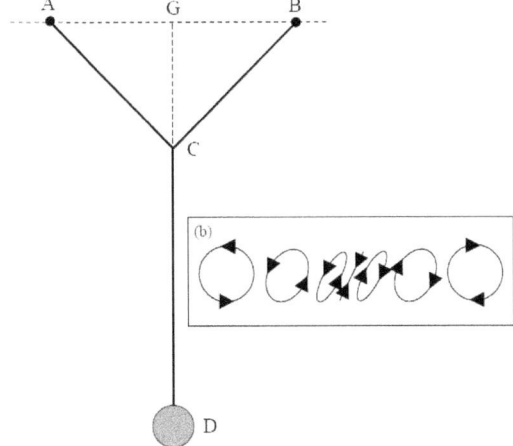

Figure 6.4. Dean's two-point pendulum. The pendulum string is fixed at points A and B, and joined at point C. By adjusting the length DC = r compared to DG = R various elliptical Lissajous or Bowditch figures can be produced. Inset (b): the sequence of oscillation shown by the pendulum, going from circular motion, to a reducing elliptical path in one direction to a reversal and expansion back out to a circular path moving in the opposite direction to the initial motion.

It is well worth the exercise of constructing a home-version of Dean's two-point pendulum since its motion is, to say the least, rather startling. Begin by releasing the pendulum in some diagonal direction. After a short while the path will change from that of a line, to that of an ellipse and eventually to that of a circle. The sequence will then reverse, changing from a circle, to an ellipse and back to a straight line again – remarkably as the sequence repeats its straight line segment the pendulum's sense of rotation switches; clockwise rotation changing to counter-clockwise rotation and *visa versa*, as illustrated in figure 6.4 inset (b).

Nathanial Bowditch, who seemed much more interested in Dean's pendulum than the actual motion of the Earth as seen from the Moon, developed a set of equations to describe the bob's motion. For small amplitude variations the (x, y) motion of the pendulum can be described by two periodic terms: $x(t) = \cos(\omega_1 t + \varphi_1)$ and $y(t) = \cos(\omega_2 t + \varphi_2)$, where it has been assumed that the amplitudes of motion in the x and y directions are scaled to unity, and where, with reference to figure (6.4), $\omega_1^2 = g/r$ and $\omega_2^2 = g/R$, with g being the acceleration due to gravity, and φ_1 and φ_2 being arbitrary phase shifts. In order to explain the observed motion of the pendulum, Bowditch introduced a time dependent phase term $h(t) = (\omega_1 - \omega_2)t + \vartheta$, which then enables the pendulum orbit to be described as $y = x\cos(h) + \sqrt{1-x^2}\sin(h)$. Essentially, what Bowditch is saying here is that the motion of Dean's pendulum can be though of in terms of adding together the motion of two *imaginary* pendulums, one swinging in the x direction and the other swinging in the y direction. The two imaginary pendulums swing backwards and forwards with constant period ($P_1 = 2\pi / \omega_1$, $P_2 = 2\pi / \omega_2$), but the phase of one pendulum slowly changes with respect to the other, with the phase change itself being periodic. In the situation where $\omega_1 = \omega_2$ and $\theta = 0$ the equation for the orbit, as would be expected, reduces to that of a circle: $x^2 + y^2 = 1$. Otherwise, with $\omega_1 > \omega_2$ the orbit shows the entire range of motion accessible to the pendulum, repeating with a period $P_h = 2\pi/(\omega_1 - \omega_2)$. In the case of the Earth's motion as seen from the Moon, P_1 and P_2 will be of order the Moon's sidereal period of 27.322 days, while the repeat period for the full motion cycle P_h is about six years.

Huygens's Odd Sympathy and the Spark-Gap Oscillator

Sometimes one simply has to be in the right place at the right time to notice an unexpected phenomenon. For pioneering Dutch horologist Christiaan Huygens, the time was February 1665 and the place was ill in his bed. Nursing a brief illness Huygens was in the right place (and presumably with time on his hands) to notice something intriguing about two pendulum clocks located upon a common shelf. Specifically he observed that the pendulums always ended-up swinging exactly opposite to each other – that is the swings were 180-degrees out of phase. No matter how he stopped and/or started the clocks their pendulum's eventually settled into an anti-phase configuration of swings. In Chapter 2 we encountered Huygens with respect to the failed sea trials of his pendulum clocks conducted in 1662 – the incessant rolling of a ship being just too much for the establishment of a constant going. Presumably, with these sea trial failures relatively fresh in his mind Huygens realized that the sympathetic, anti-phase swinging of his bedroom pendulums potentially opened-up new possibilities for stabilizing ocean-going clocks.

Writing to the newly founded Royal Society of London, Huygens described[4] his observation of the anti-phase pendulum oscillations, and while revealing it as "an odd kind of sympathy", he noted that perhaps such a system of coupled clocks could be constructed for sea-going ships. The idea essentially being that the two clocks, through their common "sympathy", might be arranged so as to compensate each other against the rolling of the ship and any associated variable going. Remarkably, for so it would appear that while pendulums cannot talk, they can communicate their presence to each other.

Following a series of experiments Huygens concluded that the anti-phase "sympathy" of his two pendulum clocks came about because of "imperceptible movements" established in the beam upon which they were supported. With this explanation, which is indeed correct, the anti-phase locking of the pendulum motion is an example of a restricted Barton's pendulum (see below) – restricted, that is, in the sense that only two pendulums are involved and that they have the same natural frequency of oscillation. More than this, however, it turns out that by adjusting the mass of the beam that connects the two clocks additional sympathetic motions can be generated. Remarkably, this latter result was only established at the turn of the 21st century, some 337 years after Huygens first wrote of his observations to the Royal Society.

Working with a pair of specially designed pendulum clocks researchers[5] at the Center for Nonlinear Science at the Georgia Institute of Technology studied the array of possible system behaviors and their relationship with the mass of the support beam. If the connecting beam was too light, then its backward and forward motion was able to rob the pendulums of their energy, resulting in what the researchers called an amplitude death mode – both clocks coming to a stop. In contrast, if the connecting beam was too massive then it was unable to transmit enough energy for synchronization of the pendulums to occur. For beam masses in between the amplitude death and no synchronization modes, however, it was possible to establish both in-phase and anti-phase synchronization of the pendulums. In-phase motion, however, soon resulted in the pendulums coming to a rest, the associated (also in-phase) lateral motion induced within the connecting beam robbing them of their energy. The Georgia Institute of Technology researchers found, just as Huygens had reported in 1665, that anti-phase synchronization of the pendulums was much longer lived than in-phase oscillations. In this latter situation, it appears that the lateral motion is effectively reduced to zero and the pendulums are accordingly allowed to oscillate without loss of energy to the connecting beam.

A hybrid demonstration devices[6], set somewhere between Dean's two-point pendulum and Huygens (Georgia Tech) double beam pendulum, was developed by Edwin Barton and Mary Browning in 1917 (figure 6.5). Interestingly, this double-cord pendulum was developed to illustrate the "phenomenon of coupled electric circuits" and specifically, the spark-gap oscillator. The Barton-Browning double-cord pendulum essentially consists of two Dean's two-point pendulums, setup side by side, with a stiff connecting rod joining the two central suspension points. The system is set in motion by pulling just one of the suspended pendulum bobs to one side (the other pendulum is left at rest). As the first pendulum swings back and forth, so the second pendulum begins to pick-up a sympathetic vibration, slowly sapping the first pendulum of its energy. After a while the first pendulum comes to a stop leaving just the second (originally stationary) pendulum swinging. The situation now reverses, with the stopped pendulum gradually increasing its amplitude of motion at the expense of the moving pendulum, which concomitantly slows to a stop. This pendulum literally oscillates between which pendulum is being excited and which pendulum is being de-excited. The phenomenon that the double-cord pendulum illustrates is a special kind of resonance motion, but to illustrate how resonances work Barton and Browning developed a new kind of multiply connected pendulum device.

Barton's Resonant Pendulum, Kirkwood Gaps and the Lorentz Atom

The harmonograph is an example of a coupled pendulum, with the motion of one pendulum being mechanically coupled to another. Likewise, Newton's Cradle, described in Chapter 1, also constitutes a set of identical pendulums that are forced to interact via direct collisions. Barton's pendulum, in contrast to the harmonograph and Newton's Cradle, however, provides us with a different and more subtle kind of multiple pendulum interaction. In this case the interaction is that of resonance excitation, with one pendulum acting to drive the motion of another. To use, albeit in an overly maffick sense, the words of 13th Century crusade leader Arnaud Amalric, *Caedite eos. Novit enim Dominus qui sunt eius* (kill them all, for God will find his own), Barton's experiment is all about the selection of identical pendulums by the excitation of their natural frequency.

Professor of Physics Edwin Henry Barton, FRS (1858 – 1925) lived his entire life in the British market town of Nottingham – famed, of course, for its link to the legendary egalitarian outlaw Robin Hood. He was a musical virtuoso, and expert on the physics of sound waves, a skilled experimenter and a great aficionado of pendulum experiments. While his early carrier focused on electrical transmission phenomena, he turned in later life to the study of analog applications of coupled and multiple pendulum systems. Writing with, first student and then colleague at University College, Mary Browning, Barton produced, between the years 1917 – 1918, a whole series of articles on pendulum experiments for the *Philosophical Magazine*. The various papers by Barton and Browning considered multiply connected pendulums having three and more linked masses as well as pendulums receiving external forcing – they also investigated the motion of variably coupled pendulums. In each case Barton and Browning not only considered the detailed mathematical theory, they also took pains to develop the appropriate pendulum apparatus. While the details of forced oscillations will be considered in the following section, here we focus on what has now become a standard high-school science experiment called – unfairly to Browning – Barton's pendulum[7]. This experiment was specifically designed to demonstrate the idea of natural frequency and resonance coupling (figure 6.6).

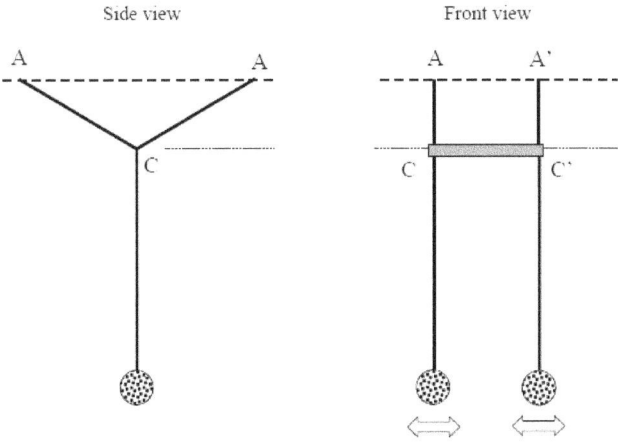

Figure 6.5. Front and side view arrangements for the double-cord pendulum. Each pendulum has two attachment points AA and A'A', and they are connected by a rigid bar CC'.

Barton's pendulum is set in motion by displacing the drive pendulum (A in figure 6.6) only – the other pendulums are all initially at rest. After a short while the amplitude of swing shown by the drive pendulum will begin to diminish, but at the same time the resonant pendulum (pendulum D in figure 6.6) begins to pick-up motion. The non-resonant pendulums will probably show some small amount of motion, but nowhere near as much as that portrayed by the resonant length pendulum. After a while the motion of the drive pendulum will come to a stop, with the resonant pendulum, however, swinging vigorously. After a short while the amplitude of the drive pendulum will once again begin to pick up, while that of the resonant pendulum starts to decrease. This back and forth switching of motion between the drive and resonant length pendulum will continue until friction eventually drains all the energy from the system.

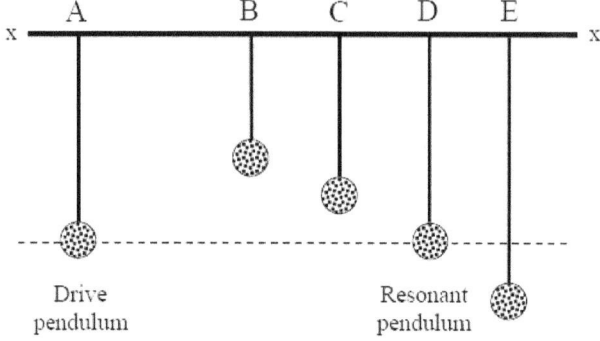

Figure 6.6. Barton's coupled pendulum. The motion of the drive pendulum (A) excites strong sympathetic (resonant) motion in only its identical length companion (pendulum D). The other pendulums (B, C, and E) have different natural frequencies to that of the drive pendulum and with no strong coupling established their motion is muted. XX is a horizontal cord to which each pendulum is attached.

The natural or fundamental frequency of any object, not just that of a pendulum, is, as its name suggests, the frequency with which it will most readily vibrate - it is the sweet spot for its oscillatory motion. If you force a specific system to vibrate, then a maximum amount of energy will be transferred to its motion if the forcing frequency is the same as the objects natural frequency. For a pendulum the natural frequency is related to its period, with $f_0 = 1 / P$, where the period, as seen earlier in equation (1.4), is related to the square root of the pendulum's length. Force a pendulum at its natural frequency and its amplitude of oscillation will get larger and larger (up to a point), but force it at some frequency other than f_0 and the response will be at best lackluster.

By continuing to excite an object at its fundamental frequency a resonance condition can be established. Resonances occur when an object is able to readily store and easily transfer its energy from one form to another – in the case of the pendulum, recall Chapter 1, this condition corresponds to the interchange between the kinetic and potential energies. Accordingly, under resonance conditions large amplitude oscillations can be established even when the driving force is very small – the driving force simply has to be applied at the appropriate natural frequency. An every day example of resonance driving is that exhibited by a person on a playground swing. By pumping their legs, energy is transferred into the swing (which of course is just a large pendulum), and by timing the leg kicks correctly (i.e. at a rate corresponding to the swings natural frequency) a large amplitude of motion can be built-up. The resonance condition is eventually broken, however, due to the frictional loss of energy from the system, and after a while the chains of the swing begin to flex and bend and this flexure will change the swings natural frequency. An example of a catastrophic, unplanned for resonance was that exhibited by the original Tacoma Narrows Bridge, built in Washington State in 1939 - 40. Even during its construction the road platform of the bridge was observed to undergo dramatic pitching oscillations if the wind chanced to blow at a speed of about 65 km/hr. It was under these very specific conditions that the small vibrations set in motion by the wind chanced to correspond to the natural frequency of the bridge platform. Just four months after the grand opening ceremony, on November 7th, 1940, the wind blew across the bridge at just the right speed and for just long enough that a large-amplitude resonant oscillation was established in the road platform and the Tacoma Narrows Bridge tumbled dramatically into the waters of Puget Sound.

The one-dimensional equation of motion for an oscillating system being periodically driven by some external force F can be written as follows:

$$m\frac{d^2x}{dt^2} + m\omega_0^2 x = F\cos(\omega t) \qquad (6.1)$$

where m is the mass, $\omega_0 = 2\pi / P$ is the system's natural frequency, and where ω is the frequency at which the forcing is being applied. The left hand side of equation (6.1) is identical to that of the simple harmonic oscillator (recall Chapter 1), while the right hand side of the equation introduces the forcing term – for the simple pendulum, of course, this term is identically zero with $F \equiv 0$. When $\omega \neq \omega_0$, equation (6.1) has the following solution

$$x = \frac{F}{m(\omega_0^2 - \omega^2)}\cos(\omega t) + C_1\cos(\omega_0 t) + C_2\sin(\omega_0 t) \quad (6.2)$$

where C_1 and C_2 are constants determined by the initial starting conditions. The propensity for large amplitude motion is contained in the denominator of the first term on the right hand side of equation (6.2). Here as $\omega \to \omega_0$ the force term F, which may be very weak in its own right, is divided by a smaller and smaller number thus leading to a numerical singularity. Indeed, if $\omega = \omega_0$ (the resonance condition for which the forcing frequency is identical to that of the system's natural frequency) then the solution to (6.1) changes to become

$$x = \frac{F}{m\omega_0} t \sin(\omega_0 t) + C_1 \cos(\omega_0 t) + C_2 \sin(\omega_0 t) \qquad (6.3)$$

In equation (6.3) we clearly see the effect of the resonance forcing in that the leading term now grows linearly with time t, resulting in an ever-increasing displacement term as time goes by. It was the great Swiss mathematician Leonhard Euler (1707 – 1783) who, in 1750, first wrote down the equation for the forced harmonic oscillator, and indeed, he also derived its time-divergent solution when the forcing frequency matched that of the natural frequency; finding thereby the first mathematical description of a resonance phenomenon. Clearly, however, equation (6.3) tells us something important about the system being considered. It tells us, in fact, that the mathematical description is no longer physically correct since no system ever displays an infinite displacement – something must happen in the 'real' world that allows resonance systems to avoid the mathematical pathology inherent in equation (6.3). Before we pick-up on this 'infinity taming' issue, however, we will first consider an example, metaphorically writ large on the landscape, in which mountains (all be they isolated ones) are literally moved by resonances.

Push something at its natural frequency and interesting things can begin to happen – be it in the motion of a child's swing, Barton's pendulum, or the Tacoma Narrows Bridge. The forcing of a naturally oscillating system, however, need not be at exactly its natural frequency for strong resonance effects to appear – indeed, within the solar system there are many examples where resonances can come about when even very weak forcing is applied at a frequency having a simple fraction $n = p / q$, where p and q are small integers, to the orbital period of a perturbing object. The quintessential such example of resonance clearing is that exhibited by Kirkwood's gaps in the main belt asteroid region between Mars and Jupiter.

First described by American mathematician and astronomer Daniel Kirkwood[8] in 1857, the diagram showing the orbital semi-major axis distribution of asteroids reveals distinct regions of avoidance or gaps (figure 6.7). In this case it is the repeated small gravitational tugs from planet Jupiter that has forced asteroid migration (technically, the jovian perturbations increase the orbital eccentricity). Indeed, the very boundaries of the main-belt asteroid region are set by resonances associated with the $n = 4 / 1$ and $n = 2 / 1$ resonances - in these locations an asteroid will make 4 orbits (inner boundary) and 2 orbits (outer boundary) for every one orbit made by Jupiter. Prominent gaps in the asteroid semi-major axis distribution are also found at locations corresponding to $n = 3 / 1, 5 / 2$ and $7 / 3$. The 3 / 1 resonance at an orbital semi-major axis of some 2.5 AU appears to be particularly strong and Jack Wisdom (MIT) has shown in a series of remarkable research papers published during the 1980s that any object located within this region is liable to evolve a chaotic orbit[9]. In this latter situation the orbit can undergo irregular transition jumps from low to high eccentricity states and this can force close encounters with the planet Mars. Mars itself can also produce resonance and strong gravitational perturbation effects upon an asteroid and this can further drive such objects into Earth crossing-orbit.

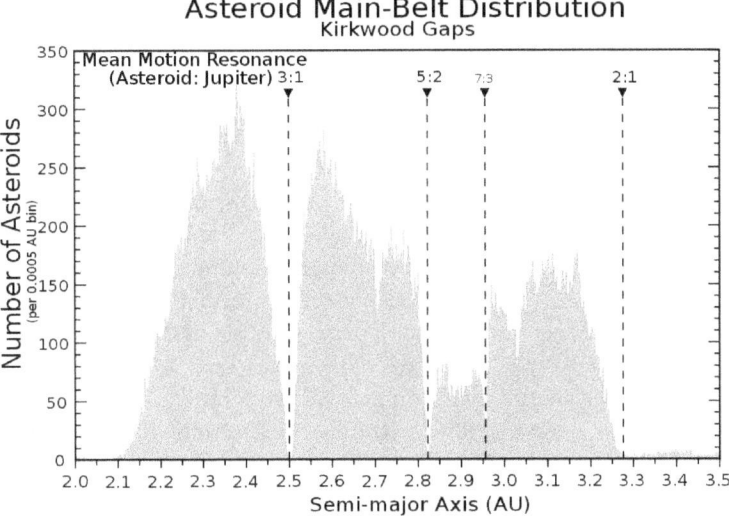

Figure 6.7. Number versus heliocentric distance for main belt asteroids. The Kirkwood gaps located at orbital resonances with Jupiter are clearly visible in terms of the reduced number of asteroids observed. The clearing of asteroid orbits with semi-major axis close to the 3/1, 5/2, 7/3 and 2/1 resonances are particularly strong. Image courtesy of NASA.

That the forcing term need not be especially large can be illustrated by considering the gravitational force acting between an asteroid and the Sun, and the same asteroid and Jupiter when it is located within the $n = 3/1$ resonance. At its closest approach to Jupiter, the ratio of gravitational forces is $F_{Jupiter} / F_{Sun} \approx M_{Jupiter} / M_{Sun} \approx 0.001$. These small magnitude, one in one thousand, gravitational tugs from Jupiter occur at intervals separated by approximately four years, and remarkably within a few tens to perhaps a few hundreds of thousands of years (a time very much shorter than the 4.56 billion year age of the solar system) the 3/1 gap can be cleared. In this situation one is reminded of the aphorism which recounts the observation: from little acorns do might oak trees grow.

Not all resonances that exist within the main asteroid belt are unstable. There is, for example, a remarkable group of minor planets, called the Trojans asteroids, that accompany Jupiter in its orbit in a $n = 1/1$ resonance.

These asteroids literally have the same orbital distance as Jupiter from the Sun, but importantly they are found in two regions, one located ahead of and the other located behind Jupiter as it moves along its orbital path. The Trojan asteroids are in fact located at two very specific stable points that exist within what is known as the restricted three-body problem, and they were first predicted and investigated by Italian-born mathematician Joseph-Louis Lagrange in 1772. The stable points are known as the L4 and L5 points and geometrically they, along with the Sun and Jupiter, form the vertices to a set of equilateral triangles. In this manner, the Trojans are located either 60° ahead of or 60° behind the Jupiter-Sun reference line. This being said, Trojan asteroids can display some remarkably complex orbital gymnastics by librating about the equilibrium (L4 or L5) point following either tadpole or horseshoe shaped trajectories. In addition to Jupiter, Earth, Mars and Neptune have their own observed sets of Trojan asteroids, while Saturn has Trojan moons.

In typical, everyday circumstances, large amplitude, resonant, oscillations do not readily occur - the essential reason being that something sooner or later happens to change the system dynamics: in the case of the Tacoma Narrows Bridge, for example, the bridge collapsed. Less dramatically, however, most oscillating systems have some form of inherent damping feature – terms, such as air resistance and friction, which results in energy dissipation and quashes large amplitude growth. The infinity lurking in equation (6.3) is removed by even the smallest, non-zero amount of energy dissipation – one might expansively say that a frictional David can easily slay a resonant Goliath.

Damping or the tendency to reduce a system's amplitude of vibration is typically introduced as a velocity dependent term, with the effect increasing linearly with the velocity and in a direction that opposes the motion. This drag-force law was first derived by British mathematician George Gabriel Stokes, who considered the motion of a sphere of radius R through a fluid (or gas) having a dynamic viscosity μ. If the sphere has a speed V, then Stokes law indicates that the drag force $F_D = 6\,\pi\mu\,R\,V$. From Stokes law it can be seen that in addition to the speed of the pendulum bob, the damping (or frictional drag) term will be dependent upon the size of the bob (through R) and the medium through which the bob is moving (through μ). In a vacuum $\mu = 0$; in air at 25 °C, $\mu = 1.983 \times 10^{-5}$ N s/m²; in water at 20 °C, $\mu = 1.002 \times 10^{-3}$ N s m².

With the damping term included, the equation of motion for a one-dimensional harmonic oscillator is written as:

$$m\frac{d^2x}{dt^2} + \gamma\frac{dx}{dt} + m\omega_0^2 x = 0 \qquad (6.4)$$

where $\gamma = 6\,\pi\,\mu\,R$ (for a pure Stokes law formulation) is a constant viscosity-like term: the smaller the viscosity, so the longer it takes for the amplitude of oscillation to decay to zero. Indeed, the solution to equation (6.4) is written as:

$$x = A\exp(-\varsigma\,\omega_0\,t)\sin\left(\sqrt{1-\varsigma^2}\;\omega_0\,t + \varphi\right) \qquad (6.5)$$

where $\varsigma = \gamma/2m\omega_0$ and φ is a constant phase offset. In equation (6.5) we now see the introduction of an negative exponential term in the equation for $x(t)$ which acts to drive the amplitude of oscillation towards zero as time increases. The characteristic decay (or e-folding) time for the damped oscillator is $\tau = 2\,m\,/\gamma$, and the damping causes a reduction in the angular frequency, compared to the undamped value of ω_0, to $\omega = \sqrt{\omega_0^2 - (\gamma/2m)^2}$. If the damping term $\gamma \ll 2m\omega_0$ so the parameter $\varsigma \to 0$, and equation (6.5) reduces to $x = A\exp(-\gamma t/2)\sin\left(\omega_0\,t + \varphi\right)$, which is the oscillatory amplitude term for the simple harmonic oscillator modified by a time-decaying exponential term [this is also the solution behind equation (2.7) in Chapter 2 where we looked at the effects of friction upon the motion of a clock's pendulum]. Clearly, if the damping term $\gamma \to 0$, so the system reverts to that of a simple harmonic oscillator – that is $\omega \to \omega_0$ and $\tau \to \infty$ indicating an infinitely slow decay time. The solution given in equation (6.5) requires that $\varsigma < 1$ in order that negative square root terms are avoided. This requirement is generally referred to as the underdamped condition. Analytic solutions to equation (6.5) can be found when $\varsigma = 1$ and when $\varsigma > 1$ but their specific form need not concern us here. The latter two solutions, however, correspond to so-called critical and overdamped systems respectively, and under these circumstances the amplitude of motion decays directly to zero without any oscillations actually occurring. Two examples of critically damped systems are those corresponding to spring-driven automatic door returns, and hydraulic dash-pots in which the motion is suppressed (damped in a literal sense) by forcing a piston to move through an oil reservoir.

Having considered driven and damped harmonic oscillators independently, the next obvious step is to combine the two effects. In doing this, however, we pass from the domain of straightforward analytic solutions to those requiring detailed numerical simulation – we also move further and further away from the domain applicable to simple pendulum experimentation.

A classical mechanical example of a forced harmonic oscillator with a damping term is that of the Lorentz atom (or oscillator). Hendrik Lorentz (who we encountered earlier with respect to the special relativistic Lorentz contraction described in Chapter 2) introduced his model atom in the early years of the 20th Century and he was interested in the question of how light (electromagnetic radiation) interacts with electrons – specifically he was interested in how the atoms (through their associated electrons) in a gas might absorb energy from an impinging electromagnetic wave of frequency ω. Lorentz was writing on these ideas in 1909, well before Ernst Rutherford discovered the atomic nucleus (in 1911), and well before the development of quantum mechanics and the introduction of the Bohr atom in 1915, and accordingly the favored atomic model of the day was the "plum-pudding" model developed by J. J. Thompson (who first demonstrated the existence of electrons in 1897). This particular atomic model pictured the atom as a spherical fluid of elemental positive charge (the pudding) containing a random distribution of electrons inside it (the plums). With these ideas in place, Lorentz wrote down the electron's equation of motion as

$$m\frac{d^2x}{dt^2} + \gamma\frac{dx}{dt} - kx = qE_0\cos(\omega t) \qquad (6.6)$$

In this representation the $\gamma\, dx\,/\,dt$ term describes the loss of energy (radiative damping) when the atom radiates energy into space; the $k\,x$ term was introduced by Lorentz on intuitive grounds and he described it as being, "a certain elastic force by which the electron is pulled back toward its position of equilibrium after having been displaced from it". The right hand side of the equation is the driving force associated with the impinging electromagnetic wave, with q being the electron charge and E_0 being the magnitude of the applied electric field. Having written down such an equation, the question that Lorentz then asked was, how does the rate of absorption of energy change with respect to the frequency of incident electromagnetic wave. The introduction of a damping term in equation (6.6) ensures that the resonant amplitude is finite when $\omega = \omega_0 = \sqrt{k/m}$ the natural frequency of the excited atom - for visible light we can expect $\omega_0 \sim$ 5 x 10^{15} radians per second. A typical excited atom will spontaneously radiate its excess energy in a time of order 10^{-8} seconds, and this sets the expected value for the typical damping or decay timescale $\tau = m\,/\,\gamma$. Given these approximations there will be something like ten million oscillations per decay time (a decidedly weak damping effect therefore). To determine the rate at which energy is absorbed by the electron we first evaluate the time derivative for the sum of its kinetic and potential energy terms, finding accordingly that

$$\frac{d}{dt}\left(\frac{1}{2}m\dot{x}^2 + \frac{1}{2}kx^2\right) = -\gamma\dot{x}^2 + \dot{x}qE_0\cos(\omega t) \qquad (6.7)$$

where we have used Newton's 'dot' notation to signify time derivatives and where, to find the right hand side, we have substitute from equation (6.6). Now, the first term on the right hand side of (6.7) corresponds to the energy dissipated through radiation, leaving, therefore, the second term to be the rate at which energy is absorbed from the impinging electromagnetic radiation. Without going through the details here, the energy absorption rate (EAR) as a function of the frequency ω, is described by the so-called Lorentzian profile, with

$$< \text{EAR} >= K \frac{(\gamma/2m)^2}{(\omega - \omega_0)^2 + (\gamma/2m)^2} \tag{6.8}$$

where the constant $K = (qE_0)^2/2\gamma$. The peak of the Lorentzian profile, naturally enough, lies at the natural frequency where $\omega = \omega_0 = \sqrt{k/m}$, and the absorption level drops by a factor of one-half of the peak when $\omega - \omega_0 = \gamma/2m$. We have looked at the Lorentzian model atom, and its interaction with an impinging electromagnetic wave, in some depth since it is an elegant example of a physical system described by a damped and forced pendulum model. Perhaps the most remarkable fact about the Lorentz's atomic model, however, is that while it has long been known to be an incorrect description of the atom, a full quantum mechanical reformulation of the interaction problem yields the exact same result for the average energy absorption rate [equation (6.8)] and its variation with the frequency of the incident electromagnetic radiation.

The planetary model of Ernst Rutherford eventually replaced the "plum-pudding" structure developed by Thompson, with the atom subsequently being pictured as a small solar system like structure, with point-like electrons (the planets) orbiting a central nucleus (the Sun). Once again, however, this model-picture is known to be in error, since such a configuration would be unstable – essentially, the electrons would radiate away their energy and rapidly 'spiral' into the nucleus. Niels Bohr solved the electron orbit problem by proposing that only very specific allowed electron orbits could exist with each orbit having a very specific associated energy[10] – Bohr's model is based upon the quantum mechanical picture of discrete energy levels that had been introduced by Max Planck (see below) in 1900. All the above being said, however, a team of researchers at Rice University, lead by Brendan Wyker, have recently succeeded in creating a Rutherford-like atom by coaxing an electron, along with its full complement of quantum mechanical uncertainties and characteristics, into a near circular orbit about the central nucleus. In addition and remarkably the orbital configurations that the Rice team have created for the electron are similar[11] in form to those exhibited by the Trojan asteroids (discussed earlier) in our solar system. The experimental trick (and difficulty of course) is to work with so-called Rydberg atoms in which the outermost electron has so much energy that its orbital is located a very long way away from the central nucleus (indeed, it can be fractions of a millimeter away from the nucleus). Targeting the outlying electron with a rapidly pulsing electric field forces it to reside somewhere in a tadpole shaped region on one side of the nucleus (this is the Trojan asteroid orbit analog). A second electric field is then applied to make the allowed electron region to rotate about the nucleus. Far from being an experiment to mimic some arcane theoretical model, however, the Rice University team hopes, eventually, to manipulate the electron configurations in any kind of atom, and to thereby engineer precise control over their associated chemical reactions, even, perhaps, producing novel compounds and new molecular structures. Once again, a remarkable threshold has been breached with the Rice University experiment – indeed, a crossover between two very different physical domains has been established, and not only are we witnessing a novel phenomenon at the very border between the quantum and classical worlds, a pleasing synergy is also established with the actions of the macrocosm being reflected within the workings of the microcosm, the gravity of the former being replaced by electric fields in the latter.

The Pendulum Goes Discrete

At the heart of quantum mechanics lies the idea of discrete energy levels. Unlike gravitational orbits, for example, which in principle can have any amount of associated energy, objects in the quantum world must have energies expressed in multiples of Planck's constant $h = 6.6262 \times 10^{-34}$ Joule seconds (to be discussed shortly). The question we ask at this stage, therefore, is it possible to transform the simple pendulum, which can in principle have any angle of oscillation, to a pendulum with only a discrete set of allowed amplitudes of swing. The answer to this is yes, and the means by achieving this result were found by Danil and Yakov Doubochinski in 1968 while undergraduate students at Moscow University[12]. The essential characteristics of the Doubochinski pendulum are shown in figure 6.8, and the distinctive additional component is that of the electromagnet solenoid placed close to the minimum height point (at $\theta = 0$). At play, in fact, in the Doubochinski pendulum are two oscillating systems. There is the simple pendulum itself, which has some natural frequency f_0, and the oscillating magnetic field with a frequency $f >> f_0$ provided by the solenoid. The two systems are made to interact by using a permanent magnet for the pendulum bob, and the system characteristics are controlled by imposing a limited range of angles over which the bob can 'sense' the solenoid's magnetic field.

The equation of motion for the Doubochinski pendulum is a slightly, but importantly modified version of the forced, damped oscillator. With the angle away from the vertical θ being the time varying quantity, we have:

$$m\ddot{\theta} + \gamma\dot{\theta} + \omega_0^2 \sin\theta = \varepsilon(\theta)\sin(\omega t) \qquad (6.9)$$

where, as seen before, γ is the damping coefficient, $\omega_0 = \sqrt{g/L}$ is the systems natural frequency, L is the length of the pendulum, g is the acceleration due to gravity, ω is the frequency modulation of the applied forcing term (from the electromagnet), and where, $\varepsilon(\theta)$ is a step function having the characteristics

$$\varepsilon(\theta) = \begin{cases} A, & |\theta| \leq \delta \\ 0, & |\theta| > \delta \end{cases} \qquad (6.10)$$

From equations (6.9) and (6.10) it can be seen that the forcing term only applies when $|\theta| < \delta$, and that outside of this range the forcing is reduced to zero. Numerical integration of equation (6.9) reveals that the restricted domain of forcing results in what are called basins of attraction forming within the system's solution set. These fixed points can be studied by constructing what is called a Poincaré map[13]; a diagram showing the repeated values of the pendulum's position θ and angular velocity $\dot{\theta}$ at some specific surface of intersection: $P_n = (\dot{\theta}, \varphi)_n$ where n corresponds to the nth intersection through the surface of section. What all this means in practical terms is that when released from any arbitrary starting point, the pendulum's motion evolves towards one of the stable oscillation modes. The number and characteristics of the pendulum's stable modes is determined by the strength (that is amplitude A) and frequency ω of the forcing term, the range $|\theta| < \delta$ over which the forcing term is applied, and the value of the damping coefficient γ. The stability of each discrete oscillation mode is maintained through a constant adjustment of the phase relationship established between the pendulum and the high frequency solenoid field, with the pendulum extracting just the right amount of energy from the magnetic field provided by the solenoid (controlled by A, ω and δ) to compensate for the frictional losses (characterized by the parameter γ) associated with a particular period of oscillation.

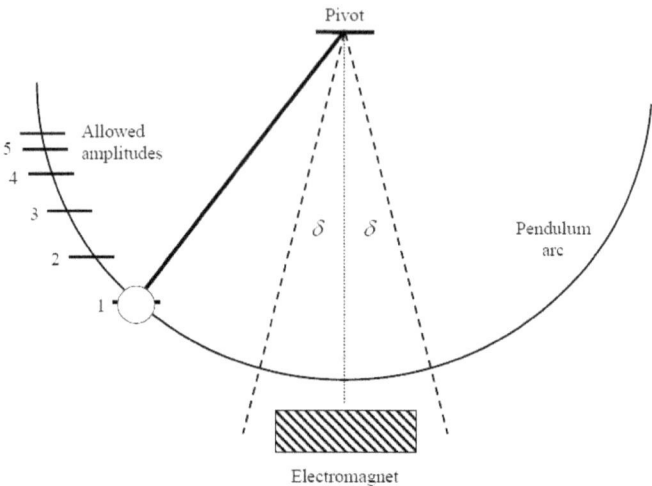

Figure 6.8. Schematic diagram of the Doubochinski pendulum. The magnet at the $\theta = 0$ location only provides forcing to the pendulum bob in the fan-like region defined by $|\theta| < \delta$. A schematic set of discrete amplitude ranges are also indicated on the left of the diagram.

Beyond the laboratory setting, the Doubochinski pendulum has a number of interesting parallels with the Fermi acceleration model for the explanation of cosmic rays – we discussed these rays in the context of time dilation experiments in Chapter 2. Although called rays[14] these entities are actually elementary charged particles, mostly protons and electrons, moving at relativistic speeds. The existence of cosmic rays is betrayed through their interaction with the Earth's atmosphere, and they were first detected while conducting balloon borne experiments by Austrian physicist Victor Hess in 1911. Indeed, Hess won the 1936 Nobel Prize for Physics as a result of his discovery of this, "extraterrestrial penetrating radiation [sic]"[14].

The astrophysical problem associated with the existence of cosmic rays is that of their extremely high velocity – literally, how is it that these particles are accelerated to speeds so close to that of light? Italian Nobel Prize winning physicist Enrico Fermi first tackled[15] the theoretical issue of cosmic ray acceleration in a 1949 publication submitted to the journal *Physical Review*. What Fermi suggested was that the charged particles attain their super-high speeds through interactions with moving magnetic field structures - such fields then already known to exist with the galaxy. The essential idea is that the charged particles behave like undamped ($\gamma = 0$) harmonic oscillators, 'bouncing' between what are effectively magnetic mirrors or walls, with more and more energy being imparted to a particle each time it encounters a highly tangled magnetic field region – such interactions effectively correspond to the

$\varepsilon(\theta)\sin(\omega t)$ forcing term in equation (6.9). Fermi later expanded upon this idea in collaboration with Polish mathematician Stanislaw Ulam, and considered the velocity change associated with a ball bouncing between two walls, one of which was fixed while the other rapidly oscillated back and forth. The Fermi-Ulam mechanism, or FUM as it is often abbreviated, has become a classical example of a nonlinear interaction process. At issue in the FUM problem is the determination of the final velocity (energy) of the ball (particle) after many collisions, and remarkably, it turns out, that under some conditions the motion and behavior of the ball (particle) is bounded or unbounded according to exactly how the moving wall oscillates – the velocity / energy remaining bound if the wall's motion varies sinusoidal, but unbound (that is tending towards infinity) if the oscillation follows a saw-tooth wave like motion.

The Quantum Oscillator – One Step Too Far

The Doubochinski pendulum, while displaying discrete modes of oscillation is not, it transpires a classical analog for a quantum mechanical oscillator. The key difference being that in the quantum mechanical world the energy levels of an oscillator must be equally spaced states with energy separations corresponding to integer multiples of Planck's constant h. It is the equally spaced energy level condition that is not satisfied by the Doubochinski pendulum.

It is in the quantum world that we begin to experience the limits of the simple pendulum. This is not to say, however, that the idea of a simple harmonic oscillator is not useful and indeed, critical to making progress in understanding quantum mechanical problems – we just need to modify some of our classical expectations. For example, the energy associated with a (classical) simple harmonic oscillator (recall the discussion at the end of Chapter 3) can be determined purely in terms of the work that is expanded in moving an object (say a mass attached to a spring – figure 6.9-A) from its starting point to its final position. In this manner expressing the work done as the integral of the force times the distance moved we find:

$$W = \int_0^x F dx = \int_0^x kx\, dx = \frac{k}{2}x^2 \qquad (6.11)$$

where we have assumed the spring obeys Hooke's law with the restoring force being directly related to the extension x and the spring constant: $F(x) = k\,x$. Figure 6.9-B illustrates the manner in which the energy of the classical simple harmonic oscillator changes with respect to the displacement x - it is a continuous, parabolic curve. Having established this result for the simple harmonic oscillator, we might now naively apply it as a model for the energy levels associated with the various vibration modes of the diatomic molecule (figure 6.9–C). We have not developed the tools or background to derive a full description of the quantum harmonic oscillator (QHO) here – although they are fairly straightforward to derive[16] – so the results will have to be stated directly. The key point at this stage is that a natural frequency ω can be associated with the QHO model of the diatomic molecule with $\omega = \sqrt{k/m_r}$, where k is the so-called bond force constant, and m_r is the reduced mass – remarkably, this natural frequency is exactly analogous to that of the classical SHO (as seen earlier). What is most surprising, however, about the QHO is that there is a zero-point, or ground state energy $E_0 = \hbar\omega/2$, where $\hbar$ is Planck's constant divided by 2π. This result is of fundamental importance in the quantum world and it indicates that, in contrast to the classical SHO, the QHO cannot have a zero energy state – there is no such thing as a stopped QHO plumb-line. Indeed, the zero point energy is a direct consequence of Heisenberg's uncertainty principle – the principle that underscores the very workings of quantum mechanics[17]. Not only this, the variously allowed energy levels of the QHO are such that $E_n = \left(n + \frac{1}{2}\right)\hbar\omega$, where $n = 0, 1, 2, 3, \dots$ is the so-called principle quantum number. Each energy level is now equally spaced with $\Delta E = \hbar\omega$ (as shown in figure 6.9-B).

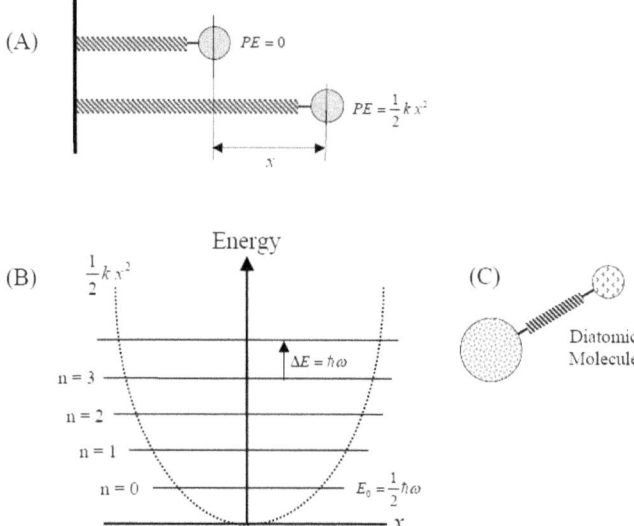

Figure 6.9. (A) The classic simple harmonic oscillator (SHO): a mass attached to a spring. (B) The continuous energy variation of the SHO plotted against displacement x. The equally spaced, discrete energy levels associated with the QHO are also shown. (C) Schematic diagram of a diatomic molecule. The vibrational modes of such molecules can be modeled in a manner akin to that of two masses joined together by a spring.

Nobel Prize winning physicists Richard Feynman poignantly wrote that, "I think I can safely say that nobody understands quantum mechanics" – by which he means, of course, the philosophical underpinning and fundamental reasons for why the world of the very small is apparently described by quantum mechanical principles is not known. The predictions derived from the theory of quantum mechanics are not only clear they have been proven to be correct time and time again. The quantum world is certainly a difficult and non-intuitive one for the non-initiated to comprehend, but it is assuredly real and it is most definitely amenable to exploration. In the quantum domain, however, we lose touch with the continuous quantities of the familiar, large-scale, everyday world, with results and experimental predictions being cast in a probabilistic rather than an absolute form. This is an alien environment to which the classical pendulum and the spring harmonic oscillator have no direct access – we have reached in the quantum world the very limits beyond which even the mighty pendulum can no longer penetrate.

A Brief Aside on Planck's Oscillators

"The whole procedure was an act of despair because a theoretical interpretation had to be found at any price, no matter how high that might be... I was ready to sacrifice any of my previous convictions about physics...", so wrote Max Planck in 1901, and such were the origins of quantum mechanics. At question was the manner in which hot objects radiate their thermal energy into space. The problem was a classic one, both in nature and within theoretical physics and it concerned the solution to what eventually became known as the ultraviolet catastrophe. According to the classical theory of thermodynamics, as developed in the later half of the 19th Century, all hot objects should radiate the bulk of their heat energy into space at very short wavelengths of light – that is at wavelengths in the ultraviolet part of the electromagnetic spectrum. The theoretical prediction was clear, undeniable and completely wrong. Laboratory experiments with analog blackbody radiators (see below) showed, in fact, the exact opposite effect to that predicted by theory actually occurred. Rather than the energy flux increasing towards shorter wavelengths, it actually decreased – a catastrophe for the theory, indeed, and it was exactly this problem that Max Planck managed to solve.

The theory of blackbody radiators was first outlined by German physicist Gustav Kirchhoff in 1862. The theory of such objects is based upon the idea that a hot object in thermodynamic equilibrium will radiate electromagnetic radiation into space in a very specific manner dependent only upon its temperature. The radiation, in fact, will be emitted over all wavelengths of light – technically from the very shortest wavelength gamma rays to the longest wavelength radio waves. As a result of laboratory experiments conducted by Jožef Stefan (in 1879) and via theoretical reasoning by Ludwig Boltzmann (in 1884) it was known that the total amount of energy radiated by a blackbody radiator per second per meter squared (the energy flux) was related to its temperature raised to the forth power (this is the Stefan-Boltzmann law: $F = \sigma T^4$). Additional experiments conducted by Wilhelm Wien, in 1893, further found that the specific wavelength at which a blackbody radiator emits its greatest energy flux varies systematically with its temperature (this is Wien's displacement law: $\lambda_{max} T = $ a constant). Even with these laws established, however, it was still not clear, by the close of the 19th Century, why the spectrum (the energy flux versus wavelength diagram) for blackbody radiators had the specific form that it did. This was the problem that Max Planck setout to investigate and he enters our story in late 1900. It was at this time that Planck was able to find a formula that correctly predicted the shape of the blackbody spectrum for a given temperature. The breakthrough was announced (to a non-impressed audience, as it turned out) at the Friday, October 19th, 1900 meeting of the German Physical Society in Berlin.

Why, it seems reasonable to ask, was Planck's initial announcement greeted with such muted interest? The answer, in fact, is straightforward and rests entirely upon the point that his solution was based upon what amounted to a mathematical trick and a seemingly *ad hoc* physical assumption. Planck's working model for a blackbody radiator was to imagine an insulated cavity containing a system of oscillators (embedded within the cavity walls) that would both radiate and interact with the electromagnetic radiation that they produced – the idea further being that after a while the whole system, oscillators and radiation, would come into equilibrium with some constant associated temperature. Planck worked with oscillators simply because the mathematics describing their interaction with radiation was essentially known and calculable (as we have seen, for example, with the Lorentz model atom described above). Planck knew from earlier research that if one had an ensemble of many oscillators then the probability $P(E)$ of any specific oscillator having an energy E was proportional to the quantity $\exp(-E/kT)$ - a rule derived from classical statistical mechanics by Ludwig Boltzmann – where k is Boltzmann's constant and T is the temperature. Using Boltzmann's distribution the average energy $<\varepsilon>$ can be calculated via the relationship $<\varepsilon> = \int \varepsilon P(\varepsilon)d\varepsilon / \int P(\varepsilon)d\varepsilon$. From this result, Planck reasoned that the intensity $I(f, T)$ of radiation, at a specific frequency f, associated with a blackbody radiator of temperature T would be proportional to $N(f) <\varepsilon(f, T)>$, where $N(f)$ corresponds to the number of oscillators having frequency f. If one now assumes, as indeed Planck did, that the various oscillators cannot have just any energy (as allowed for by classical theory) but must have some discrete amount of energy given as $\varepsilon = nhf$, where n is some integer and h is a constant (Planck's constant), then it turns out that $<\varepsilon> = hf / (\exp(hf/kT) - 1)$. With his quantized energy rule, Planck was essentially imposing a differential tax (or weighting function) upon the number of oscillators with a specific frequency f that could contribute to the overall energy. Since short wavelength radiation has the highest associated frequency, and therefore the highest associated energy, so the number of such oscillators must be relatively small (this is the high tax bracket). Long wavelength radiation, in contrast, has a small associated frequency and therefore it contributes a miniscule amount of energy, and accordingly the number of oscillators in this range can be high (this is the low tax bracket). Performing a detailed accounting for the number of oscillators with a specific frequency Planck was finally able to show that the intensity of radiation for a blackbody radiator is $I(f,T) = \left(2hf^3/c^2\right)/\left(\exp(hf/kT)-1\right)$, and the key point, as far as Planck was concerned, is that this formula, for a given temperature T provides an exact fit to the experimentally observed spectrum of a blackbody radiator. Further, by integrating Planck's intensity function over all frequencies one can derive the Stefan-Boltzmann law, and by differentiating Planck's function one can find the frequency (or wavelength) corresponding to the maximum energy flux and hence Wien's law. Planck's formula hit, as it were, all the right buttons – it explained the profile of the blackbody spectrum and it provided a theoretical understanding of the experimental laws deduced by Wien, Stefan and Boltzmann. The problem for physicists in general, however, was that while this was all mathematically consistent, who had ever heard of an oscillator having a discrete, that is quantum energy distribution? Essentially, what Planck had done, and this was in part his act of despair, was to find a way of fixing the mathematics of the result, that is he assumed quantized energies for the oscillators, without worrying about what the physical implications of that assumption were.

History reveals that although Max Planck was awarded the Nobel Prize for physics in 1918 for his work on blackbody radiators, he grew increasingly dissatisfied with the continued developments relating to its application. Indeed, after Planck's death in 1947 his former student James Franck commented that, "I watched his hopeless struggle to avoid quantum theory and make its influence as small as possible". Planck was the reluctant hero of the new quantum world order and it fell upon the shoulders of other researchers to take his radical ideas more seriously. Indeed, it was Albert Einstein who first saw the extended value of the quantum hypothesis and it was in 1905 that he was able to explain the photoelectric effect – a phenomenon first observed in 1887 by Heinrich Hertz (the discoverer of radio waves). Einstein explained this specific phenomenon in terms of the energy of light being quantized[18] with units of $E = hf$, and for his efforts he duly received the 1921 Nobel Prize for physics. Later, of course, it is also well known that Einstein himself grew to dislike the way in which some areas of quantum theory were developing, arguing against its inherently probabilistic nature through the now immortal lines, "I am convinced that He [God] does not play dice". In rather prophetic terms, Max Planck is also remembered for once arguing that, "an important scientific innovation rarely makes its way by gradually winning over and converting its opponents: what does happen is that the opponents gradually die out". Indeed, in spite of the philosophical angst felt by Planck and Einstein (and many others) in their later careers, quantum mechanics had not only arrived by the close of the first quarter of the 20th Century, it was very much alive and kicking.

Trevelyan's Hot Rocker

A number of mechanisms designed to counteract heat induced changes in the length of a pendulum were discussed in Chapter 3. The idea behind such devices, recall, was to compensate for the expansion, or contraction, of the pendulum material, with the aim of keeping the overall length of the pendulum constant. In the case of Trevelyan's rocker, however, it is a heat-induced expansion that drives a mechanical pendulum. The workings of this particular oscillator were first described in detail[19] by Arthur Trevelyan in the *Transactions of the Royal Society of Edinburgh* in 1834. He writes that, "in the month of February 1829, I discovered accidentally that a bar of iron, when heated and placed one end on a solid block of lead, in cooling vibrated considerably, and produced sounds similar to those of an Aeolian harp". While less commonly encountered in the experimental laboratories of today, Trevelyan's rocker, as his oscillator became known, was once an object of great interest, controversy and speculation[20]. Generally, the rocker is composed of a triangular bar of brass that has one longitudinal ridge filed down to accommodate a groove (figure 6.10).

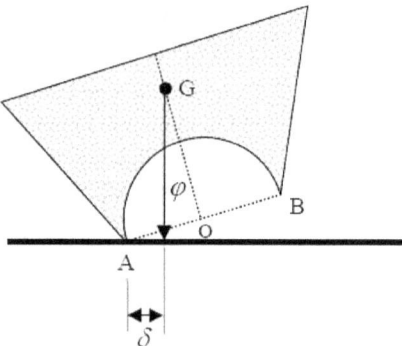

Figure 6.10. Cross-section through Trevelyan's rocker. G indicates the center of mass, and the distance AB = Δ corresponds to the width of the longitudinal groove.

Once the brass bar has been heated, it is the grooved edge that is placed upon a lead block which is at room temperature. Upon contact the rocker begins to vibrate producing a buzzing sound. The essential explanation for this phenomenon is that after making contact with the heated edges of the rocker the lead beneath the ridges expands rapidly and this expansion produces a kick which slightly offsets the balance of the triangular prism. As each heated edge of the rocker makes contact with the lead base plate it receives a small rotary impulse from the expanding lead underneath it and it is this effect that drives the continued oscillations. A lead base plate is used primarily because lead has a high coefficient of expansion (twice that of copper and steel in fact – see table 3.1) but a low thermal conductivity (half that of steel and one tenth that of copper) – this ensures a rapid expansion response but a slow overall heating effect.

The equation of motion for Trevelyan's rocker can be cast in terms of the free-mechanical motion of the triangular bar under the action of gravity[21]. In this sense the forcing of the motion, by the expanding lead underneath each edge in turn, is essentially a benign excitation mechanism that just compensates for any energy losses in the otherwise freely oscillating rocker. Using figure 6.10 as a guide, the equation of motion for the rocking behavior is written as

$$- mk^2 \ddot{\varphi} = mg \, \delta = mg \left(\frac{\Delta}{2} - h \sin \varphi \right) \cos \varphi \qquad (6.12)$$

where mk^2 is the rocker's moment of inertia, and h = OG. Since the motion being considered is for very small angles of φ so equation (6.12) can be reduced to the expression $\ddot{\varphi} = (hg / k^2)(\varphi - \Delta / 2h)$, which upon the substitution $x = \varphi - \Delta / 2h$, reduces to $\ddot{x} = \omega^2 x$, where $\omega = \sqrt{hg/k^2}$. This second order differential equation, while looking deceptively similar to that describing simple harmonic motion (only the minus sign on the right hand side is missing), does not, in fact, have a time-dependent oscillatory solution; rather it has a solution expressed as two time-dependent exponential terms[22]. In order to establish the boundary conditions for the equation of motion, let the rocker start at time t = 0, while resting on ridge A (figure 6.10), with some initial displacement φ_{max}. Since the rocker starts from rest, so $d\varphi / dt$ = 0 will also hold true at the start time. Now, at a later time t = $T/4$, where T is the period of oscillation, edge B (figure 6.10) will contact the lead base plate and at this time the angular displacement will be zero: φ = 0. The process is now imagined to repeat with the expansion impulse from the base being applied at A (figure 6.10), reversing the direction of motion, and so on. In fits and starts, therefore, the rocker is set in motion, and for small angles of oscillations the period is[23]: $T = 8k\sqrt{\varphi_{max}/g\Delta}$. The sound made by the rocker will have a frequency $f = 2\pi / T$, and this will vary according to the inverse square root of the initial angular displacement φ_{max} - all the other terms, for a given rocker, being held constant. This prediction concerning the sound frequency and the maximum angle of offset was actually tested experimentally, and confirmed, by Bhargava and Ghosh[24] (Allahabad University, India) in 1922, some 93 years after Arthur Trevelyan first noticed the phenomenon that now bears his name.

The theoretical description of Trevelyan's rocker, along with its (apparent) experimental verification seems, at first glance, appealing and complete – and yet, it is clearly not the whole story. The actual motion of the rocker and the sound that it produces must be much more complicated than presented. The first hint that the analysis must be incomplete is encapsulated in the result that the frequency of the sound produced by the rocker varies with the inverse square root of the maximum angular displacement. Given this condition, as φ_{max} decreases towards zero so the sound frequency must become higher and higher, and this is not something that is observed. Clearly, therefore, the equation of motion should include a damping term that accounts for mechanical friction, as well as heat exchange, between the rocker and the base plate. There should also be some cut-off condition which stipulates a cessation of the rocker's motion once φ_{max} falls below some critical value. In addition, while the expansion and impulse imparted to the rocker's ridges by the lead base plate has been acknowledge, it has not been directly included in the analysis – Trevelyan's rocker is clearly a forced oscillator, but no actual description of the forcing term has been provided. Indeed, the reasons for this are simple enough and relate to the fact that a full analysis, including heat transfer and material deformation effects, would be extremely complicated and certainly not amenable to producing an analytic solution. To the authors knowledge no physically complete analysis of Trevelyan's rocker has ever been published. Remarkably, we have the situation where what appears to be a straightforward oscillating phenomenon has a deep, complex, and entirely hidden underbelly; and while up to a point its essential behavior can be described by a straightforward equation of motion (equation 6.12) we have only managed to catch but an approximation of the full theory – we have seen only the tip of the proverbial iceberg, or to continue our metaphors, we have but glimpsed the ghost in the machine. With the quantum oscillator we lost contact with the physical (everyday experiential world) pendulum; with Trevelyan's rocker we begin to lose contact with straightforward analytic techniques and we enter the realm within which only approximate numerical solutions exist.

Duffing's Non-linear Oscillator
The restoring force in the harmonic oscillator, simple pendulum, equation is linear with respect to the displacement (recall Chapter 3 and above) and in this manner the equation of motion is just $m\ddot{x} + k_1 x = 0$, where we have used Newton's 'dot' notation to signify the time derivative and $k_1 > 0$ is a constant. There is no physical reason, however, to restrict our attention to just linear restoring terms, and accordingly the next logical step is to intro-

duce a non-linear variation, with the restoring force now being expressed as a sum of terms involving various powers of the displacement x.

The first person to study in detail the effects of an additional restoring term, making the equation non-linear with respect to the displacement, was the British physicist John William Strutt (3rd Lord Rayleigh). In his extensive two-volume work relating to *The Theory of Sound* (1894) Strutt considered the situation where the deflection characteristics of an oscillator were non-linear and symmetrical, finding thereby solutions to the equation $m\ddot{x} + k_1 x + k_3 x^3 = 0$. The German physicist Herman von Helmholtz soon extended Strutt's analysis to include a forcing term, and in his 1895 work *On the Sensations of Tone as a Physiological Basis for the Theory of Music*, he considered solutions to the equation $m\ddot{x} + k_1 x + k_2 x^2 = F\sin(\omega t)$. Helmholtz was specifically interested in finding solutions to this equation since he thought (incorrectly as it turned out) that the human eardrum could be modeled as an asymmetric oscillator (hence the additional x^2 term - which is always positive even when x becomes negative). Indeed, Helmholtz suggested that it was the asymmetric oscillation characteristics of the ear that caused additional harmonics to be imposed upon any input tone.

Neither Strutt nor Helmholtz considered damping terms in their specific equations, and this was a topic first analyzed in detail by George Duffing, a German engineer, in a now well referenced but rarely read (at least in its original form) monograph published in 1918. Entitled (in English) *Forced Oscillations with variable natural frequency and their technical significance*, Duffing's monograph was based upon his own mathematical investigations and his detailed knowledge of engineering systems. Duffing's equation adds a damping term to Strutt's equation and is written as: $m\ddot{x} + k_4\dot{x} + k_1 x + k_3 x^3 = F\sin(\omega t)$, where the k terms are all constants. An example of a system that obeys Duffing's equation is that shown in figure 6.11. The key point about this system is that it has a flexible wire pendulum that can move between two attracting magnets – it is the presence of these two magnets that makes the oscillator non-linear. To see how the system can show many different kinds of behaviors, let us consider the non-driven situation ($F \equiv 0$) and split the equation into two parts: $y = \dot{x}$ and $\dot{y} = x - x^3 - k_4 y$ - where we have assumed $m = k_3 = 1$, $k_1 = -1$ and simply assume $k_4 > 0$. The zero velocity condition can now be written as $y = \dot{x} = 0$, and using the second equation this indicates that $\dot{y} = x(1 - x^2) = 0$. The latter zero condition is satisfied for $x = -1$, 0 and 1, and these correspond to the so-called fixed points of the system at which the velocity will be zero. It can be further shown, for the parameter set chosen, that the fixed point at $x = 0$ is unstable, while the fixed points at $x = -1$, and $+1$ are stable. In this manner, as time goes by, the position of the flexible pendulum wire will converge, that is remain bent towards, one magnet or the other – which magnet the beam will eventually converge (bend) towards depends entirely upon the initial velocity and initial displacement of the beam. If we had chosen the parameter set $m = 1$, $k_1 = 1$, $k_3 = 1$, and $k_4 > 0$, however, there is just one fixed point at $x = 0$, and in this case, over time, the flexible beam will always converge to the origin, hanging vertically, mid-way point between the two magnets. By this simply demonstration, of just allowing k_1 to be either positive or negative, remarkable different end point configurations are seen to occur, and this suggests that the general Duffing oscillator (with a full set of none-zero parameters) is likely to realize a large range of possible behaviors.

The solutions to Duffing's equation are clearly going to be sensitive to the values adopted for the k_i (i = 1, 2, 3), F and ω terms. If F and k_4 are very small then the solutions will follow those of the Strutt's oscillator. If F, k_3 and k_4 are very small the motion will follow that of the harmonic oscillator; if F and k_3 are very small then the motion will follow that of Euler's resonance oscillator [equation (6.4)]. In the (normal) situation where none of the parameters are especially small the numerical solutions to Duffing's equation show a whole range of periodic, aperiodic and convergent stable orbit solutions. Most remarkably of all, however, there are some combinations of the parameters that result in highly complex, indeed, chaotic oscillations (figure 6.12). While the meaning of the word chaos is reasonably well understood in the everyday life sense, mathematically it is a more difficult beast to define. From a physical perspective we have to separate out just complex-looking, but entirely predictable behavior from that of truly chaotic and entirely unpredictable motion. Fortunately, there are several indicators that can reveal the true presence of chaos. One relates to sensitivity in the starting conditions, with even the smallest of changes resulting in dramatically different system configurations over time. Another condition follows from the topological properties of a system's attractors. In simple systems, the attractors represent the outcome of long-term trajectory evolution, and these can be described as points, lines, and closed loops. The central location $x = 0$ is the attractor point towards which all damped simple pendulums will evolve, whereas an un-damped, un-forced Duffing oscillator can evolve to a limit cycle trajectory around one of the stable points located at $x = \pm 1$. Points, lines and limit cycles have well behaved topologies of dimension 0, 1 and 2 respectively, but when chaos is present so the attractor morphs into a strange attractor described by having a fractal or non-integer dimension. French-American mathematician Benoit Mandelbrot (1924 – 2010) first discussed the properties of fractal objects[25] in detail in the late 1960s. At this stage, however, all we really need to appreciate is that a fractal can be thought of as a structure that is infinitely self-similar – that is, no matter how fine a slice or section that we might take through a fractal, it will always look pretty much the same. The three-dimensional, non-linear Lorenz oscillator, to be discussed below, can develop chaotic solutions for which the strange attractor has a fractional dimension of 2.06. The fractal dimension of the strange attractors encountered in Duffing's oscillator can vary, it turns out, between 1.3 and 1.7.

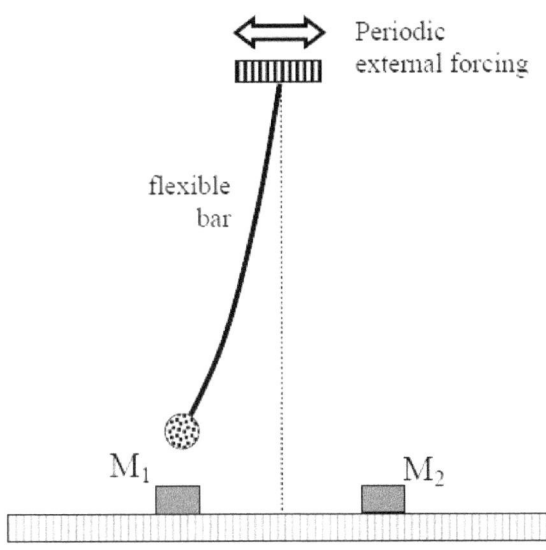

Figure 6.11. Schematic design for a Duffing oscillator. The suspension point of the flexible bar is made to undergo periodic oscillations. M_1 and M_2 correspond to the locations of two fixed magnets.

With the introduction of non-linear terms, such as in Duffing's oscillator, we have clearly passed an important threshold in pendulum behavior, and we can no longer rely on simple everyday intuition and analytic formula to describe resultant behaviors. Indeed, recourse must be made to numerical computational methods, as well as deeply subtle mathematics, in order to make any headway in understanding what such systems might do, and even then, with the appearance of a strange attractor we can never be fully certain what the long term behavior of a system will be. We have passed beyond the realm of certainty and fully predictable phenomena – the pendulum has gone where physical computation cannot follow.

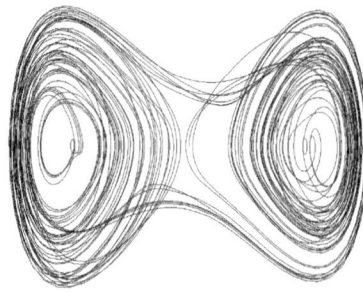

Figure 6.12. Phase diagram for one specific solution set to Duffing's equation. This diagram shows the velocity $\dot{x}$ (vertical axis) against position x (horizontal axis). The locations of the two fixed points are situated at the centers of the two shell-like lobes either side of the origin. In this particular simulation the motion is chaotic. Image courtesy of Wikimedia Commons.

Constancy meets Chaos

The Anglo-Austrian mathematician Sir Herman Bondi began a 1989 essay on chaos as follows, "chaos is old. The interesting point is not that chaos was recently 'discovered,' but that for so long it was systematically ignored"[26]. This is a remarkable statement, and it is an arguably correct one. Indeed, Bondi suggests, greatly overstating his case, however, that chaos was long ignored because of the, "absurd philosophical doctrine of determinism that should never have entered physics". While we would contend that determinism has its place (a very important place) within physics, Bondi is largely correct in seeing it as the historic cause for the marginalization of chaos – at least as an area for systematic study. At the heart of determinism is the rationalistic philosophy of Pierre Simon Laplace, who famously argued that given the initial starting conditions (at any one specific instant) of a particular dynamical system then the entire and complete future of that system was determined and could be mapped out into the indefinite future. Even entities as complex as the solar system or a spiral galaxy containing hundreds of billions of stars, indeed, the very universe itself might, in principle, according to Laplace, be described deterministically. Once a set of initial conditions are known then all future behavior is entirely expressed through the equations of motion and their solution. Laplace placed his beliefs upon the solid gearing of the Newtonian clock-work heavens, and the sturdy mechanics of Newtonianism were further reinforced by the mathematicians (specifically Charles Picard, Ernst Lindelöf, Rudolf Lipschitz and Augustin-Louis Cauchy) who in the late 19th Century proved the existence and uniqueness theorems for first order differential equations.

The uniqueness theorem for differential equations is a powerful result. By requiring only that the function $f(x, y)$ and the derivative of y be continuous in a rectangular region of (x, y) values, then the equation $dy / dx = f(x, y)$ has one, and only one, solution that will uniquely satisfy the boundary condition $y(x_0) = y_0$. As with all powerful theorems, the uniqueness theorem is limited in practice. It doesn't tell you, for example, what the solution to a specific differential equation is - rather, it only says that if the equation satisfies specific conditions then a unique solution must exist. Such results, however, are sufficient to bolster a strong belief in determinism. If I have, for example, a legitimate set of differential equations to describe the motion of a given system of objects, and a complete set of initial boundary conditions at time t_0, then I also known that a unique set of solutions to those equations must exist for the system at any other time $t > t_0$. Here is mathematical power indeed. This power, however, as Herman Bondi pointed out in his 1989 review, is a chimera.

The doctrine of determinism as espoused through Newtonianism and the grandiloquence of Laplace are a fine example of Plato's ideal realm – a universe that we can only encounter in our minds-eye and through our intellect – indeed, a universe that our own, the tactile one that we actually live in, is but a crude approximation. The great power of the uniqueness theorem is crippled in the real world by the overwhelming diversity of the small. First of all, we can never hope to know and/or measure with enough precision all of the starting parameters, and second the uniqueness theorem does not save us from divergent results. Simply put, the uniqueness theorem does not say that a small change in the initial conditions must produce a small change in final solutions. Indeed, this is entirely what chaos is all about, with even exquisitely small changes in the starting parameters producing substantially different behavior outcomes in a system. It is this behavior of the solutions to differential equations that gives the fallacy to the rationalistic work of Laplace and the clockwork of Newtonianism, and forces us to accept, in fact, that the future is not deterministic. The point being that the long-term behavior of a chaotic system is inherently unknowable and that the mystic veil thrust over our future gaze cannot be lifted by simply finding out more about the system and/or measuring the starting conditions with greater and greater precision.

Early signs of the chaos that can lurk within what appear to be stable dynamical systems were found by French mathematician Henri Poincaré in the late 19th Century. Indeed, he found that the orbits of only slightly perturbed objects (moving under gravitational forces alone) while often stable could, under some initial conditions, become divergent over time. The chaos was not only hidden within the equations, it was also, for all appearances, temperamental. The Russian mathematician Aleksandr Lyapunov used the idea of divergence of orbits to establish a condition that described the timescale of chaos. If two particles, on two very similar orbits, have an initial separation $d(0)$ then the presence of chaos can be detected by following their separation

$d(t)$ over time. If the system is regular, and not chaotic then it is found that $d(t)$ typically increases linearly with time: $d(t) \sim d(0)\, t$. If, however, the system is chaotic then the separation increases exponentially with time: $d(t) = d(0) \exp(\lambda\, t)$, where λ is the Lyapunov exponent. With this condition established, the characteristic timescale over which the chaos, that is divergence of orbits, will begin to manifest itself will be of order $1 / \lambda$. It is important to remember what such results actually tell us. They do not say that after several Lyapunov timescale intervals that the system under investigation will fly apart and/or be destroyed. What they do reveal, however, is the limit to our knowledge of where the system components will be in the future – remember the equations themselves are fully deterministic. In the case of our solar system, the orbits of the planets have clearly been stable for at least some 4.5 billion years, and yet, as we saw in Chapter 2, the orbit of Pluto is chaotic. Indeed, the Lyapunov timescale for Pluto's orbit is about 10-20 million years. In similar fashion, the Lyapunov timescale for planetary orbits within the inner solar system are found to be of order 4-5 million years, while those for the outer solar system planets are of order several hundred million years. With the appearance of chaos we learn that while the solar system may well be stable (almost forever – and at least on a timescale of tens of billions of years) we cannot know precisely where the planets will be in their orbits, based upon any single set of calculations, on timescales in excess of a few hundred million years into the future.

The Lorenz Attractor

Complex, non-periodic behaviors (or trajectories) resulting from the apparently simple combination of various sets of equations was a mathematical topic that could only flourish once the electronic computer came into existence. It was not that mathematicians were unaware of chaos or aperiodic functions (as discussed above); it was simply that there was no quick and straightforward way to generate the diagrams to show the complex behavior. The development and availability of affordable table-top, electronic computers solved this problem, however, and since the early to mid 1970s the field of chaos has blossomed. One of the now classic icons of chaos is that of the Lorenz strange attractor (figure 6.13) and appropriately its origin lay within the domain of that most un-predictable and chaotic element that surrounds and affects us all – the weather.

First studied by American mathematician and meteorologist Edward Lorenz (Massachusetts Institute of Technology) in the early 1960s, the three equations that describe the Lorenz oscillator are a minimal, stripped-down simplification of the so-called Navier-Stokes equations of fluid flow[27]. Without explanation or development the three, deceptively simple looking, equations are:

$$\frac{dx}{dt} = \sigma(y - x), \qquad \frac{dy}{dt} = x(\rho - z) - y, \qquad \frac{dz}{dt} = xy - \beta z$$

These coupled, non-linear, first-order differential equations describe the time variation of the quantities x, y and z (these terms are not to be thought of as spatial coordinates, but are in fact related to the convective properties and temperature of the fluid). The equations can only be solved for numerically, hence the requirement for a fast electronic computer, and the resultant time variation in the (x, y, z) positions will change according to the values chosen for the parameters σ, β and ρ. The remarkable feature of the Lorenz oscillator is that the trajectory of the (x, y, z) points map out a surface with fractal characteristics (hence the name strange attractor). Indeed, the trajectory followed by the Lorenz oscillator is highly sensitive to the starting conditions and the specific parameter set chosen, with the long-term behavior, that is, the (x, y, z) position at some given time t, being entirely unpredictable. The equations are deterministically chaotic in that they allow a solution to be generated at any time t, but the solution set obtained for earlier times provides no hint as to what the future behavior might be. For weather forecasting this is all bad news, and Lorenz noted in a research paper published in 1961 that, "in view of the inevitable inaccuracy and incompleteness of weather observations, precise very-long-range forecasting would seem to be non-existent". Such detailed mathematical results only help to underscore the commonly held perception that, the only constant thing about the weather is that it is not constant.

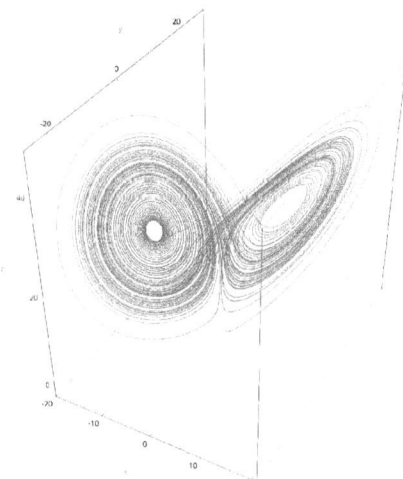

Figure 6.13. The Lorenz strange attractor. In this plot the (x, y, z) trajectory begins, at time $t = 0$, at the origin $(0, 0, 0)$ and then follows a path that unpredictably switches between the two 'buttery fly wing' loops.

The Double Compound Pendulum

Chaos is inherent in all compounded things.
Strive on with diligence.
Gautama Buddha

Chaos, the very antithesis of order, has become, like the pendulum, another great cultural icon; the world is in chaos; the stock markets are chaotic and unpredictable; and our lives (occasionally) are in a chaotic mess. Remarkably, however, there is a bridge between the seemingly disparate domains of chaos and constancy, and the double beam pendulum is our mechanical guide. The double compound pendulum is a remarkable beast, and at rest it hides its unpredictability well (figure 6.14). Externally the double compound pendulum looks to be a harmless enough device; two rigid arms joined so that the lower arm can rotate freely about the upper arm, which in turn can rotate freely about a fixed support point. Setting such a pendulum in motion, however, reveals its Janus nature. For small angles of offset, with the lower beam below the upper one, the motion is reassuringly stable and the behavior is just like that of a double string (two-dimensional harmonograph-like) pendulum – rhythmic and stable. Add a little more initial height and velocity to the lower beam, however, and the pendulums hidden chaos is released, with the spatial and temporal locations of the lower arm becoming unpredictable after just a matter of seconds[28] – indeed, the motion of the lower arm varies from being almost stalled to something akin to a whirling dervish (Figure 6.15). The equation of motion for the double pendulum, like its physical appearance, shows little sign of the chaos that lurks within it, and we have (with reference to figure 6.14)

$$m_2 \, L_2 \, \ddot{\eta} + m_2 \, L_1 \, \ddot{\varphi} \, \cos(\varphi - \eta) - m_2 \, L_1 \, \dot{\varphi}^2 \, \sin(\varphi - \eta) + m_2 \, g \sin\eta = 0$$

where L_1 and L_2 are the lengths of each pendulum arm. The equation of motion for the simple pendulum can be found within its double-armed cousin by simply setting φ equal to some constant value; it seems remarkable that by adding just two additional terms into the equation of motion, when φ is not constrained to be some constant, that the demon of chaos can be set loose.

The double compound pendulum is a mechanical metaphor for the so-called Butterfly Effect[29]; a phenomenon that encapsulates the extreme sensitivity of some systems to their initial starting conditions. Indeed, this effect accounts for the phenomena whereby the very smallest of differences at the beginning of a particular experiment can result in extremely large variations in its long-term behavior. Such chaotic motion is often called pathological, and even the most refined super-computer can make no accurate, long-term prediction of the exact configuration of a double compound pendulum just minutes after it has been set in motion.

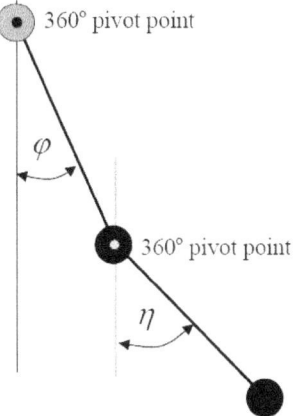

Figure 6.14: The double compound pendulum. Each arm is allowed to rotate freely about the pivot points, with the angles φ and η being allowed to vary between 0 to 360°.

With the double compound pendulum we come full circle – from the inherently predictable to the inherently unpredictable, and yet the pendulum, indeed, the mighty pendulum, provides us with a physical demonstration of these two extremes. For all of our modern-world sophistication it is perhaps comforting to know that there is still room for an age-old simple device that can challenge our perceptions, provide insight into new mathematics, and redirect our concepts of reality.

Figure 6.15: Long exposure photograph of a double compound pendulum having a light attached to its lower arm. Image by George Loannidis and courtesy of Wikimedia Commons.

Nowhere is the pendulum synergy of certainty and doubt better physically demonstrated than at the church of St. Mary Redcliffe in Bristol, England. Located in the 15th Century transept of this beautiful church is a chaotic pendulum (figure 6.16). Based upon a design by Brian Pippard[30] (formerly Cavendish Professor of Physics at Cambridge University), this simple hollow-bar pendulum pivots about the central point of a Christian cross – it is a water-fed pendulum, with the water continuously entering into and along the hollow bar. It is absolutely fascinating to watch. Without the water feed the bar would simply oscillate up and down in a normal and entirely predictable manner; this predictability, however, is removed by the water that flows in the bar's interior - the amount of water at any one instant varying according to the angle of the bar and the input flow rate. It is truly an unpredictable pendulum and from one minute to the next it is not clear how the oscillations will proceed – indeed, an exact equation of motion for the system is not known[30]. When he first introduced it, Pippard knew that there was more, much more in fact, to his water-fed 'seesaw' pendulum than its mere mechanical motion, and he commented that, through it one might, "recognize how physics can open a little window on to a reality far more multifarious than itself, perhaps even put into your hand a clue to guide your first steps into a labyrinth where your teacher can no longer lead you, and often enough dare not even follow". If ever there was a mechanical metaphor for the various ups and downs of life, our emotions, our faith and indeed, the very workings of science, then it is assuredly encapsulated in the behavior of the wonderful pendulum at St. Mary Redcliffe church – if you get the chance go and see it for yourself.

Afterwords

The stately swing of the pendulum has become a universally recognized icon. Across all cultures the steady back and forth motion of the pendulum is a metaphor for cyclical change; renewal and decay, life and death. It encapsulates in miniature the rhythmical flow of time, the change of our moods and the dance of the seasons; it is the steadfast swing of the incense-bearing thurible; it mimics our lives and it provides us with a standard of ideal predictability. When we set a pendulum in motion, we know what it should do, and it is by this standard that astronomers, physicists, geologists and mathematicians have, throughout history, been able to tease from nature some of its most profound secrets. Carefully crafted, simple pendulum experiments, such as the one conducted by Leon Foucault, have enabled a definitive proof that the Earth must be spinning, while the six-monthly and yearly pendulum-like oscillations of the stars in the sky tells astronomers that the Earth must be in motion about the Sun, and that even the closest stars are at immense distances away from us. Near and far, to and fro, the simple pendulum has enabled prospectors to find oil-fields and it has allowed geologists to measure the heights and depths of Earth's crust, interior and oceans. The pendulum has regulated human time, it has helped weigh the Earth, and it has enabled architects to design earthquake-proof buildings.

Figure 6.16. The chaotic pendulum installed at St. Mary Redcliffe church in Bristol, England. The hollow pendulum bar (positioned at a 45° angle in the picture) rotates about the center of the supporting wooden cross. Water enters the bar at the central pivot point and is collected, after running through the pendulum bar, in the semi-circular trough seen in the lower part of the picture. Photograph by Janice Lane (Trelewis, South Wales).

Long ago poet John Donne complained that the new philosophy of Copernicus "put all in doubt". What was previously predictable and understood was cast into question and new ideas and new observations were called upon to explain the new cosmos. More recently, indeed, just over one century ago, the deterministic world of classical physics was shattered by the necessary introduction of quantum mechanics, whereby absolute determinism was replaced by the probabilistic properties of collapsing wave functions and the time dependent Schrödinger equation. And again, in the past quarter century, the foundations of classical science were further shaken by the new discoveries relating to chaos, strange attractors, complexity and fractal geometry. Throughout all this change, however, the mighty pendulum has remained our resolute guide; embracing the new within the comforting fold of the old.

A clock aeonian, steady and tall,
With its back to creation's flaming wall,
Stands at the foot of a dim, wide stair.
Swing, swang, its pendulum goes,
Swing—swang—here-there!
Its tick and its tack like a sledge-hammer blows
Of Tubal Cain, the mighty man!
But they strike on the anvil of never an ear.
On the heart of man and woman they fall,
With an echo of blessing, and echo of ban;
For each tick is hope, each tack is a fear,
Each tick is a kiss, each tack is a blow,
Each tick says "why", each tack "I don't know".
Swing, swang the pendulum!
Tick and tack, and "go" and "come",
With a haunting, far-off, dreamy hum,
With a tick, tack, loud and dumb,
Swings the pendulum.

"The Clock of the Universe"
George MacDonald

FURTHER NOTES AND REFERENCES

Chapter 1

1. The approach appears in Newton's *Principia* (Book II, section 7, proposition 32). Newton specifically talked in terms of what he called 'dynamical similarity', but the effect is the same as matching common units. A very readable and detailed book on this topic is that by Rudolf Kurth, *Dimensional Analysis and Group Theory in Astrophysics* (Pergamon Press, Oxford, 1972). For a very readable and informative discussion on dimensional analysis, as well as an introduction to the closely related topic of "Fermi Problems", see John Adam's book *Mathematics in Nature: modeling patters in the natural world* (Princeton University Press, 2003).

2. There are many texts that have explored the epic struggle to map-out the geodesic arc stretching across France and the determination the meter. A recommended reading on this topic is Ken Alder's, *The Measure of All things: the seven year odyssey and hidden error that transformed the world* (The Free Press, New York, 2002). The 1820 definition for the second was later refined to correspond to $1/86,400^{th}$ of the mean solar day January 1, 1900. The length of day was eventually removed from the definition of the second in 1967, when it was agreed that one second corresponds to 9,192,631,770 transitions between the hyperfine ground state of the caesium-133 atom. As a point of interest, the meter was redefined at the 17th General Conference on Weights and Measures, in October of 1983. The meter is now taken to be the distance traveled by a light ray in vacuum during a time interval of $1/299,792,458$ of a second. Even in the modern era the less than exciting issue of official units can be a contentious political problem, as exemplified by the letter by Vivian Linacre (British Weights and Measures Association) to the *Daily Mail* newspaper for Friday, May 18, 2012. Linacre is critical of the May 16th comments by Lord Howe of Abervan (British Parliamentary Under-Secretary of State) that British weights and measures are, "in a mess", and that a switch to the metric system should be engineered, "as swiftly and cleanly as possible". Indeed, Britain is the only European country to hold onto an imperial weights and measures system. Linacre comments that an "Act of Parliament in 1897 legalized the optional use of metric measures for all trade purposes", and suggests that the only way that a full metric change over might be accomplished is to make the use of, "any customary [imperial] units a criminal offence".

3. In *SI* temperature is measured on the thermodynamic scale with the Absolute Zero point (where molecular energy is at is lowest possible value) being the lower fixed point and the triple point of water (the temperature at which ice, water and water vapor are in equilibrium) being the upper fixed point. The Kelvin unit, named in honor of William Thomson, 1st Baron Kelvin, corresponds to $1/273.16$ of the triple point of water.

4. The rational behind the Avogadro Project is to define the kilogram in terms of Avogadro's constant N_A which is itself defined according to the number of atoms in exactly 12 grams of carbon-12. For a recent comprehensive review see Robert Crease's web-article, Metrology in the balance (*Physicsworld.com*, May 22, 2011).

5. Einstein, A. Elementare Betrachtungen über die thermische molekularbewegung in festen körpern (*Annalen der Physik*, 340, 679–694, 1911).

6. The term order of magnitude simply means that we are not so much interested in the exact value of the number but whether the number is of order 1, or 10, or 100, or 1000, and so on.

7. It seems fair to generalize that Leibniz and Newton did not see eye-to-eye on most, if not all topics. Principally, however, they became the foci of an acrimonious debate concerning the discovery and first use of differential calculus with neither party being able to accept the possibility of independent invention. Neither Newton's nor Leibniz's calculus had the rigorous form of modern theory, which was largely developed in the 19th Century, but the most commonly used d-symbol notation for the differential is that presented by Leibniz. Newton's 'dot' notation for the time derivative, however, is still commonly used in calculations relating to dynamics. For the history of mathematics we recommend the web site http://www-history.mcs.st-andrews.ac.uk/Indexes/HistoryTopics.html.

8. Given the expression $\varphi = \cos(a\,t)$, where a is a constant, then the first and second time derivatives are: $d\varphi / dt = -a \sin(a\,t)$, and $d^2\varphi / dt^2 = -a^2 \cos(a\,t)$. Substituting these two expressions into the differential equation then gives the result that $-L\,a^2 \cos(a\,t) = -g \cos(a\,t)$, which yields the relationship $L\,a^2 = g$. By setting $a = 2\pi / P$ so equation (1.4) is obtained.

9. There is a slight mathematical point worth making at this stage, π is an irrational number and its exact value cannot be written down (although many millions of its digits have been evaluated). In a practical sense, however, this is not a problem since there is a very definite limit to the accuracy with which any time interval can be measured.

10. The standard gravitational acceleration is actually an arbitrary value, but an internationally accepted one, loosely based upon the average measured value for the acceleration due to gravity at different locations on Earth's surface.

11. A French Republican Calendar was also designed and consisted of twelve, thirty-day months, with each month being divided into three, ten-day weeks. Five additional holiday days were added every year (six in a leap year) to make up a 365 day (or 366 day in a leap year) cycle. The calendar was never popular and saw but fleeting employment during the last few years of the 18th century.

12. See the book by John Keay, *The great arc: the dramatic tale of how India was mapped and Everest named* (Harper Collins, London, 2000). The trigonometric survey and the first successful ascent of Mount Everest are additionally described in Wade Davis's engaging book, *Into the Silence: the Great War, Mallory, and the conquest of Everest* (Alfred Knope, New York, 2011).

13. Henry Kater, An account of experiments for determining the length of the pendulum vibrating seconds in the latitude of London. *Philosophical Transactions of the Royal Society*, 108, 33-102 (1818). A version of equation (1.5) when $P_1 = P_2$ can be readily derived by simply considering the pendulum equations when the suspension points are from knife edge 1 and knife edge 2 (figure 1.3). Accordingly we have $P_1 = 2\pi\sqrt{I_1/mgh_1}$ and $P_2 = 2\pi\sqrt{I_2/mgh_2}$, where I_1 and I_2 are the moments of inertia. Now we have to use what is called Steiner's theorem to link the moment of inertia values I_1 and I_2 to the moment of inertia through the pendulum's center of mass I_C. Accordingly, we have $I_1 = I_C + mh_1^2$ and $I_2 = I_C + mh_2^2$. Equating the two expressions for the period reveals that the equality condition requires that $(h_1-h_2)(I_C - mh_1h_2) = 0$.

The interesting case for the zero condition is that when $I_c = mh_1h_2$, since the $h_1 = h_2$ zero condition just gives us two identical pendulums with the associated problem of determining the exact values for the pendulum lengths. So, with the pendulum adjusted to give equal periods when suspended from each of the knife edges, we have $P_1 = P_2 = 2\pi\sqrt{(mh_1h_2 - mh_2^2/mgh_2)} = 2\pi\sqrt{(h_1 + h_2)/g}$. This equation now expresses the experimentally measured period in terms of the more accurately determined distance between the two knife edges $(h_1 + h_2)$, and is indeed equivalent to equation (1.5) when $P_1 = P_2$. In terms of experimental procedure it is more straightforward to measure the two periods of swing and use equation (1.5) rather than adjust the masses so that $P_1 = P_2$. The full expression of Equation (1.5) was first derived by the German mathematician and astronomer Friedrich Bessel in 1826.

14. John Bird (1709 – 1776) was celebrated for his great skill in dividing and graduating instrument scales. It was a transit telescope and scale devised by Bird that Charles Mason and Jeremiah Dixon used in their four year survey, starting in 1763, to determine the boundary (the Mason-Dixon Line) between the States of Maryland and Pennsylvania.

15. Hooke and Newton had a very public and rather acrimonious row about who first suggested the inverse square law for gravity. There seems little doubt that Hooke made the suggestion first in several articles written between 1660 and 1680. It was Newton, however, who was able to show in his *Principia* (published 1687) that this law accounted for Kepler's laws and the observed planetary orbits. Newton later argued that even if he had heard of the inverse law from Hooke (which he continued to claim he had not) then Hooke had only guessed at the rule, while he had shown that it must be true by mathematical construction.

16. The term h / R is always going to be a small number; even for Chomolungma (Mount Everest), the highest mountain on Earth, $h / R = 1.39 \times 10^{-3}$. Specifically, however, provided $h / R << 1$, so the expression $(1 + h / R)^{-2}$ can be expanded in terms of the infinite series: $1 - 2 (h / R) - 3(h / R)^2 - 4(h / R)^3 +$ terms of higher order. Now, since h / R is small, so this term squared, cubed and so on will be even smaller, and to a good order of approximation $(1 + h / R)^{-2} \approx 1 - 2 (h / R)$.

17. The gravitational acceleration due to the mountain will be of order $g_M = G M_M / (h / 2)^2$, where M_M is the mass of the mountain, and $G = 6.67 \times 10^{-11}$ N m^2 / kg^2 is the universal gravitational constant (the formula for g_M is described in Chapter 4). If it is assumed that the mountain is made of rock having a density of 2500 kg/m^3, then $M_M = 1.05 \times 10^{13}$ kg. Substituting these numbers into the formula gives the gravitational attraction of the mountain as $g_M = 7 \times 10^{-4}$ m/s^2.

18. The Bouguer correction due to an infinite slab of material of thickness h is given by the conveniently simple formula $\delta_B = 2\pi G \rho h$, where ρ is the density of material and h corresponds the height of the mountain. This expression is actually three times larger than that resulting from the sphere approximation used in the text – but its derivation is more complicated. The Bouguer correction for Chomolungma (Mount Everest) amounts to 9.91 x 10^{-3} m/s^2, while the Bouguer correction for the Challenger Deep located at the southern end of the great Mariana Trench, with a depth of 10.991 km, is -7.64 x 10^{-3} m/s^2.

19. The properties of the tautochrone are described by the author in the article, The Catenaric Oscillator (*The Bulletin of the Institute of Mathematics and its Application* **27**, 152-154, 1991). Additional mathematical details can be found in John M. McKinley, Bra-

chistochrones, tautochrones, evolutes and tessellations (*American Journal of Physics*, 47, 81-86, 1979), and Ramesh Chander's article, Gravitational fields whose brachisto-chrones and isochrones are identical curves (*American Journal of Physics*, 45, 848-850, 1977). For a brief historical review, see the article by R. Gómez *et al.*, An alternative solution to the general tautochrone problem (*Revista Mexicana De Fisica*, E54, 212-215, 2008).

20. Dark matter is presently a profound mystery. The observations of galaxy rotation curves and the dynamics of galaxies within clusters clearly indicates that there is more gravitational mass than can be accounted for in the terms of visible matter – that is stars, gas and dust (or more generally any matter that can interact with or emit elec-tromagnetic radiation). Many objects and particles no doubt contribute to the dark matter reserves. Some amount of dark matter will be baryonic, and located in the form of very faint stars, brown dwarfs, planets, asteroids and cometary nuclei, but this component cannot account for more than perhaps a few percent of the total mass required. The general consensus at the present time is that the bulk of dark matter mass must reside in the form of some as yet unidentified subatomic particle. The key characteristics of any dark matter particle are that it must have a gravitational mass, of course, but only interact very rarely with ordinary matter (hence the difficul-ty in detecting such particles in terrestrial experiments). Alternative solutions to the dark matter problem look to modify the law of gravity. One of the most studied non-standard models in recent times is that of MOND – Modified Newtonian Dynamics – introduced by Mordehai Milgrom in the article, A Modification of the Newtonian dynamics as a possible alternative to the hidden mass hypothesis (*The Astrophysical Journal*, 270, 365-370, 1983). A generally more accessible account of MOND is given by Milgrom in the article, Does Dark Matter Really Exist? (*Scientific American*, 52, 42-50, 2002). Although MOND was developed to explain the large scale motion of stars within a galaxy, effects local to the Sun should also be measurable. The author, for example, has investigated the orbital motion of Proxima Centauri about α Centauri AB which are currently separated by about 15,000 AU (The orbit of Proxima Centau-ri: a MOND versus standard Newtonian distinction – *Astrophysics and Space Science*, 333, 419-426, 2011), while, Lorenzo Iorio has investigated the dynamics at the very edge of the solar system (MOND orbits in the Oort Cloud – *The Open Astronomy Journal*, 3, 156-166, 2010).

21. The historical fame of Kepler's laws and Newton's interpretation of them in terms of gravitational attraction tend to make us forget that Kepler's result only applies to a two-body system. Technically speaking, none of the planets within the solar system actually move along perfectly formed and constant elliptical orbits. There is a continual gravitational interaction between the Sun and all of the planets. The long term orbital change in Earth's orbit is discussed under the topic of the Croll-Melankovitch cycle which is described in the author's book, *Terraforming: the creation of habitable worlds* (Springer Publishing, New York, 2009).

22. Richard Baum and William Sheehan have written a comprehensive historical review of this topic in their book, *In Search of Planet Vulcan: the ghost in Newton's clock-work machine* (Plenum Press, New York, 1997). While it is now clear that no intra-Mercurial planet exists, there is still every chance that a class of asteroids, appropri-ately called the vulcanoids, does exist within the orbit of Mercury. Systematic search-es have been made to find these objects, but their location close to the Sun and their

expected small size (less than ~10-km diameter) makes them a very difficult observational target.

23. Asaph Hall, A suggestion in the theory of Mercury (*The Astronomical Journal*, 14, 49-51, 1894). Hall made use of a remarkable theorem developed by French mathematician Joseph Bertand in 1873. Bertrand's theorem shows that in the two body problem only two types of force potentials can produce stable, closed orbits – namely the inverse square central force (e.g., such as that provided by gravitational and electrostatic potentials) and the linear harmonic oscillator potential (e.g., such as that provided by a spring or wire obeying Hooke's law). If the orbit is governed by a central force of the form $F = k\ r^n$, where k and n are constants and r is the radius, the angle between the radius vectors for the maximum and minimum radii (that is the furthest and closest approach distances) will be $\theta = \pi/\sqrt{n+3}$ and for $n = $ -2, we see that the standard Newtonian solution is recovered. For Mercury, the observations indicate that the perihelion (closest approach) radius advances by 43 arc seconds per century and accordingly from Bertrand's formula $n = $ -2.00000016.

24. The connection between spacetime dimensionality and the ability of observers to make meaningful predictions about future events in a stable universe have been discussed in detail by Max Tegmark in his article, On the dimensionality of spacetime (*Classical and Quantum Gravity*, 14, L69-L75, 1997). John Barrow reviews the cosmological implications of multiple-dimensionality in his article, Cosmology: a matter of all and nothing (*Astronomy and Geophysics*, 43, 4.8-4.15, 2002). See also the review by J. Barrow and F. Tipler in their highly recommended (and highly comprehensive) book *The Anthropic Cosmological Principle* (Clarendon Press, Oxford, 1986, section 4.8).

25. P. Ehrenfest, In what way does it become manifest in the fundamental laws of physics that space has three dimensions? (*Koninklyke Akademie van Wetenschappen*, 20, 200-2009, 1918). See also J. Barrow and F. Tipler, *The Anthropic Cosmological Principle* (Clarendon Press, Oxford, 1986, pp.260-262).

26. Numerous excellent books and articles have been written about the life, works and times of Robert Hooke. A very good place to begin any study on Hooke, however, is Allan Chapman's essay, England's Leonardo: Robert Hooke (1635-1703) and the art of experiment in Restoration England (*Proceedings of the Royal Institution of Great Britain*, 67, 239-275, 1996). Also recommended are Lisa Jardine's book, *The Curious Life of Robert Hooke: the man who measured London* (Harper Collins, New York, 2003); Stephen Inwood's book, *The Man Who Knew too Much: the strange and inventive life of Robert Hooke* (Pan MacMillan, London, 2002). Hooke was the first person to receive a salary to do actual scientific work; beginning with his appointment as Curator of Experiments to the Royal Society of London in 1662. He was later, in 1664, appointed Professor of Geometry at Gresham College in London and Cutlerian Lecturer in Mechanics. Gresham College was founded by Sir Thomas Gresham in 1597 and continues to host public lectures in all areas of science to this very day. The Cutlerian Lectures were founded by merchant and Member of Parliament Sir John Cutler in 1664. Hooke received a lifetime's annual bursary to deliver several lectures per year, but after falling out with Cutler he had great difficulty in actually collecting his fee.

27. J. C. Maxwell, On governors (*Proceedings of the Royal Society*, 16, 270-283, 1868). The characteristics of governors are also explored by Mark Denny in his article, Watt steam governor stability (*European Journal of Physics*, 23, 339-351, 2001). See also, Denny's wonderful book, *Ingenium: Five machines that changed the world* (John Hopkins University Press, Baltimore, 2007).

28. All three of Allais's (English language) articles were provocatively entitled, *Should the laws of gravitation be reconsidered?* In order to test such remarkable claims, Andreas Heck (Institute for Gravitational Physics, University of Strasbourg) conducted a series of experiments with a paraconical pendulum from 2005 to 2007, and found no evidence to indicate that any astronomical influences acted upon the pendulum's motion – details are available at http://www.goede-stiflung.org/uk/experiment/E-Allais-Pendel-homepage.pdf. Contrasting conclusions to those by Heck are offered by I. A. Savrov in the article, [The] Paraconical pendulum as a detector of gravitational effects during solar eclipses (*Measurement Techniques*, 38, 253-260, 1995). See also, Michele Caputo's article, On new limits of the coefficient of gravitational shielding (*Journal of Astrophysics and Astronomy*, 27, 439-441, 2006).

Chapter 2

1. The key cosmological observations that indicate a Big Bang origin for the universe are (1) the microwave background, (2) the universal expansion of space and (3) the overwhelmingly predominant abundance of hydrogen and helium. The cosmic microwave background radiation was discovered by Arno Penzias and Robert Wilson in 1964 and is the remnant glow of the thermal radiation that existed at the time of decoupling, which took place some 380,000 years after the Big Bang. Universal expansion was first observed in detail and articulated by Edwin Hubble in 1929. Building upon his own as well as data from other observers Hubble found that the speed V(km/s) with which distant galaxies are moving away from us is proportional to their distance D(Mpc) measured in mega-parsecs: V(km/s) = H D(Mpc), where H is Hubble's constant (now known to have a value of 72 km/s/Mpc). Hubble's constant actually changes with the age of the universe such that $H = \dot{R}/R$, where R is a scale factor describing the expansion of the universe, and $\dot{R}$ is the rate at which R changes with time. The variation of R is determined by the equations of general relativity and the specific matter/energy content of the universe. The cosmic abundance of hydrogen and helium is determined by the conditions that prevailed in the universe after the first three minutes of its existence. The essential theory describing cosmological nucleosynthesis was developed by George Gamow, Ralph Alpher and Robert Herman in 1948. The only elements to be produced during the epoch of cosmological nucelosynthesis were hydrogen (about 75 percent by mass fraction), helium-4 (about 25 percent by mass fraction), and small amounts (less than 0.01 percent by mass fraction) of deuterium, lithium and beryllium. The entire epoch of cosmological nucleosynthesis lasted for about seventeen minutes and was entirely controlled by the rate at which the temperature and density of the universe decreased with age.

2. Technically speaking speed and velocity are different quantities. Speed is a so-called scalar quantity and is accordingly ascribed some numerical value and has the associated units of distance per unit time. Velocity is a vector quantity that has both magnitude (that is a speed associated with it) and direction characteristics.

3. The modern day theory of statistical mechanics technically allows the reverse heat flow process to take place – it is just a highly improbable occurrence, and effectively, therefore, an event never likely to be observed. A very readable introduction to the physics of this problem is given by P. Davies, The Arrow of Time (*Astronomy and Geophysics*, 46, 1.26-1.29, 2005). See also the highly recommended book by Roger

Penrose, *The Road to Reality: a complete guide to the physical universe* (Alfred Knopf, New York, 2004).

4. Tunneling is one of the defining characteristics of quantum mechanics and it is a direct consequence of the wave-particle duality of matter. The phenomenon breaks the classical constraint that in order for a particle to overcome a barrier of energy E_B it must have an energy $E_P > E_B$ – under the rules of quantum mechanics, however, there is a finite probability that a particle will successfully penetrate a barrier even if $E_P < E_B$. The tunneling effect was first developed in the 1930s in order to explain the process of nuclear decay and the emission of particles from an unstable nucleus. Quantum tunneling is vital with respect to the Sun's energy source, provided through the proton-proton chain of fusion reactions, since the thermal energies of the nuclei within its interior are much smaller than the coulomb barrier (the energy due to the electrostatic interaction of two nuclei) that must be overcome for nuclear fusion to take place. Not only must quantum tunneling take place for the proton-proton chain to work, but in the first step, when two protons first approach each other, at the very instant of the tunneling and fusion process one of the protons must also undergo a beta-decay into a neutron. In this manner the first step of the chain is the creation of a stable deuterium nucleus rather than a highly unstable diproton or helium-2 nucleus. Quantum tunneling can also be thought of in terms of imaginary time. In spite of its name, imaginary time is taken to be a 'real', second time dimension, with the imaginary label corresponding to the concept of complex numbers. Imaginary time is described by the transform $t \rightarrow it$, where $i = \sqrt{-1}$. In real time, that is, the time domain that we physically experience, events take place in either the past, the present instant or the future; in imaginary time multiple events, that we cannot directly experience, may take place within any given instant.

5. The speed of light in vacuum is a universal physical constant. Within the context of special relativity it is the maximum speed at which energy, matter and information can be transmitted. All zero mass particles, such as the photon, must travel at the speed of light, and in terms of time dilation (see note 6) all such particles are timeless. In principle an object can travel faster than the speed of light, but such hypothetical (tachyon) particles would need to have extremely bizarre physical properties and would be entirely inconsistent with the presently known laws of physics – they would also violate the fundamental relationship of causality and they could never drop to a speed below that of light.

6. A time dilation effect can also be associated with regions of different gravitational potential, and it has been shown experimentally, for example, that atomic clocks run at different rates according to their altitude above Earth's surface. Gravitational time dilation was described by Albert Einstein in 1907 and comes about as a consequence of his special theory of relativity. The extreme gravitational time dilation effect is that associated with the passage through a black hole's event horizon. At the event horizon boundary the speed required to escape from the black hole's gravitational pull is equal to that of the speed of light - see note 5. Once an observer has crossed the event horizon they may never thereafter escape from the black hole. An external viewer, however, watching a hapless voyager approach an event horizon will observer that the traveler's clock will tick more and more slowly. Indeed, to the external observer, the traveler will never actually appear to cross the even horizon since from their perspective it will take an infinite amount of time for the traveler to reach it.

7. The Rossi-Hall experiment is nicely described on the HyperPhysics website: http://hyperphysics.phy-astr.gsu.edu/hbase/relative/muon.html. The original publication appeared as, Variations of the rate of Decay of Mesotrons with Momentum (*The Physical Review*, 59, 223-228, 1941).

8. The muon is a short-lived elementary atomic particle with characteristics similar to that of the electron. Unlike the electron, however, the muon is unstable and will decay into either an electron or positron (and an associated neutrino) in a typical life time of 2.2 microseconds.

9. Rutherford introduced the analogy between the solar system and the atom in 1911 after reviewing the results of the Geiger-Muller experiment in which helium nuclei (alpha particles) were fired at a gold foil target. Unexpectedly (at the time) some of the alpha particles were scattered back towards the emission aperture. Rutherford famously commented that this observation was akin to firing "a 15-inch shell at a piece of tissue paper and it came back and hit you". The important point, however, that Rutherford deduced from the Geiger-Muller experiment was that the central positively charged nucleus of the atom must be concentrated into a very small region of space. Rutherford was less clear about how the (negatively charged) electrons might be distributed, but built upon the idea of Japanese physicist Hantaro Nagaoka who had suggested in 1904 that the electrons might orbit a massive central nucleus in a manner analogous to that in which the ring-system particles orbit around planet Saturn.

10. The relativistic pendulum is described in Cahit Erkal's article, The simple pendulum: a relativistic revisit (*European Journal of Physics*, 21, 377-384, 2000). See also the articles by Robert Penfield and Henry Zatzkis, The relativistic linear harmonic oscillator (*Journal of the Franklin Institute*, 262, 121-125, 1956), and Louis Gold, Note on the relativistic harmonic oscillator (*Journal of the Franklin Institute*, 264, 25-27, 1957). P. J. Torres, Periodic oscillations of the relativistic pendulum with friction (*Physics Letters A*, 372, 6386-6387, 2008) has further shown the somewhat surprising result that unlike the classical non-relativistic pendulum the externally forced (relativistic pendulum can support periodic solutions even when a velocity dependent friction term is present. The equation studied by Torres has the form $d(mV)/dt + k_1V + k_2 \sin x = F(t)$, where $V = dx/dt$, k_1 and $k_2 > 0$, m is the appropriately Lorentz transformed mass, and $F(t)$ is a periodic forcing term.

11. The Armillary Sphere is essentially a model of the Earth-centered cosmos. The outer shell represents the celestial sphere and various circular hoops designate the celestial equator, the ecliptic and the tropics. The Earth is set at the center of the sphere and some models contained a movable plate to designate the observer's horizon. Armillary spheres were typically used as teaching aids, rather than observing instruments, and they were not intended to show the locations of the stars and constellations on the sky.

12. Equation (2.5) is often referred to as Beverloo's law, after the work by Wim Beverloo (Agricultural University of Wageningen, in the Netherlands). See, W. A. Beverloo, H. A. Leniger and J. van de Velde. The flow of granular solids through orifices (*Chemical Engineering Science*, 15, 260-269, 1961).

13. Kepler introduced the first two of his laws of planetary motion in his *Astronomia Nova* published in 1609. The first law states that planets move along elliptical orbits with the Sun at one focus. The second law states that the planet-Sun line sweeps out

equal areas in equal amounts of time. Kepler's third law was introduced in 1918 in *Harmonius Mundi*. This law states that the square of the planets orbital period is proportional to the cube of its orbital semi-major. Isaac Newton eventually showed in his *Principia* that the second law is a consequence of the conservation of angular momentum, and that the third law is related to the inverse square law of gravitational attraction.

14. Much has been written on the Antikythera Mechanism, the author provides a summary of some recent writings in the article, The Antikythera Mechanism: still a mystery (*Journal of the Royal Astronomical Society of Canada*, 101, 93-94, 2007). A recent, detailed analysis and possible reconstruction of the Antikythera Mechanism is provided by Tony Freeth *et al.*, Decoding the ancient Greek astronomical calculator known as the Antikythera Mechanism (*Nature*, 444, 587-591, 2006). A broader historical account is given in Jo Marchant's book *Decoding the Heavens: solving the mystery of the world's first computer* (William Heinemann, London, 2008). Further up-to-date details on research can be found at the Antikythera Mechanism Research Project web page: http://www.antikythera-mechanism.gr/.

15. *God's Clockmaker: Richard of Wallingford and the invention of time*, by John North (Hambledon, London. 2005), is a highly readable (and highly recommended) account of the life and times of Richard of Wallingford, and the development of the first mechanical clocks. The details of a replica reconstruction of Wallingford's clock at St. Albans Cathedral can be found at http://wallingfordclock.talktalk.net/

16. See, Bedini, S. A., and Madison, F. R., Mechanical Universe: the astrarium of Giovani de'Dondi. *Transaction of the American Philosophical Society*, new. ser. 56, pt. 5 (1966). Images of a modern day reconstruction of Dondi's *astrarium* can be seen at http://www.clockmaker.it/ingle37astrario.htm.

17. The Dresden Planetenlaufuhr is described, with numerous color pictures and line diagrams, by Fortunat Mueller-Maerki in *Antiquarian Horology*, 31 (6), 795-809 (2009). See http://www.skd-dresden.de/de/museen/math_phys_salon/sammlung.html.

18. Ironically, Newton would have only partially agreed with the clock analogy that is accredited to his mechanics. Indeed, Newton felt that the presence of a creator, who was actively interested in the universe, was required at all times. The universe according to Newton may well have had the mechanical beauty of a fine watch, but it still needed God to wind it. As with many of Newton's proclamations, Leibniz was horrified by the idea that God hadn't created a perfect universe which would run unattended for ever. See also Chapter 1, note 7.

19. Irrespective of era, the orrery provides a powerful symbol for the passage of celestial time. Indeed, the future-looking Long Now Foundation (http://www.longnow.org/about/) has recently commissioned and had built a near 3-meter high orrey made of long-wearing monel and stainless steel. The author discusses the symbolic role of the clock and the orrey in the contemplation of humanities future in, The Clock of the Long Now – a reflection (*The Journal of the Royal Astronomical Society of Canada*, 101, 4–5, 2007). See also notes 33 and 34 below.

20. The cometarium was never an especially common lecture device, and was never intended to be a predictive machine. The heyday of these devices was probably located in the later half of the 18th Century at which time a number of London instrument makers produced them to coincide with the 1758 return of Halley's Comet. Desagulier's original device was designed to illustrate the orbit of Mercury, and it appears that the name 'cometarium' was introduced by London-based instrument maker

Benjamin Martin circa 1740. The author has described the early history of such device in, The Mechanics and Origin of Cometaria (*Journal of Astronomical History and Heritage*, 5(2), 155-163, 2002). Additional historical comments can also be found in an article by the author, On Ptolemy's equant, Kepler's second law, and the non-existent empty-focus Cometarium (*Journal of the Royal Astronomical Society of Canada*, 99, 120-123, 2005).

21. In August 2006 the International Astronomical Union (IAU) adopted the following conditions for an object to have planetary status: (1) it must be in orbit about the Sun, (2) it has sufficient mass to be spherical due to its own self gravity, and (3) has cleared the neighborhood around its orbit of smaller objects. A non-satellite body satisfying conditions (1) and (2) is classified as a dwarf planet. The distinction between planets and dwarf planets is deemed to be real and that the categories distinguish between two distinct classes of objects. Under the new IAU definitions Pluto along with the largest of the main belt asteroids, Ceres were reclassified as being dwarf planets. The classification scheme remains controversial and will undoubtedly undergo future revision (see note 22).

22. Part of the controversy surrounding the 2006 IAU definition was that it approached the definition of a planet from a purely astronomical point of view. Geologists, for example, might reasonably say that any object (spherical or not, and in any orbit) that has undergone internal differentiation (that is density sorting) could be classified as a planet. Likewise, the definition is rather parochial since as of this writing nearly 1000 planets [sic] have been found to orbit stars other than the Sun. The term exoplanet and extrasolar planet is often used to describe these latter objects.

23. Kuiper Belt Objects (KBOs) move along circular orbits in the outer solar system and are found within the region extending from beyond the orbit of Neptune (30 AU from the Sun) to about 50 AU from the Sun at the location of the so-called Kuiper cliff where there numbers drop-off dramatically. The existence of a population of objects beyond the orbit of Pluto was suggested by amongst others, Gerard Kuiper (in 1951) and Kenneth Edgeworth (in 1943). The region is also called the Edgeworth- Kuiper belt but increasingly the objects in the Kuiper belt region are simply being called trans-Neptunian objects (TNOs). A sparsely populated scattered disc region extends beyond the Kuiper belt for many hundreds to thousands of AU away the Sun, and the objects within this region have highly eccentric orbits.

24. The key article is G. J. Sussman and J. Wisdom, Numerical evidence that the motion of Pluto is chaotic (*Science*, 241, 433-437, 1988). For details concerning the digital orrery see, J. H. Applegate *et al.*, The outer solar system for 200 million years (*The Astronomical Journal*, 92, 176-194, 1986); and J. H. Applegate *et al.*, A Digital Orrery (*IEEE Transactions on Computers*, C34, 822-831, 1985). A more recent review of this topic is given by Wayne Hayes, Is the Outer Solar System Chaotic? (*Nature Physics*, 3, 689-691, 2007).

25. Boys, C. V. A new astronomical clock. *Monthly Notices of the Royal Astronomical Society*, 38, 74-78 (1877). Perhaps more famous than the clock developed by Boys is the version invented, in 1921, by British engineer William Shortt. The Shortt clock had, like that by Boys, a separate master pendulum that oscillated in a vacuum cylinder. A second clock pendulum, however, was slaved to the master pendulum via an electromagnetic circuit enabling the system to maintain an accuracy of about 1 second per year. It was in 1926 that a Shortt clock was used to demonstrate the non-isochronal spin of the Earth – see, for example, the article by John Jackson (then at

the Royal Observatory at Greenwich), Shortt Clocks and the Earth's rotation (*Monthly Notices of the Royal Astronomical Society*, 89, 239-250, 1929). The effect of the medium through which the pendulum bob moves (air, water, vacuum and so on) was studied in great detail, both theoretically and experimentally, in the classic paper by George Gabriel Stokes, On the effect of the internal friction of fluids on the motion of pendulums (*Transactions of the Cambridge Philosophical Society*, 9, 8-106, 1851).

26. A brief review of the research by Flambaum and co-workers is provided in the article, Proposed Nuclear Clock May Keep Time with the Universe (www.ScienceDaily.com web page for March 8, 2012). For a more general overview of the historical development of timekeeping accuracy see the article by Matthew Chalmers, Super clocks: more accurate than time itself (*New Scientist*, issue 2694, 39-41, February 7, 2009).

27. A very readable introduction to this topic is provided by Bryan E. Penprase in his book *The Power of Stars: how celestial observations have shaped civilization* (Springer, New York, 2011). See also the comprehensive review by Humphrey M. Smith, International Coordination and Atomic Time (*Vistas in Astronomy*, 28, 123-128, 1983).

28. For a general review, see F. R. Stephenson's article, Historical eclipses and Earth's rotation (*Astronomy and Geophysics*, 44, 2.22-2.27, 2003). The eclipse circumstances are specifically used to estimate the quantity $\Delta T = TT - UT$, where TT corresponds to an idealized constant terrestrial time (used to calculate planetary ephemeredes), and where UT corresponds to universal time. The determination of ΔT is based upon the historical accounts describing the time of onset or even observation of an eclipse and the computational predictions using TT and assuming a constant Earth spin rate.

29. Writing in 1687 [see Waller, R. (ed)., *The Posthumous Works of Robert Hooke*, Johnson Reprint Corporation: New York, 1969], Hooke explains how to align a telescope to the NCP, commenting that he uses a helpful group of stars, "consisting of six stars in the Rose itself, and several others in the Leaves and Branches, one of these in the Center of the Rose, and five in the five green leaves and the Knob. This I have somewhere described about ten years since, but have mislaid them at the present…". To the authors knowledge Hooke's [1677] description of the English Rose has never been found, but I have speculated upon which group of stars he might have been refereeing to in the article, In Search of the English Rose, Robert Hooke's lost Constellation (*Journal of the Royal Astronomical Society of Canada*, 98, 183-186, 2004).

30. The manner in which the human brain 'sees' and 'asses' intrinsic patterns in point data sets has been studied by J. D, Barrow, S. P. Bhavsar, and D. H. Sonoda, in the research paper, Minimal spanning trees, filaments and galaxy clustering (*Monthly Notices of the Royal Astronomical Society*, 216, 17-35, 1985). The computational method that the authors employ involves the construction of a minimal spanning tree, which is the shortest length graph that connects together the data points with each branch of the tree having a length falling between two upper and lower pre-set limits.

31. An optical binary is simply a pair of stars that appear close together when seen on the sky from the Earth. They are not physically interacting or indeed necessarily at similar distances from the Sun.

32. Mizar is the 6th brightest star (ζ Ursae Majoris) in the constellation of the Great Bear (Ursa Major). Its position corresponds to the second star from the end of the 'bears' tail. The name Mizar is derived from the Arabic for waistband or girdle. Alcor (derived from the Arabic word for forgotten or neglected) is the faint companion star

to Mizar, and its visibility has long been used as a measure of an individual's eye sensitivity. The pair is easily resolved into two distinct stars with low power binoculars, but detailed telescopic studies have revealed that Mizar is in fact a quadruple system, composed of a binary pair of binary stars; Alcor is also a binary star system. The two star systems are located about 25 pc away from the Sun and separated by about 1/3rd of a parsec.

33. Stewart Brand is co-founder of the Long Now Foundation – Details relating to project can be found at http://longnow.org/about/. See also Brand's book, *The Clock of the Long Now: time and responsibility – the ideas behind the world's slowest computer* (Basic Books, New York, 1999).

34. This quotation is from the article, Measuring Our Escape Velocity as we Exit the 20th Century: http://longnow.org/press/articles/ArtStrategicFinance.php. Also see the descriptive article by Danny Hillis and co-authors, *Time in the 10,000 Year Clock* (http://arxiv.org/pdf/1112.3004.pdf).

Chapter 3

1. I have previously described my own journey to determine the Earth's radius in the article, The Measure of the Earth – a Saskatchewan diary (*Journal of the Royal Astronomical Society of Canada*, 99, 7–9, 2005).

2. In all but the modern physical interpretation, this tendency of solid matter to seek out its 'final destiny' is exactly what we now call gravitational attraction. Indeed, being spherical through their own gravity is now part of the official IAU definition for planetary status (see Chapter 2, note 21).

3. One of the best introductions to this topic is *Sundials: history, theory and practice*, by René R. J. Rohr (Dover Publications, 1970). To find north and to determine the time of year you need to know the direction and length of the sundials gnomon at noon. The shadow at noon points due north, and the time of noon is indicated when the Sun is at its greatest altitude above the horizon (when it is on your meridian and due south). The shadow length at noon is at its longest on the day of the winter solstice, and at its shortest on the day of the summer solstice.

4. The day of the summer solstice corresponds to the day on which the noon time Sun (see note 3) attains its greatest altitude above the horizon (corresponding to the shortest noon time shadow length). On any other day of the year, the Sun at noon is always lower in the sky.

5. In modern terms we would simply say that the latitude of the two locations differs by 7 degrees.

6. The units are 'odd' in the sense that in the modern era we don't use stadia to measure distances and because the Greeks didn't use degrees to measure angles. The angle of the Sun's offset from the vertical at Alexandria was quoted by Eratosthenes as being 1/50th of a circle. A scholarly discussion of the measurements assumed by the ancients is presented by Irene Fischer's article, Another Look at Eratosthenes' and Posidonius' Determinations of the Earth's Circumference (*Quarterly Journal of the Royal Astronomical Society*, 16, 152-167, 1975). See also the detailed discussion in James Evans's highly recommended book, *The History and Practice of Ancient Astronomy* (Oxford University Press, Oxford, 1998. pp. 63-66).

7. The radius can be determined from the well known equation for the circumference of a circle $C = 2\pi R$, where R is the radius and $\pi = 3.14159$. Accordingly, R =

252,000 / 6.28319 = 40,107 stadia. The standard conversion for the stadia is that it equals $1/6^{th}$ of a kilometer, and hence Eratosthenes estimate for the radius of the Earth is 6684.5 km. The Earth's mean radius is actually 6378.14 km, so the value determined by Eratosthenes is about 5% too big.

8. A star (or planet) culminates when it is due south of an observer and situated on their meridian, an imaginary line (great circle in fact) which runs from the north celestial pole through the observer's zenith point and on over the celestial sphere to the south celestial pole.

9. The life and times of Gerard de Cremer (Mercator is the Latinized form of his name, meaning *merchant*) are described in Andrew Taylor's book, *The World of Gerard Mercator: the mapmaker who revolutionized geography* (Harper Collins, London, 2004).

10. Although often called a cylindrical projection, what Mercator actually did is more complex (and is technically a transform rather than a projection). Mercator, in fact, did not explain his methods clearly and it fell upon the shoulders of English mathematician Edward Wright to correctly describe, in 1599, the mathematical details of the transform. The key point about the Mercator 'projection' is that while it distorts areas and coastlines it conserves angles (it is technically what is called a conformal transformation). This, with respect to navigation, is the important point since it gives the true direction in which to travel – the so-called great circle path. The co-ordinate transforms for taking the latitude and longitude (φ, λ) to the Mercator (x, y) points are: $x = R(\lambda - \lambda_0)$ and $y = R \ln((1 + \sin\varphi)/\cos\varphi)$, where R is the radius of the Earth corresponding to the map scale being used and λ_0 is the longitude corresponding to the $x = 0$ point. An inordinate amount of nonsense, mostly driven by politicians with no sense of history or mathematics, has been written about map projections in recent decades, especially by Arno Peters (1916-2002) and his followers. But then, of course, there is nothing more political than a set of imaginary lines and boundaries projected onto a piece of flat paper that can never be more than a 2-dimensional distortion (in one form or another) of a 3-dimesnional world. A good introduction to the history of map making is provided by Mike Parker in his book, *Map addict: a tale of obsession, fudge and the Ordnance Survey* (Collins, London, 2009).

11. See Chapter 2, note 13. Kepler's third law links together the period P and semi-major axis a of the orbit so that $P^2 = K a^3$, where K is a constant. If the units of years for the orbital period and astronomical units for the semi-major axis are used then, conveniently, the constant $K = 1$. In general, however, $K = 4\pi^2/G(M_{planet} + M_{Sun})$, where now the *SI* units of seconds, meters and kilograms will apply to the various quantities.

12. A detailed history and review of the determination of the astronomical unit is provided by David W. Hughes in his article, Six stages in the history of the astronomical unit (*Journal of Astronomical History and Heritage*, 4, 15-28, 2001).

13. A very detailed history of the Peruvian expedition has been presented by Larrie Ferreiro in his book, *Measure of the Earth: the enlightenment expedition that reshaped the Earth* (Basic Books, New York, 2011).

14. By way of a mathematical example, let us take a network of three connected bars - as shown in figure 3.7. Let the central bar be made of iron and have a length L_0, which is the desired length of the pendulum. If the temperature increases by an amount ΔT, then the central iron bar will expand, according to our earlier formula, to

a length of L_0 $(1 + \alpha_{Fe} \Delta T)$. Now, attached to the base of the central iron rod is a second brass rod of length L_{B0}, which after the temperature increase will have a length L_{B0} $(1 + \alpha_{Brass} \Delta T)$. Finally, attached to the top of the brass rod is a second iron bar of length L_1, which after a temperature change of ΔT will have a length of L_1 $(1 + \alpha_{Fe} \Delta T)$. The trick now is to fix L_{B0} and L_1 so that the final length of the network is L_0, ensuring therefore that the total length of the pendulum hasn't changed. Let us set L_0 = 10-m (this is a big pendulum) and take a temperature change of ΔT = 10 degrees. The coefficients of expansion for brass and iron are given in table (3.1). The equation to satisfy is $L_0 = L_1$ $(1 + \alpha_{Fe} \Delta T)$ - L_{B0} $(1 + \alpha_{Brass} \Delta T)$ + L_0 $(1 + \alpha_{Fe} \Delta T)$. Putting numbers into the formula we find, that L_{B0} = [(1.00012) L_1 – 0.0012] / (1.0002), if we further set L_1 = 1-m (since it is a length we can choose) then L_{B0} = 0.9987-m. In other words, for this particular example the brass rod must be cut about 1 millimeter shorter than the second iron rod in order that the total length of the pendulum remains constant, irrespective of any surrounding temperature variations.

15. These comments were made by John Herschel, then President of the Royal Astronomical Society, during the 28th Annual General Meeting held in February 1848 (*Monthly Notices of the Royal astronomical Society*, 8, 116-118, 1848). Herschel also directed high praise towards Everest's then recent publication, *An account of the measurements of two sections of the meridional arc of India* (J & H Cox, London, 1847).

16. Amongst Himalayan explores there is the well known, and apparently true, story that the first person to place "two feet" on top of Mount Everest was Indian mathematician Radhanath Sikdar in 1856. This, of course, is a play on words and the reference is to a measure of 2 feet rather than the placing of two actual feet on the summit of Everest. Sikdar rose to the rank of Chief Computer within the Trigonometric Survey of India, and was assigned in 1854 to the task of determining the height of Peak XV (later to be named Mt. Everest) in the Himalayan range. After two years effort he eventually determined that the height of Peak XV was exactly 29,000 feet. The survey officials, specifically Surveyor-General Andrew Waugh, thinking that such an exact number would be viewed suspiciously arbitrarily added on an additional 2 feet.

17. While Pluto is the closest large KBO to the Sun, the largest object presently known in the Kuiper belt is the Dwarf Planet Eris, discovered in 2005. Eris is about 25% more massive and about 5% larger than Pluto and it is accordingly the 9th largest (currently known) object in orbit about the Sun.

18. Technically this formula also includes a latitude dependent centrifugal correction due to Earth's rotation – this effect is described in Chapter 5.

19. The quintessential hollow Earth story is Jules Verne's 1864, *A Journey to the Center of the Earth*. In this story the journey begins with the adventurers descending a series of volcanic lava tubes in Iceland and ends with them being lofted back to Earth's surface upon a rising lava column through Mt. Etna in Sicily. For more detailed historical perspectives on the hollow as well as flat Earth misconceptions see, David Standish, *Hollow Earth* (Da Capo Press, Cambridge, MA, 2006) and Christine Garwood, *Flat Earth: the history of an infamous idea* (Pan Books, London, 2007). That Earth's subterranean regions might be inhabited was not an especially controversial point for Halley to make, since it was generally believed at that time that all of the planets and even the Sun were inhabited.

20. Hooke's law relates to the mechanical stretching and elasticity of materials – especially wires and springs. The key point is that the extension is directly proportional to the applied load, up to, that is, a certain limit (corresponding to the yield strength) at which point the material becomes deformed and the linear extension law breaks down. Hooke appears to have been particularly proud of his law of extension, but initially announced his discovery in the form of a Latin anagram "ceiiinossssttuv", the solution to which he only released in 1678 as "ut tension, sic vis", meaning "as the extension, so the force". While the anagram established Hooke's priority of discovery, the secrecy concerning its form was partly inspired by interests in developing spring-drive timepieces. An acrimonious dispute broke out between Hooke and Christiaan Huygens (and various members of the Royal Society – especially its Secretary, Henry Oldenburg) in the 1670s concerning priority over the development of coiled-spring driven watches. While Hooke appears to have demonstrated the possibility of using a spring to power a timepiece in 1664, it was Huygens, in 1674/5, who first constructed such an actual device. For further reference details see note 26 in Chapter 1.

21. Of the various Mars Landers and Rovers only the two Viking Landers (operational from 1976 to 1980/2) included seismograph instrument packages. The seismometer on Viking 1 failed to work, however, and that aboard Viking 2 suffered from a high noise background (probably due to wind vibration and thermal expansion and contraction effects upon the Lander itself). A review paper on the Viking 2 results is given by D. L. Anderson *et al.*, in the article, Seismology on Mars (*Journal of Geophysical Research*, **82**, 4524-4546, 1977). The French Space Agency (CNES) is currently preparing to build a network of four seismometers to travel to Mars (at some future date) to specifically map-out the planet's interior structure. Current theoretical models suggest that something like 1000, magnitude 3 on the Richter scale, marsquakes should take place per year.

22. The Apollo missions to the Moon established a network of five seismometers, and these were operational until the end of 1977. An attempt to revive the network in 1986 failed due to insufficient power to run the various system packages. A brief summary of the Apollo Passive Seismic Experiment (PSE) and the requirements for the establishment of a new seismic network are given by C. R. Neal *et al.*, in the article, The Lunar Seismic Network: mission update (*Lunar and Planetary Science XXXV* (2004), paper 2093.pdf).

23. See L. M. Burko's paper, Effect of the spherical Earth on a simple pendulum (*European Journal of Physics*, **24**, 125-130, 2003). We can simplify the potential energy term in equation (3.5) by factoring out an R^2 term from the square root expression to obtain $gR^2 / \sqrt{R^2\left[(1 + L/R)^2 + (L/R) - 2(L/R)(1 + L/R)\cos\phi\right]}$ which, when L/R is taken to be vanishingly small gives $gR^2 / \sqrt{R^2} = gR = G\,M/R$.

24. The spherical Earth correction does not apply in the infinite pendulum case since the pendulum wire is taken to move as if $\phi = 0$ at all times, and the pendulum bob sweeps out a straight line path - being a segment of arc in an 'infinite' radius circle. In this situation the entire square root term in equation (3.5) reduces to a value of R since $\cos\phi = 1$ at all times. In addition, note that to be consistent with figure 1.6, we must also make the transformation that $R \rightarrow R + h$, where h is the instantaneous vertical height of the pendulum bob above the Earth's surface.

25. See Richard Berg and Todd Marshall's paper, Wilberforce pendulum oscillations and normal modes (*American Journal of Physics*, 59, 32-38, 1991). See also the original paper by L. R. Wilberforce, On the vibrations of a loaded spiral spring (*Philosophical Magazine*, 38, 386-392, 1894).

26. See, for example, the text by Kurt Magnus, *Vibrations* (Blackie and Son Ltd., London, 1965. Section 4.2).

27. See the informative paper by P. Arun, the moving center of mass of a leaking bob (arXiv:1002.3956v1). See also the article by Rafael Digilov, M. Riener, and Z. Weizman, Damping in a variable mass on a spring pendulum (*American Journal of Physics*, 73, 901-905, 2005).

28. There is no such physical thing as negative friction, of course. Rather, the amplitude increases as a consequence of the conservation of energy. If we just think about the potential energy at the extremity of the pendulum's arc, we know that the initial energy will be PE $(t = 0) = m_0gh_0$, where m_0 is the initial mass, g is the acceleration due to gravity, and h_0 is the initial release height. Conservation of energy requires that when the bob, at the extreme of its arc, is at rest so PE $= $ PE$(t = 0)$, and so when the bob mass is $m_t < m_0$ we must have $m_0gh_0 = m_tgh_t$, which dictates that $h_t = (m_0/ m_t) h_0 > h_0$, and this increase in height is accommodated for by a larger angle of swing by the pendulum.

29. There are, in fact, deep philosophical issue at play with respect to this idea of passing from the ideal realm to that of the real world - not least in Plato's philosophy. Given the very un-real assumptions in the derivation of equation (1.4) – a mass less, none-extensible string, no-friction, constant gravitation field, point mass for the bob, no air-resistance to motion and so on - it is remarkable that the formula describes anything useful at all. Indeed, we should always remember that fundamental equations do not govern the motion of objects in reality; they only describe how idealized objects move in an idealized model. For further reading on this topic a good place to begin is with Nancy Cartwright's book, *How the Laws of Physics Lie* (Oxford University Press, 1994).

Chapter 4

1. The summit of Mt. Chimborazo is, in fact, the highest point on the Earth's surface. It has this distinction even though its sea-level elevation is only 6267-m making it, therefore, apparently smaller than Chomolungma (Mt. Everest) by several kilometers. The claim holds for Mt. Chimborazo, however, provided one measures the summit to Earth's center distance. In this measure the equatorial bulge places the tip of Chimborazo further from Earth's center than of Chomolungma. The first ascent of Chimborazo was claimed by famed British mountaineer Edward Whymper (who was also the first person to reach the summit of the Matterhorn) in 1880. Whymper actually climbed Chimborazo twice in 1880 as part of an investigation into the effects of reduced pressure and altitude sickness on the human body.

2. For detailed experimental investigations, see: J. Thomas *et al.*, Testing the inverse square law of gravity on a 465-m tower (*Physical Review Letters*, 63, 1902-1905, 1989); Mark Ander, *et al.*, Test of Newton's inverse-square law in the Greenland ice cap (*Physical Review Letters*, 62, 985-988, 1989); Mark Zumberge *et al.*, Submarine measurements of the Newtonian gravitational constant (*Physical Review Letters*, 67, 3051-3054, 1991). For an historical review of the topic see Luigi Foschini's paper, Short

range gravitational fields: rise and fall of the fifth force (*Annales de la Foundation Louis de Broglie*, 23, 156-160, 1998).

3. Nevil Maskelyne, A proposal for measuring the attraction of some hills in the Kingdom by astronomical observations (*Philosophical Transactions of the Royal Society of London*, 65, 495-499, 1775). The actual results of Maskelyne's expedition were presented by Charles Hutton in, An account of calculations made from the survey and measures taken at Schehallion [*sic*, see note 5 below] in order to ascertain the mean density of the Earth (*Philosophical Transactions of the Royal Society of London*, 68, 689-788, 1778).

4. Following figure 4.4(B) the tangent of angle φ is given by the ratio $\tan(\varphi) = F_{Mountain} / F_{Earth}$. Now, these two force terms can be expressed in terms of Newton's formula such that for a small pendulum bob of mass m, $F_{Mountain} / F_{Earth} = (G M_M m / d^2) / (G M_\oplus m / R_\oplus^2) = (M_M / M_\oplus) (R_\oplus / d)^2$. If the mass is now expressed in terms of the bulk density of the material out of which the mountain and the Earth are made then $\tan(\varphi) = \frac{1}{2} (\rho_M / \rho_\oplus) (d / R_\oplus)$, where ρ_M and $\rho_\oplus$ are the bulk densities of the material that constitutes the mountain and the Earth respectively. The factor of $\frac{1}{2}$ enters at this stage since the mountain is assumed to be a hemisphere. When it is assumed that the densities are the same then $\tan(\varphi) = \frac{1}{2} (d / R_\oplus)$.

5. There are many different spellings for Schaehallion of which Schiehallion and Schehallion appear to be the most common alternatives. Schaehallion is one of the most visible of mountains in the Scottish Grampians – the same range that contains Ben Nevis, the highest mountain in the United Kingdom. Maskelyne comments that the mountain was known as the *Maiden-Pap* by the local inhabitants, while the word Schaehallion itself is usually translated to mean *The Fairy Hill of the Caledonians* – which is a lot of meaning to fit into just one word!

6. A general account of Airy's work at Harton Colliery is provided in his book, *Lecture on the pendulum experiments at Harton Pit* (Longman and Co. London, 1855). This text is based upon a talk delivered at the Central Hall in South Shields on 24 October, 1854. A modern day equivalent to the Airy experiment is described in a research paper by Mark Ander and co-workers, Test of Newton's inverse-square law in the Greenland ice cap (*Physical Review Letters*, 62, 985-988, 1989). In this latter experiment the measurements were made with La Coste-Romberg gravimeters at various depths within a 2033-m borehole.

7. The pendulums used in Airy's experiment were essentially measuring a difference in the gravitational attraction at two vertically separated points within the Earth. With reference to figure 4.5, the gravitational acceleration at point Q will be $g_Q = G M_\oplus m / R_\oplus^2$, where m is the mass of the pendulum. At point P, on the other hand, $g_P = G (M_\oplus - M_{shell}) m / (R_\oplus - r)^2$. The values of g_Q and g_P are found by determining the periods of the pendulum at locations P and Q, and then, by estimating r and M_{shell} so the Earth's mass can be found. That the gravitational force due to the shell of material above the pendulum cancels out is one of the most important points of Newton's gravitational theory. In order to illustrate this result, imagine the shell to be very thin and composed of material having a mass density ρ kg/m^2. Now for any point P interior to the shell, we can construct two cones radiating outwards – the center line of each cone lying along a common diameter cutting through the center point of the shell. One cone will have a side of length r_1 and the other a side of r_2, and each cone will have the same vertex angle of φ radians at P. The base area of each cone swept out upon

the surface of the shell will then be $A_1 = \pi r_1^2 \varphi / 4$ and $A_2 = \pi r_2^2 \varphi / 4$. Further, the gravitation force due to each area at point P will be $F_1 = G\rho A_1 / r_1^2$ and $F_2 = G\rho A_2 / r_2^2$. Now, upon substituting for the area terms we find, $F_1 = F_2 = G\rho\pi\varphi / 4$ = constant. This is the result we needed in the sense that it tells us that the gravitational forces from the two surface areas are equal and opposite irrespective of the location of P inside to the shell. Generalizing and expanding this argument across the entire surface area of the shell gives us the final result that the gravitational force at any point interior to a shell is exactly zero.

8. Some of the details relating to the Dolcoath (a name derived from the Cornish for Old Ground) pendulum experiments are given in the Presidents Address (Airy was actually the President of the RAS at the time) to the Royal Astronomical Society for 10 November, 1854 (*Monthly Notices of the Royal Astronomical Society*, 15, 35-36, 1854). The Dolcoath mine became the largest and deepest tin and copper mine in Cornwall having a history that stretched from the early 1700s to its final closure in 1920.

9. Henry Cavendish was apparently an extremely shy man and, for example, he found it unbearable to look at or even talk to women – the maids of his household could in fact be dismissed if he even so much as caught a glimpse of them. Of his character, it was once commented that 'Cavendish probably uttered fewer words in the course of his life than any man who ever lived to fourscore years, not at all excepting the monks of La Trappe'. Independently wealthy Cavendish lived frugally (some would have said shabbily) in central London and dedicated his life to experimental research. While his researched spanned many fields of inquiry Cavendish published very few of his results and researchers are still sifting through and exploring the great number of manuscripts that he produced but never saw into press. Cavendish features amongst many other famous scientists (including Einstein, Feynman, Maxwell, and Newton) in Clifford A. Pickover's book, *Strange Brains and Genius: the secret lives of eccentric scientists and mad men* (Harper Collins, New York, 1998).

10. Let us assume that the spheres envisioned by Newton are made of rock with a density of 4500 kg/m³. In this case, each will have a mass m = 66.72 kg. If we now imagine one sphere to be fixed, then the gravitational force acting upon its freely moving neighbor will be $F = G\, m^2 / (0.31115)^2$, where G is the universal gravitational constant and the initial distance between centers is 12.25 inches (0.31115-m). From Newton's second law the initial acceleration will be given by $a = F / m$, and accordingly the time T taken for the second sphere to move ¼ of an inch (0.635 cm) will be given by the formula $T^2 = (2 \times 0.00635) / a$. Accordingly we find T = 525.6 seconds or about 9 minutes.

11. The author has studied the oscillation times for small objects (mini black holes – see note 22 below) moving within the interiors of spheres with varying density distributions. See, Oscillations and settling times for black holes placed within planetary and stellar interiors (*Journal of the British Interplanetary Society*, 60, 257-262, 2007. The crossing time T_{cross} is equivalent to the free-fall collapse time T_{ff} under gravity. That is, if by some strange quirk of physics all the pressure support inside of the Earth (or a planet or a star) suddenly vanished then gravity would crush all the material down to a black hole singularity is a time T_{ff}. A dimensional analysis argument can be employed to find the equation for T_{ff} – assume that the collapse time depends upon the gravitational constant G, and the density ρ, and radius R of the object undergoing collapse. Then, we have $[T_{ff}] = k[\rho]^A[R]^B[G]^C$. Substituting for the base units of length

l, mass *m* and time *s*, we find the following identity: $l^0 \, m^0 \, s^1 = l^{B-3A+3C} \, m^{A-C} \, s^{-2C}$, which yields the identities: $A = C = -\frac{1}{2}$ and $B = 0$, and accordingly $T_{ff} = k \, / \sqrt{G\rho}$. This expression is the same as that for T_{cross}, but emphasizes the point that the collapse time is inversely proportional to the density (more dense objects collapse more quickly) and that the collapse time is independent of the size of the object undergoing collapse. While the collapse time for the Earth is about 42 minute, the free-fall collapse times for the Sun and Jupiter are about the same and of order 27 minutes.

12. Physicist and renowned science fiction writer Robert Lull Forward, along with D. Berman, discussed this particular experiment at the 14th Annual Meeting of the American Astronautical Society, in Dedham, Massachusetts in 1968. The formula shown in the text follows from the harmonic oscillator crossing time equation when the mass of the sphere is expressed in terms of its density and volume. Interestingly, it turns out that while the sphere must have a uniform density it doesn't matter that the borehole passes directly through the center of the sphere – indeed, it can be at any angle through the interior and the oscillation time will be exactly the same. The author has discussed the history of the mathematical problem associated with the Earth tunnel in *The Observatory Magazine*, December issue, 2013.

13. See for example, Airy's argument in his article, The Interior of the Earth (*The Observatory*, 2, 45-52, 1878). For a more contemporary picture see James Jackson's article, Mountain roots and the survival of cratons (*Astronomy and Geophysics*, 46, 2.33-2.35, 2005).

14. Galileo's early biographer Vincenzo Viviani recounts the story of two objects of different mass being dropped from the top of the Tower in Pizza in an attempt to disprove Aristotle's dictate that more massive objects fall more rapidly than low mass ones. As a biographer, however, Viviani has been found to be less than reliable, and since Galileo never made any mention of this supposed highly public experiment it more than likely never took place. That this story continues to be found in many present day introductory texts is testament to the enduring power of history being perpetuated by rumor. The equivalent of Galileo's experiment was actually conducted by astronaut David Scott during the Apollo 15 lunar mission. Scott simultaneously dropped a feather and a hammer and the recorded video sequence clearly shows them falling at the same rate.

15. Edward Teller, On the changing of physical constants (*Physical Review Journal*, 73, 801-802, 1948). A more detailed account than the one page article published by Teller is given by P. Pochoda and M. Schwarzschild in their article, Variation of the gravitational constant and the evolution of the Sun (*Astrophysical Journal*, 139, 587-593, 1964). For a more recent review of cosmological models in which various universal constants are allowed to change with time, see John Barrow's article at http/arxiv.com/abs/0912.5510. See also, Fred Adams, Stars in other universes: stellar structure with different fundamental constants (*Journal of Cosmology and Astroparticle Physics*, 8, article 010, 2008).

16. See the biography by Graham Farmelo, *The Strangest Man: the hidden life of Paul Dirac, mystic of the atom* (Basic Books, New York, 2009). Although Farmelo's book is highly readable, it completely fails to explain why Dirac should be considered "the strangest man". This being said, Dirac's upbringing was certainly unusual, and his general behavior and later life demeanor could be considered especially eccentric, but

no more (or less) than many other historical and even contemporary genii – see note 9 above.

17. Dirac first introduced the idea of the large number hypothesis in a brief, one-page paper entitled The Cosmological Constants (*Nature*, 139, 323, 1937). Later in life he came back to discuss the topic again in Cosmological models and the large number hypothesis (*Proceedings of the Royal Society A*, 338, 439-446. 1974).

18. We note here that while G may not change with the age of the universe it has not been shown experimentally that α_G is constant. Indeed, m_p, h and c may yet be found to vary over time.

19. See section 5.6 of John Barrow and Frank Tipler's, *The Anthropic Cosmological Principle* (Clarendon Press, Oxford, 1986).

20. The supernova disruption of massive stars is required to generate all of the chemical elements beyond that of hydrogen and helium. Clearly, therefore, the existence of interesting chemistry, planets, life and us is entirely dependent upon the historical supernova rate within the galaxy.

21. See, for example, the author's book, *The Large Hadron Collider – unraveling the mysteries of the universe* (Springer Publishing, New York, 2011).

22. The idea of primordial black holes was first explored in detail by Stephen Hawking in the 1970s. The essential idea is that such black holes would form in the very earliest moments of cosmic creation (the Big Bang), rather than through the collapse of supermassive stars. It has been suggested that some (all ?) of the dark matter in our galaxy might be composed of primordial black holes in the $10^{14} - 10^{23}$ kg mass range. Due to the emission of Hawking radiation, primordial black holes with masses smaller than 10^{12} kg will have evaporated by now. See the author's book, *The Physics of Invisibility* (Springer Publishing, New York, 2012) for a discussion on analog black hole laboratory experiments.

23. See Yang Luo, Shravan Hanasoge, Jeroen Tromp and Frans Pretorius, Detectable seismic consequences of the interaction of a primordial black hole with Earth (arXiv:1203.3806v1). Luo and co-workers find that the energy deposited by a 10^{12} kg PBH is comparable to that associated with a magnitude 4 earthquake.

24. The ratio of the Coulomb force to the gravitational force for an electron-proton pair is $F_C / F_g = (k/G)(e^2/m_e m_p)$, where k is the Coulomb constant.

25. String theory is one of the present frameworks under which physicists are hoping to unite the theories of quantum mechanics and general relativity. It is a contender for a theory of everything (TOE) in that it sets out to explain all of the fundamental forces and all of the known forms of atomic matter. A mathematically gentle outline of the theory is given by Steven Gubser in his book *The Little Book of String Theory* (Princeton University Press, Princeton, 2010).

26. For a detailed review see John Barrow's article, Cosmology: a matter of all and nothing (*Astronomy and Geophysics*, 43, 4.8-4.15, 2002).

27. See the comprehensive review by John Webb, Are the laws of nature changing with time? (*Physics World*, 33-38, April 2003). A review of the experimental methods for testing the constancy of the fundamental constants over time is given by J. C. Berengut and V. V. Flambaum in their article, Manifestations of a spatial variation of fundamental constants on atomic clocks, Oklo, meteorites, and cosmological phenomena (arXiv:1008.3957v1).

28. Lyman absorption lines correspond to transitions taking the electron from the ground state of the hydrogen atom. The Lyman alpha line corresponds to the $n = 1$ to $n = 2$ electron transition. The Lyman series of emission lines were first investigated by American spectroscopist Theodore Lyman in 1906.

29. See J. Chaste *et al.*, A nanomechanical mass sensor with yoktogram resolution (*Nature nanotechnology*, 2012.42). See also, K. Jensen, K. Kim and A. Zettl, An atomic-resolution nanomechanical mass sensor (*Nature nanotechnology*, 2008.200).

Chapter 5

1. Jack Repcheck provides an excellent biographical account in his book, *Copernicus' Secret: how the scientific revolution began* (Simon and Schuster, New York, 2008).

2. The formative members and formative years of the Royal Society of London are eloquently described by John Gribbin in his book, *The Fellowship: the story of a scientific revolution* (Overlook Press, New York, 2007). An early history is provided by Thomas Sprat in his, *The history of the Royal Society of London for the improving of natural knowledge* (Samuel Chapman, London, 1722).

3. See note 26 in Chapter 1.

4. No known portrait of Robert Hooke has survived to the modern era, and while the story may well be apocryphal it is often reported that one of Newton's first actions upon becoming President of the Royal Society was to order the burning of Hooke's portrait. Recently, artist Rita Greer produced a portrait of Hooke for the Institute of Physics in London. The portrait is based upon a number of written descriptions of Hooke; his friend John Aubrey, for example, described him thus, "he is but of middling stature, something crooked, …. his head is large, his eie full and popping, and not quick. He haz a delicate head of haire, browne, and of an excellent moist curl. He is and ever was temperate and moderate in dyet".

5. *The Diary of Robert Hooke: 1672-1680*. H. W. Robinson and W. Adams (Eds.), Taylor and Francis, London.

6. The time t_{fall} for the weight to fall through a distance $h = 8.23$-meters is given by the equation $h = \frac{1}{2} g (t_{fall})^2$, where g is the acceleration due to gravity. Accordingly, $t_{fall} = 1.296$ seconds. The relative velocity between the Earth's surface and the tangential velocity of the weight will be ωh, where $\omega = 7.272 \times 10^{-5}$ is the Earth's spin rate in radians per second. Accordingly at the latitude of London (taken to be 50º) we have a relative velocity of 0.385 mm/s, and the displacement of the falling weight will be about $\frac{1}{2}$ a millimeter.

7. The early history of the Greenwich Observatory, Flamsteed's study on Earth rotation and Tompion's clocks are described in great detail by Derek Howse in his article, The Tompion clocks at Greenwich and the dead-beat escapement. The article is published in two parts; Part I: 1675-1678 (*Antiquarian Horology*, 7, 18-35, 1970); Part II: 1678-1971 (*Antiquarian Horology*, 7, 114-134, 1970).

8. The 'pendulum mania' that followed in the wake of Foucault's experiment is documented by Michael Conlin in, The popular and scientific reception of the Foucault Pendulum in the United States (*Isis*, 90, 181-204, 1999). A general biographical account of Foucault is provided by Amir Aczel in his book, *Pendulum: Leon Foucault and the triumph of science* (Washington Square Press, New York, 2003). An accessible mathematical account of Foucault's pendulum is presented by John Hart, R. Miller and R.

Mills in their article, A simple geometric model for visualizing the motion of a Foucault pendulum (*American Journal of Physics*, 55, 67-70, 1987).

9. The Amundsen-Scott South Pole Station is situated very close to the Earth's actual spin axis and is an ideal, albeit inhospitable place to perform the Foucault pendulum experiment. The experiment used a 33-meter pendulum. See: http:www.physastro.sonoma.edu/graduates/baker/southpolefoucault.html.

10. See the article by V. M. Babović and S. Mekić, The Bravais pendulum: the distinct charm of an almost forgotten experiment (*European Journal of Physics*, 32, 1077-1086, 2011). For a comprehensive historical review of the various attempts to measure the rotation of the Earth see the article by William Rigge, Experimental proofs of the Earth's rotation (*Popular Astronomy*, 21, 208-216, and 267-276, 1913).

11. The currently favored theory for the formation of the Moon is that of the giant impact hypothesis. Indeed, this model nicely explains the observational data with respect to the Moon's bulk composition and it fits naturally into the solar nebula hypothesis describing the formation of the inner solar system. In the giant impact hypothesis it is reasoned that the newly formed Earth suffered a final grazing impact from a Mars-sized proto-planetary body. This impact ejected material from the Earth's outer mantle into space. Some of the collision debris was captured into Earth orbit and then aggregated into the Moon.

12. Angular momentum describes the rotational state of a physical system. If, for example, an object of mass m rotates with velocity V a distance r away from a fixed origin then its angular momentum is given by the product $h = m\,V\,r$. The key point about the angular momentum (in a closed system that experiences no external torques) is that it is a conserved quantity. This means that the numerical value of h must remain the same at all times, and that if the radius r, for example, is changing, then the velocity V must also vary in sympathy so that the product $r\,V$ remains constant (it being assumed that the mass is constant). For rigid spinning bodies, such as the Earth, the spin angular momentum is give by the product $L = I\,\omega$, where I is the so-called moment of inertia and $\omega = V/r$ is the angular velocity. The moment of inertia will vary according to the shape and mass distribution of the spinning body. A uniform density sphere of mass m and radius R, for example, will have a moment of inertia $I = (2mR^2)/5$, while a uniform density beam of mass m and length L suspended about its mid-point (as might be used in a torsion balance experiment) will have $I = (mL^2)/12$. For systems such as the Earth and Moon the sum of the spin and the orbital angular momentum is conserved, and this results in the fact that as the Earth spins down so the Moon's orbital radius increases.

13. Seth Carlo Chandler (1846 – 1913) was an American astronomer who studied variable stars and worked on improving estimates for the constant of aberration (as discovered by James Bradley in 1725). He is perhaps best known today, however, for the discovery of Chandler Wobble which is the small, semi-regular movement in the location of the Earth's spin axis relative to the Earth's surface - the variation typically amounts to no more than 15 meters on a timescale of years.

14. For a recent detailed study on this topic see J. J. Lissauer, J. W. Barnes and J. E. Chambers, Obliquity variations of a Moonless Earth (*Icarus*, 217, 77-87, 2011).

15. As the Sun's luminosity increases, so, for a given heliocentric distance, the energy flux at the top of Earth's atmosphere (the insolation) must increase. For details, see

the author's book, *Terraforming: the creating of habitable worlds* (Springer Publishing, New York, 2009).

16. The stage show counterpart to this statement is, of course, the great Indian rope trick, in which a person is seemingly able to climb an erect, unsupported rope. Apparently, the Magic Circle in London established a (still unclaimed) prize of 500 guineas in 1934, to be collected by the first person to successfully perform an outdoor version of the rope trick. In the sense of science being much more miraculous, but perhaps less entertaining, than magic, however, a rope (or multiply linked chain of rods) can be made to stand vertically by fixing its base point to a rapidly vibrating platform. This stabilization phenomenon was first described mathematically by Andrew Stephenson (University of Manchester) in a series of articles (both titled, On induced stability) in the *Philosophical Magazine* (15, 233-236, 1908, and 17, 765-766, 1909).

17. The Segway Personal Transporter uses an active torque control upon its wheel drive to maintain stability, and its motion is controlled by tracking variations in the center of mass as the passenger shifts their weight. In some, hair-splitting, sense the active torque pendulum is really an articulated arm, rather than a free-pendulum. Mark Spong has described the design and control theory behind a two-armed motor driven arm in his article, The swing-up control problem for the Acrobat (*IEEE Control Systems Magazine*, 15, 49-55, 1995). Spong and Daniel Block have also described a variant to the Acrobot in their article, The Pendubot: a mechatronic system for control research and Education (*34th IEEE Conference on Decision and Control*, New Orleans, 1995, pp. 555).

18. See D. ter Harr (Ed.), *Collected Papers of P.L. Kapitza* (Pergamon Press, Oxford, 1986). Although generally called the Kapitza pendulum, the theory and experimental verification of the stable vertical pendulum was first provided by Andrew Stephenson in 1908 (see note 19).

19. An alternative way of describing the inverted pendulum is to make use of the imaginary time concept [recall note 4 in Chapter 2]. Starting with the equation of motion for a simple pendulum $d^2\theta/dt^2 + (g/L)\theta = 0$, we introduce the transform $t \rightarrow it$, which, since $i^2 = -1$, gives $-d^2\theta/dt^2 + (g/L)\theta = 0$. The equation for the simple pendulum can be recovered, however, if we further introduce the transform $g \rightarrow -g$. In other words, the inverted simple pendulum is the mirror-image equivalent of the simple pendulum in an imaginary time universe in which gravity is repulsive.

20. For a general account see David Acheson and Tom Mullin's article, Ropy Magic (*New Scientist magazine*, 21 February, 1998). The mechanical demonstration of a 3-linked vertical pendulum is described in D. Acheson and T. Mullin, Upside-down pendulums (*Nature*, 366, 215-216, 1993). See also the detailed mathematical paper by T. Mullin *et al.*, The Indian rope trick by parametric excitation (*Proceedings of the Royal Society* A, 459, 539-546, 2003). See also note 19. For a detailed account of Daniel Bernoulli's contribution to the theory of vibrating strings, see Olivier Darrigol's article, The acoustic origins of harmonic analysis (*Archive for history of exact sciences*, 61. 343-424, 2007).

21. M. W. Spong., P. Cole, and R. Lozano, Nonlinear control of the inertia wheel pendulum (*Automatica*, 37, 1845-1851, 2001). See also the highly comprehensive ebook by Daniel Block, Karl Ånström and Mark Spong, *The Reaction Wheel Pendulum* (Morgan and Claypool Publishers, 2007).

22. K. Furuta, M. Yamakita, and S. Kobayashi, Swing-up control of inverted pendulum using pseudo-state feedback (*Journal of systems and Control Engineering*, **206**, 263-269, 1992). The full set of equations describing the motion of the Furuta pendulum are presented by Benjamin Cazzolato and Zebb Prime in their article, On the dynamics of the Furuta pendulum (*Journal of Control Science and Engineering*, article ID 528341, 2011).

Chapter 6

1. The design for a basic harmonograph is given by H. N. Cundy and A. P. Rollett in their classic and highly recommended book *Mathematical Models* (Tarquin Publications, Norfolk, 1989). An applet for visualizing Lissajous figures can be found at the history of mathematics website maintained by St. Andrews University, Scotland: http://www-history.mcs.st-and.ac.uk/Curves/Lissajous.html.

2. Nathanial Bowditch, On the motion of a pendulum suspended from two points (*Memoirs of the American Academy of Arts and Science*, 3, 413-436, 1815).

3. James Dean, An investigation of the apparent motion of the Earth, viewed from the Moon, arising from the Moon's librations (*Memoirs of the American Academy of Arts and Science*, 3, 241-245, 1815). See also, A. D. Crowell, Motion of the Earth as viewed from the Moon and the Y-suspended pendulum (*American Journal of Physics*, 49, 452-454, 1981).

4. Huygens letter is reported in Thomas Birch's, *The history of the Royal Society of London for improving natural knowledge* (Johnson Reprint Corporation, New York, 1968).

5. Matthew Bennett, M. F. Schatz, H. Rockwood and K. Wiesenfield, Huygens's Clocks (*Proceedings of the Royal Society of London* A, 458, 563-579, 2002). See also the article by Erica Klarreich, Huygen's clocks revisited (*American Scientist*, July/August, 2002). A press release on the work by Bennett *et al.* can be found at http://gtresearchnews.gatech.edu/reshor/rh-foo/time.html.

6. E. H. Barton and H. M. Browning, Vibrations under variable couplings quantitatively elucidated by simple experiments (*Philosophical Magazine*, 34, 246-270, 1917).

7. E. H. Barton and H. M. Browning, Forced vibrations experimentally illustrated (*Philosophical Magazine*, 36, 169-178, 1918).

8. Daniel Kirkwood, On the theory of Meteors (*Proceedings of the American Association for the Advancement of Science*, 1866, pp. 8-14). Kirkwood also pioneered the idea that annual meteor showers are produced when the Earth passes through the debris trail associated with the decay of a cometary nucleus. I have discussed Kirkwood's early contributions to meteor physics (and the urban myth associated with them) in, The Makings of Meteor Astronomy: Part XII (*WGN, Journal of the International Meteor Organization*, 24, 89-91, 1996).

9. The mathematics associated with the appearance of resonances within the solar system is formidable, but see, for example, J. Wisdom, The resonance overlap criterion and the onset of stochastic behavior in the restricted three-body problem (*The Astronomical Journal*, 85, 1122-1133, 1980); Chaotic behavior and the origin of the 3/1 Kirkwood gap (*Icarus*, 56, 51-74, 1983); and, Meteorites may follow a chaotic route to Earth (*Nature*, 315, 731-733, 1985). See also note 24 to Chapter 2.

10. Bohr assumed that the electron orbits of the hydrogen atom are circular with quantized angular momentum. In this manner $m_e V_n r_n = n h$, where m_e is the electron mass, V_n is the electron velocity and r_n is the orbital radius of the $n = 1, 2, 3,...$ al-

lowed orbit. Motion in a circle further provides a relationship between the centripetal force and the Coulomb force between the electron and proton: $F = m_e V_n^2 / r_n = e^2 / 4\pi\varepsilon_0 r_n^2$, where ε_0 is the permittivity of free space, and e is the elementary charge. Combining these two expressions and inserting constant values gives $r_n(\text{nm}) = 0.052918 n^2$ for the radii of the allowed electron orbits in the hydrogen atom. The energy associated with the various electron orbits is further given by $E_n(\text{Joules}) = -(1.602 \times 10^{-19}) / n^2$. The veracity of Bohr's relationship for the energy levels is established by studying the emission lines produced when electrons move between allowed electron orbits in a heated hydrogen gas; the measurable energy of the emitted photon being exactly equal to the energy difference between the electron's starting and final orbital. While quantum mechanics destroys the absolute determinacy of the classical world, it importantly explains why matter is (predominantly) stable and why chemistry is an understandable and predictive science. The quantization of electron orbits, for example, means that all atoms (with the same total number of electrons and protons) are identical. In the classical world, where electron orbits could have any energy value, no identical atoms would exist, and there would be no constancy between chemical reactions.

11. B. Wyker *et al.*, Creating and transporting Trojan wave packets (*Physical Review Letters*, 108, 043001(5), 2012). Technically, the atom developed by Wyker *et al.* is not an analog for a Trojan asteroid orbit in the solar system. The atomic Trojan configuration has just one central nucleus, while the solar system Trojans require the presence of two gravitating masses (the Sun and Jupiter) in order to come about.

12. The history and mathematical theory behind the Doubochinski pendulum is nicely described by Jonathan Tennenbaum in his article, Amplitude quantization as an elementary property of macroscopic vibrating systems (*21st Century Science and Technology*, winter 2005-6, 50-63).

13. The Poincare map is sometimes called the first return map. In the case where there is a known forcing frequency ω, the points plotted on the Poincare map (phase plane) correspond to the position and velocity at times $t = 0$ and integer multiples of $2\pi / \omega$. For a periodic system, such as the simple pendulum, the Poincare map will consist of just one point (the so-called fixed point) within the phase plane; the Poincare map for a damped pendulum will be a series of points with a trajectory moving towards the origin.

14. The term cosmic ray was introduced by American physicist Robert Millikan in a classic paper entitled, High frequency rays of cosmic origin (*Proceedings of the National Academy of Sciences*, 12, 48-55, 1926). Millikan and collaborators reasoned that their experiments were best interpreted by thinking of cosmic rays as a high energy gamma ray (a high frequency form of electromagnetic radiation), and it was only later found, in the 1930s and 40s, that they were, in fact, relativistic charged particles. For an historical review, see Per Carlson's, A Century of Cosmic Rays (*Physics Today*, February 2012, pp. 30-36).

15. Enrico Fermi, On the origin of the cosmic radiation (*Physical Review*, 75, 1169-1174, 1949). The Fermi-Ulam (FUM) process has been discussed in great detail since it was first introduced in 1961. The essential idea of the problem, however, is to find the velocity of a particle after multiple interactions with the oscillating wall. If after the nth collision the particle has a velocity $V(n)$ then the equation to be studied is: $V(n + 1) = -|V(n) + 2 U(n)|$ where $U(n) = \varepsilon \cos[t(n) + \varphi(n)]$ is the velocity of the

oscillating wall at time $t(n) = t(n - 1) + 2/V(n - 1)$ and where $\varepsilon < 1$ is a scale factor. The $\varphi(n)$ term is a random variable in the interval between 0 and 2π. A few years prior to studying the FUM process, Fermi was involved (in the early-1950s) in another famous project with Ulam and John Pasta. Generally referred to as the Firmi-Pasta-Ulam (FPU) problem this particular work was one of the very first electronic computer studies to be made of a nonlinear process. The FPU problem essentially considers the energy partition within a one-dimensional chain of atoms 'linked' by 'springs' obeying Hooke's law with an additional nonlinear perturbation term. The atoms and springs were essentially deemed, therefore, to be slightly perturbed simple harmonic oscillators (recall Chapter 3). Here the idea was to model how the atoms exchanged energy as the vibrations were passed along the chain of oscillating components. The FPU problem was first described in Los Alamos report, *Studies on non linear Problems* (LA-1940, May 1955).

16. To determine the energy levels of a quantum harmonic oscillator one must solve the time-independent Schrödinger equation for an appropriately defined wave function – such examples of this procedure can be found in any introductory textbook on quantum mechanics.

17. The Heisenberg Uncertainty Principle lies at the very heart of quantum indeterminacy. In its most common form the uncertainty principle is written in terms of the position and momentum such that if Δx and Δp are the uncertainties in the measurements of the position and momentum of an atomic particle, then $\Delta x\, \Delta p \geq \hbar/2$. Niels Bohr, under the guise of the Copenhagen interpretation, used the idea of Uncertainty Principle to argue that should the precise position of, say, an electron be determined ($\Delta x = 0$) then this measurement would so affect the electron that its momentum would be entirely unknowable ($\Delta p \to \infty$).

18. The photoelectric effect is observed through the emission of electrons from matter as a consequence of their absorbing electromagnetic radiation. The key part of the quantum mechanical explanation to this phenomenon is that the energy with which the electrons are emitted varies with the frequency of the impinging electromagnetic radiation and not its intensity. The term photon, now used to describe the quantized packets of energy invoked by Einstein in order to explain the photoelectric effect, was coined by American chemist Gilbert Newton Lewis in a letter written to the journal *Nature* (118, 874-875) in 1926. Interestingly, Lewis was proposing in his letter an altogether different (and as it turns out wrong) interpretation of Einstein's ideas on light quanta.

19. A. Trevelyan, Notice regarding some experiments on the vibration of heated metals (*Transactions of the Royal Society of Edinburgh*, 12, 137-146, 1834).

20. See, for example, S. H. Frisbee, Mechanical analysis of the Trevelyan rocker (*Nature*, 17, 242-243, 1878); W. H. Eccles, Note on an electrical Trevelyan rocker (*Proceedings of the Physical Society of London*, 23, 204-208, 1910); and I. M. Freeman, What is Trevelyan's rocker? (*The Physics Teacher*, 12, 348-349, 1974).

21. See, S. Bhargava and R. N. Ghosh, Note on Trevelyan's rocker (*Physical Review*, 22, 517-521, 1923).

22. The general solution to an equation of the form $\ddot{x} - \mu^2 x = 0$ is given by two exponential terms, with $x(t) = C_1 \exp(\mu t) + C_2 \exp(-\mu t)$, where the constants C_1 and C_2 are determined by the boundary conditions.

23. For the small angle approximation solution to equation (6.12), it turns out that the constant terms in the general solution (see note 22 above) are such that $C_1 = C_2 = C = \vartheta_{max} - \Delta/2h$, and hence $-\Delta/2h = C(e^{\mu T/4} + e^{-\mu T/4})$ when $t = T/4$. The final result depends upon expanding the two exponential expressions in terms of power series, keeping only linear and quadratic terms in $\mu T/4$, which yields $C^{-1} = (1 - 2h\vartheta_{max}/\Delta)^{-1} \approx 1 + 2h\vartheta_{max}/\Delta$.

24. Experimental studies have been conducted by, S. Bhargava and R. N. Ghosh (note 21 above), and by E. G. Richardson, The theory of the Trevelyan Rocker (*Philosophical Magazine*, 45, 976-989, 1923)

25. The study of fractals, literally fractional dimensional objects, began in the early 1980s following the publication of Benoit Mandelbrot's somewhat quixotic book *The Fractal Geometry of Nature* (W. H. Freman and Co. New York, 1977). Within his book Mandelbrot describes a fractal as being a self-similar structure, meaning that it is a "geometric shape that can be split into parts, each of which is (at least approximately) a reduced-size copy of the whole". It is often said that Fractals are the geometry of Nature – but this cannot possibly be true. Fractals are a mathematical (infinite regression) construct; Nature is made up of small, but very much finite sized building blocks. As an investigative methodology fractal concepts have not proven especially useful – it is a descriptive (at some level) tool rather than a physically predictive one.

26. H. Bondi, The prehistory of Chaos (*Bulletin of the Institute of Mathematics and its Applications*, 25, 107-108, 1989). It is probably not unreasonable to argue that the study of chaos actually began in the 1880s when French polymath Henri Poincare found non-periodic solutions to the so-called three-body gravitational problem. While modeling the dynamics relating to the gravitational interaction of three or more bodies is a decidedly complex matter, it turns out that even very simple systems of equations can display chaotic behavior. French mathematician and astronomer Michel Hénon, for example, introduced the Hénon map composed of the points (x_n, y_n), $n = 1, 2, 3\ldots$ where $x(n+1) = y(n) + 1 - a[x(n)]^2$ and $y(n) = bx(n)$, and where a and b are constants. According to the values chosen for a and b, the map will diverge to infinity, converge to a periodic orbit or become chaotic. When $a = 1.4$ and $b = 0.3$, for example, the Hénon map is chaotic, producing a strange attractor with a fractal dimension of 1.26.

27. There are numerous websites dedicated to the explanation and graphical description of the Lorenz oscillator. A good place to start is http://mathworld.wolfram.com/LorenzAttractor.html. The paper, Attractor sets and quasi-geostrophic equilibrium, by Lorenz (*Journal of the Atmospheric Sciences*, 37, 1685-1699, 1980) affords a detailed introduction to the mathematics involved in deriving the Lorenz oscillator equations.

28. See e.g., http://scienceworld.wolfram.com/physics/DoublePendulum.html.

29. The Butterfly Effect has been a greatly misunderstood and misused term in the modern era. It encapsulates, however, the idea of the sensitive dependency upon initial conditions in chaos theory. The name originates from the visual appearance of the Lorenz attractor (see figure 6.13) - which looks superficially like a pair of butterfly wings.

30. The water-driven chaotic pendulum was first described in A. B. Pippard's wonderful little book entitled *Reconciling Physics with Reality* (Cambridge University Press, 1972). Describing an experimental demonstration of the water-fed pendulum, Pip-

pard notes that, "the number of swings it makes on each side before tilting right over is highly variable". The pendulum was installed at St. Mary Redcliffe Church in 1997 as part of the *Journey into Science* project of the Clifton Scientific Trust, see http://www.clifton-scientific.org/journey_chaos.htm). See also the study by Wanjiao Lee at: http://www.enm.bris.ac.uk/teaching/projects/2007_08/wl4457/index.html.

INDEX

www.ingramcontent.com/pod-product-compliance
Lightning Source LLC
Chambersburg PA
CBHW071401170526
45165CB00001B/142